# Power Systems

Xiao-Ping Zhang, Christian Rehtanz, Bikash Pal
Flexible AC Transmission Systems: Modelling and Control

Xiao-Ping Zhang, Christian Rehtanz,
Bikash Pal

# Flexible AC Transmission Systems: Modelling and Control

With 156 Figures

Dr. Xiao-Ping Zhang
University Warwick
School of Engineering
Coventry CV4 7AL
United Kingdom
x.p.zhang@warwick.ac.uk

Bikash Pal
Imperial College London
Dept. of Electrical & Eelctronic Engineering
Exhibition Road
London SW7 2BT
United Kingdom
b.pal@imperial.ac.uk

Dr. Christian Rehtanz
ABB Corporate Research China
Universal Plaza, 10 Jiuxianqiao Lu
Chaoyang District
Beijing, 100016
P.R. China
christian.rehtanz@ieee.org

Library of Congress Control Number: 2005936513

ISBN-10   3-540-30606-4 Springer Berlin Heidelberg New York
ISBN-13   978-3-540-30606-1 Springer Berlin Heidelberg New York

This work is subject to copyright. All rights are reserved, whether the whole or part of the material is concerned, specifically the rights of translation, reprinting, reuse of illustrations, recitation, broadcasting, reproduction on microfilm or in other ways, and storage in data banks. Duplication of this publication or parts thereof is permitted only under the provisions of the German Copyright Law of September 9, 1965, in its current version, and permission for use must always be obtained from Springer. Violations are liable to prosecution under German Copyright Law.

**Springer is a part of Springer Science+Business Media**
springer.com

© Springer-Verlag Berlin Heidelberg 2006
Printed in Germany

The use of general descriptive names, registered names, trademarks, etc. in this publication does not imply, even in the absence of a specific statement, that such names are exempt from the relevant protective laws and regulations and therefore free for general use.

Typesetting: Digital data supplied by editors
Final processing by PTP-Berlin Protago-T$_E$X-Production GmbH, Germany
Cover-Design: deblik, Berlin
Printed on acid-free paper        89/3141/Yu – 5 4 3 2 1 0

# Foreword

The electric power industry is undergoing the most profound technical, economic and organisational changes since its inception some one hundred years ago. This paradigm change is the result of the liberalisation process, stipulated by politics and followed up by industry. For many years the electric power industry was characterized by a vertically integrated structure, consisting of power generation, transmission/distribution and trading. The liberalisation process has resulted in the unbundling of this organizational structure. Now generation and trading are organised in separate business entities, subject to competition, while the transmission/distribution business remains a natural monopoly. Since the trading of electric energy happens on two levels, the physical level and the contractual level, it has to be recognized that these two levels are completely different. However for understanding the electricity market as a network based industry both levels have to be considered and understood. The fundamental properties of electric energy are as follows:

- Electricity always needs a network for transportation and distribution.
- Electricity cannot be stored in a substantial amount, hence production and consumption have to be matched at each instant of time.
- The physical transport of electricity has nothing to do with the contracts for trading with electricity

The role of the electric network is of prime importance within the electric energy business. Its operation is governed by physical laws. The electric network has a fixed structure consisting of different voltage levels; the higher levels are for transmission purposes whereas the lower levels are used for the distribution tasks. Each network element has a finite capacity, limiting the amount of electricity to be transported or distributed. As a consequence of the liberalisation process the operation of the networks has been pushed closer towards its technical limits. Hence the stress on the system is considerably bigger than in the past. The efficient use of all network elements is of prime interest to the network operator because the cost constraints have also become much tighter than in the past. Recognizing that the operation of a large electric network is a complex and challenging engineering task, it becomes evident that the cost constraints increase the operational complexity considerably. The bigger the interconnected network becomes the more flexibility is required with respect to the cross border trading of electricity. Simultaneously the complexity of operational problems increases due to voltage, angle and frequency stability problems.

The traditional planning approaches for power networks are undergoing a reengineering. The long lasting experience with the power flowing purely from the generation plants to the customers is no longer valid. Growing volatility and increasingly unpredictable system behaviour requires innovative equipment to handle such situations successfully. Keeping in mind that the interconnected power networks have been designed such that each network partner may contribute with reserve power in case of emergency, the trend is now towards extensive cross border energy trading. Another fundamental development is the construction of micro grid on the distribution level. The introduction of dispersed generation close to the customers changes the functionality and the requirements of the distribution networks. The grid operator is requested to provide network access to any interested stakeholder in a transparent and non-discriminatory manner. So, while in the past the power flow in distribution networks was unidirectional, now the system must handle bidirectional power flows. This allows the distribution network to take on more and more the function of a balancing network. At the same time, the capacity of individual elements may not be sufficient to cope with the resulting power flow situations.

Summarizing the current developments, it must be noticed that both planning and operation of electric networks are undergoing fundamental and radical changes in order to cope with the increased complexity of finding economic and reliable network solutions. The operation of the transmission and distribution networks will be closer to their physical limits. The necessity to design electric power networks providing the maximal transmission capacity and at the same time resulting in minimal costs is a great engineering challenge. Innovative operational equipment based on power electronics offers new and powerful solutions. Commonly described by the term 'Flexible AC Transmission Systems' or 'FACTS-devices', such equipment has been available for several years, but has still not been widely accepted by all grid operators for several reasons.

The introduction of innovative equipment has a great impact on the operation. A more flexible transmission or distribution system may cause new problems during normal or disturbed operating states. Furthermore, the proper understanding of innovative equipment is also an educational problem because there is not much experience reported so far with this innovative equipment.

On the other hand, the opportunities for new solutions are substantial and important. FACTS-devices can be utilized to increase the transmission capacity, improve the stability and dynamic behaviour or ensure better power quality in modern power systems. Their main capabilities are reactive power compensation, voltage control and power flow control. Due to their controllable power electronics, FACTS-devices always provide fast control actions in comparison to conventional devices like switched compensation or phase shifting transformers with mechanical on-load tap changers.

This book offers a concise and modern presentation of the timely and important topic of flexible AC transmission networks. There is no doubt that these innovative FACTS-devices will find a definite place in transmission and distribution networks. The complete description of the functionality of such devices is supported with extensive mathematical models, which are required when planning the use of this type of equipment in electrical networks. The first part of the book deals with the modeling of single and multi-converter FACTS-devices in single and three-phase power flow studies and optimal power flow solutions.

The in depth discussion of the operational and controlling aspects in the second part of the book makes it a most valuable compendium for the design of future electric networks. Without a complete and powerful solution of the control problems, the FACTS-devices will not find their application in power systems because they have to operate in normal and contingency situations in a reliable and economic way. System security must not be weakened by the FACTS-devices, even if the system is operated closer to its limits. The control speed of the FACTS-devices can only be utilized, if they are first given higher priority from the operator, then designed to react in a coordinated but autonomous manner in dynamic or even contingency situations. A novel and original control strategy based on the autonomous control theory fulfilling these requirements is presented in the book.

Due to the influence of FACTS-devices on wide system areas, especially for power flow and damping control, an exchange of system information with the FACTS-controllers is required. A wide area control scheme is introduced and applied for power flow control. The dynamics of FACTS-devices provide effective damping capability. Inter-area oscillations require wide area system supervision and a wide area control scheme. For this application time delays in the wide area control loop play a significant role in the controller design. Based on detailed modeling, an innovative approach is presented considering this time delay, making wide area damping control feasible. Only with such a control scheme, FACTS-devices can be applied beneficially in the future.

Based on the authors' extensive experience, this book is of greatest importance for the practical power engineers for both planning and operational problems. It provides a deep insight into the use of FACTS-devices in modern power systems. Although the technology of modern power electronics will change very quickly, the results presented in this book are sustainable and long lasting. The combination of theoretical and practical knowledge from the international team of authors from academia and industry provides an invaluable contribution for the future application of FACTS-devices. I am convinced that this book will become a standard work in modern power engineering. It will serve equally as a text book for university students as well as an engineering reference for planning and operation of modern power systems.

Prof. Dr.-Ing. Edmund Handschin                    Dortmund, Germany, 2005

# Preface

Electricity market activities and a growing demand for electricity have led to heavily stressed power systems. This requires operation of the networks closer to their stability limits. Power system operation is affected by stability related problems, leading to unpredictable system behavior. Cost efficient solutions are preferred over network extensions. In many countries, permits to build new transmission lines are hard to get, which means the existing network has to be enforced to fulfill the changing requirements.

Power electronic network controllers, the so called FACTS-devices, are well known having several years documented use in practice and research. Several kinds of FACTS-devices have been developed. Some of them such as the Thyristor based Static Var Compensator (SVC) are a widely applied technology; others like the Voltage Source Converter (VSC) based Static Compensator (STATCOM) or the VSC-HVDC are being used in a growing number of installations worldwide. The most versatile FACTS-devices, such as Unified Power Flow Controller (UPFC), although still confined primarily to research and development applications, have the potential to be used widely beyond today's pilot installations.

In general, FACTS-devices can be utilized to increase the transmission capacity, the stability margin and dynamic behavior or serve to ensure improved power quality. Their main capabilities are reactive power compensation, voltage control and power flow control. Due to their controllable power electronics, FACTS-device provide always a fast controllability in comparison to conventional devices like switched compensation or phase shifting transformers. Different control options provide a high flexibility and lead to multi-functional devices.

To explore the capabilities of FACTS-devices, a specific operation and control scheme has to be designed. Fundamental to their operation and control is their proper modeling for static and dynamic purposes. The integration of FACTS-devices into basic tools like power flow calculation and optimal power flow (OPF) is mandatory for a beneficial system operation. Due to the wide area and dynamic impact of FACTS-devices, a pure local control is desired, but is not sufficient in many cases. The requirements for normal and emergency operation have to be defined carefully. A specific control design has to address these different operational conditions. This book introduces the latest results of research and practice for modeling and control of existing and newly introduced FACTS-devices.

## Motivation

This book is motivated by the recent developments of FACTS-devices. Numerous types of FACTS-devices have been successfully applied in practical operation. Some are still in the pilot stage and many are proposed in research and development. From practical experience it has been seen that the investment into FACTS-devices, in most of the cases, only pays off by considering their multi-functional capabilities, particularly in normal and emergency situations. This requires a three-phase modeling and a control design addressing both normal and emergency conditions which, in most of the cases, uses wide area information. The recent results and requirements for both modeling and control have motivated this book.

## Focus and Target

The focus and target of this book is to emphasize advanced modeling, analysis and control techniques of FACTS. These topics reflect the recent research and development of FACTS-devices, and foresee the future applications of FACTS in power systems. The book comprehensively covers a range of power system control problems like steady state voltage and power flow control, voltage and reactive power control, voltage stability control and small signal stability control using FACTS-devices.

Beside the more mature FACTS-devices for shunt compensation, like SVC and STATCOM, and series compensation, like TCSC and SSSC, the modeling of the latest FACTS-devices for power flow control, compensation and power quality (IPFC, GUPFC, VSC HVDC and Multi-VSC-HVDC, etc.) is considered for power system analysis. The selection is evaluated by their actual and future practical relevance. The multi-control functional models of FACTS-devices and the ability for handling various internal and external operating constraints of FACTS are introduced. In addition, models are proposed to deal with small or zero impedances in the voltage source converter (VSC) based FACTS-devices. The FACTS-device models are implemented in power flow and optimal power flow (OPF) calculations. The power flow and OPF algorithms cover both single-phase models and especially three-phase models. Furthermore the unbalanced continuation power flow with FACTS is presented.

The control of FACTS-devices has to follow their multi-functional capabilities in normal and emergency situations. The investment into FACTS is normally justified by the increase of stability and primarily by the increase of transmission capability. Applications of FACTS in power system operation and control, such as transfer capability enhancement and congestion management, are used to show the practical benefits of FACTS devices.

A comprehensive FACTS-control approach is introduced based on the requirements and specifications derived from practical experience. The control structure is characterised by an autonomous system structure allowing, as far as possible, control decisions to be taken locally, but also incorporating system wide information where this is required. Wide Area Measurement System (WAMS) based control methodologies, which have been developed recently, are introduced for the first time in a book. In particular, the real-time control technologies based on Wide Area Measurement are presented. The current applications and future developments of the Wide Area Measurement based control methodologies are also discussed. As a particular control topic, utilizing the control speed of FACTS-devices, a special scheme for small-signal stability and damping of inter-area oscillations is introduced. Advanced control design techniques for power systems with FACTS including eigenvalue analysis, damping control design by the state-of-the art Linear Matrix Inequalities (LMI) approach and multiple damping controller coordination is presented. In addition, the time-delay of wide area communications, which is required for a system wide damping control, is considered.

These aspects make the book unique in its area and differentiate from other books on the similar topic. The work presented is derived both from scientific research and industrial development, in which the authors have been heavily involved. The book is well timed, addressing current challenges and concerns faced by the power engineering professionals both in industries and academia. It covers a broad practical range of power system operation, planning and control problems.

## Structure

The first chapter of the book gives an introduction into nowadays FACTS-devices. Power semiconductors and converter structures are introduced. The basic designs of major FACTS-devices are presented and discussed from a practical point of view. The further chapters are logically separated into a modeling and a control part. The modeling part introduces the modeling of single and multi-converter FACTS-devices for power flow calculations (Chapter 2 and 3) and optimal power flow calculations (Chapter 4). The extension to three phase models is given in chapter 5. This is fundamental for proper system integration for steady state balanced and unbalanced voltage stability control or the increase of available transmission capacity.

Chapter 6 and 7 present the steady state voltage stability analysis for balanced and unbalanced systems. The increase of transmission capacity and loss reduction with power flow controlling FACTS-devices is introduced in chapter 8 along with the financial benefits of FACTS. From these results it can be seen, that the benefits of FACTS can be increased by utilizing the fast controllability of FACTS together with a certain wide area control scheme.

The control part of the book starts with chapter 9 introducing a non-intrusive system control scheme for normal and emergency situations. The chapter takes the view, that a FACTS-device should never weaken the system stability. Based on this condition, the requirements and basic control scheme for FACTS-devices are derived. Chapter 10 introduces an autonomous control system approach for FACTS-control, balancing the use of local and global system information and considering normal and emergency situations. Due to the influence of FACTS-devices on wide system areas, especially for power flow and damping control, an exchange of information with the FACTS controllers is required. A wide area control scheme for power flow control is introduced in chapter 11.. Only with wide area system information can the benefits of power flow control be achieved.

The control options available with FACTS-devices can provide effective damping capability. Chapter 12 and 13 deal with small signal stability and the damping of oscillations, which is a specific application area utilizing the control speed of FACTS. The coordination of several FACTS damping controllers requires a formally introduced wide area control scheme. This approach has to consider communication time delays carefully, which is a specific topic of chapter 13.

## Acknowledgements

The authors would like to thank Prof. Edmund Handschin at the University of Dortmund, Germany for his support and encouragement to write this book. Significant progress was made in the modeling of FACTS in power flow and optimal power flow analysis when Dr. Zhang was working in Prof. Handschin's Institute at the University of Dortmund, sponsored by the Alexander van Humboldt Foundation, Germany. Subsequent work has been sponsored by the Engineering and Physics Sciences Research Council (EPSRC), UK. Therefore, Dr Zhang would like to take the opportunity to acknowledge the support from the Alexander van Humboldt Foundation and the EPSRC.

Dr. Rehtanz would like to thank the following researchers for their contributions to some of the chapters. Chapter 8 is based on collaborative work with Prof. Jürgen Haubrich, Dr. Feng Li of RWTH, and Dr. Christian Zimmer and Dr. Alexander Ladermann of CONSENTEC GmbH, Aachen, Germany. Dr Christian Becker, who was working with the University of Dortmund, and is now working with AIRBUS Deutschland GmbH, has contributed to chapter 10. Dr. Mats Larsson, Dr. Petr Korba, and Mr. Marek Zima, ABB, Switzerland have contributed with their work to chapter 11. Special thanks are given to Prof. Dirk Westermann of the Technical University Ilmenau, Germany for his useful contributions, inputs and comments to chapters 9 to 11.

Dr. Bikash Pal would like to thank Dr. Balarko Chaudhuri of GE Global Research Lab, Bangalore and Mr Rajat Majumder, a PhD student at Imperial College for supporting him for the preparation of chapter 13 through simulation results. The control design techniques presented in this chapter primarily comes from the research conducted by them under the supervision of Dr. Pal at Imperial College. Dr. Pal also expresses his gratitude to EPSRC (UK) and ABB for sponsoring this research at Imperial College. Dr. Pal is also thankful to Dr. John McDonald of the Control and Power research group at Imperial College for proof reading chapters 12 and 13.

The challenging task of writing and editing this book was made possible by the excellent co-operation of the team of authors together with a number of colleagues and friends. Our sincere thanks to all contributors, proofreaders, the publisher and our families for making this book project happen.

| | |
|---|---|
| Xiao-Ping Zhang | University of Warwick, Coventry, UK, 2005 |
| Christian Rehtanz | ABB China Ltd, Beijing, China, 2005 |
| Bikash Pal | Imperial College London, London, UK, 2005 |

# Contents

**1 FACTS-Devices and Applications** ................................................................. 1
    1.1 Overview ........................................................................................................ 2
    1.2 Power Electronics ......................................................................................... 5
        1.2.1 Semiconductors ................................................................................... 5
        1.2.2 Power Converters ............................................................................... 8
    1.3 Configurations of FACTS-Devices ............................................................ 10
        1.3.1 Shunt Devices .................................................................................... 10
        1.3.2 Series Devices ................................................................................... 15
        1.3.3 Shunt and Series Devices ................................................................. 19
        1.3.4 Back-to-Back Devices ...................................................................... 24
    References ........................................................................................................... 25

**2 Modeling of Multi-Functional Single Converter FACTS in Power Flow Analysis** ............................................................................................................ 27
    2.1 Power Flow Calculations ............................................................................ 27
        2.1.1 Power Flow Methods ........................................................................ 27
        2.1.2 Classification of Buses ..................................................................... 27
        2.1.3 Newton-Raphson Power Flow in Polar Coordinates ........................ 28
    2.2 Modeling of Multi-Functional STATCOM ................................................ 28
        2.2.1 Multi-Control Functional Model of STATCOM for Power Flow Analysis .............................................................................................. 29
        2.2.2 Implementation of Multi-Control Functional Model of STATCOM in Newton Power Flow ...................................................................... 35
        2.2.3 Multi-Violated Constraints Enforcement ......................................... 37
        2.2.4 Multiple Solutions of STATCOM with Current Magnitude Control . 39
        2.2.5 Numerical Examples ......................................................................... 40
    2.3 Modeling of Multi-Control Functional SSSC ............................................ 44
        2.3.1 Multi-Control Functional Model of SSSC for Power Flow Analysis. 44
        2.3.2 Implementation of Multi-Control Functional Model of SSSC in Newton Power Flow ........................................................................ 48
        2.3.3 Numerical Results ............................................................................. 51
    2.4 Modeling of SVC and TCSC in Power Flow Analysis .............................. 54
        2.4.1 Representation of SVC by STATCOM in Power Flow Analysis ........ 55
        2.4.2 Representation of TCSC by SSSC in Power Flow Analysis .............. 56
    References ........................................................................................................... 56

**3 Modeling of Multi-Converter FACTS in Power Flow Analysis** ..................... 59

3.1 Modeling of Multi-Control Functional UPFC ........................................... 59
    3.1.1 Advanced UPFC Models for Power Flow Analysis .......................... 60
    3.1.2 Implementation of Advanced UPFC Model in Newton Power Flow . 66
    3.1.3 Numerical Results .............................................................................. 67
3.2 Modeling of Multi-Control Functional IPFC and GUPFC ........................ 70
    3.2.1 Mathematical Modeling of IPFC in Newton Power Flow under Practical Constraints ................................................................................... 71
    3.2.2 Mathematical Modeling of GUPFC in Newton Power Flow under Practical Constraints ................................................................................... 75
    3.2.3 Numerical Examples ......................................................................... 78
3.3 Multi-Terminal Voltage Source Converter Based HVDC ........................ 82
    3.3.1 Mathematical Model of M-VSC-HVDC with Converters Co-located in the same Substation ................................................................................. 83
    3.3.2 Generalized M-VSC-HVDC Model with Incorporation of DC Network Equation ....................................................................................... 88
    3.3.3 Numerical Examples ......................................................................... 91
3.4 Handling of Small Impedances of FACTS in Power Flow Analysis ......... 95
    3.4.1 Numerical Instability of Voltage Source Converter FACTS Models . 95
    3.4.2 Impedance Compensation Model ...................................................... 95
References ........................................................................................................ 97

**4 Modeling of FACTS-Devices in Optimal Power Flow Analysis ................. 101**
4.1 Optimal Power Flow Analysis ................................................................. 101
    4.1.1 Brief History of Optimal Power Flow ............................................. 101
    4.1.2 Comparison of Optimal Power Flow Techniques ........................... 102
    4.1.3 Overview of OPF-Formulation ........................................................ 104
4.2 Nonlinear Interior Point Optimal Power Flow Methods ......................... 105
    4.2.1 Power Mismatch Equations ............................................................. 105
    4.2.2 Transmission Line Limits ................................................................ 106
    4.2.3 Formulation of the Nonlinear Interior Point OPF ........................... 106
    4.2.4 Implementation of the Nonlinear Interior Point OPF ...................... 109
    4.2.5 Solution Procedure for the Nonlinear Interior Point OPF ............... 112
4.3 Modeling of FACTS in OPF Analysis ..................................................... 112
    4.3.1 IPFC and GUPFC in Optimal Voltage and Power Flow Control ..... 113
    4.3.2 Operating and Control Constraints of GUPFC ................................ 113
    4.3.3 Incorporation of GUPFC into Nonlinear Interior Point OPF .......... 116
    4.3.4 Modeling of IPFC in Nonlinear Interior Point OPF ........................ 121
4.4 Modeling of Multi-Terminal VSC-HVDC in OPF .................................. 123
    4.4.1 Multi-Terminal VSC-HVDC in Optimal Voltage and Power Flow . 123
    4.4.2 Operating and Control Constraints of the M-VSC-HVDC .............. 123
    4.4.3 Modeling of M-VSC-HVDC in the Nonlinear Interior Point OPF .. 124
4.5 Comparison of FACTS-Devices with VSC-HVDC ................................ 126
    4.5.1 Comparison of UPFC with BTB-VSC-HVDC ................................ 126
    4.5.2 Comparison of GUPFC with M-VSC-HVDC ................................. 128
4.6 Appendix: Derivatives of Nonlinear Interior Point OPF with GUPFC.... 131
    4.6.1 First Derivatives of Nonlinear Interior Point OPF .......................... 131

    4.6.2 Second Derivatives of Nonlinear Interior Point OPF ...................... 133
    References............................................................................................ 136

## 5 Modeling of FACTS in Three-Phase Power Flow and Three-Phase OPF Analysis ........................................................................................................ 139
    5.1 Three-Phase Newton Power Flow Methods in Rectangular Coordinates 140
        5.1.1 Classification of Buses ...................................................................... 140
        5.1.2 Representation of Synchronous Machines ....................................... 141
        5.1.3 Power and Voltage Mismatch Equations in Rectangular Coordinates
        ......................................................................................................... 142
        5.1.4 Formulation of Newton Equations in Rectangular Coordinates ....... 143
    5.2 Three-Phase Newton Power Flow Methods in Polar Coordinates ........... 149
        5.2.1 Representation of Generators ........................................................... 149
        5.2.2 Power and Voltage Mismatch Equations in Polar Coordinates ........ 149
        5.2.3 Formulation of Newton Equations in Polar Coordinates.................. 151
    5.3 SSSC Modeling in Three-Phase Power Flow in Rectangular Coordinates
    ............................................................................................................... 152
        5.3.1 Three-Phase SSSC Model with Delta/Wye Connected Transformer
        ......................................................................................................... 153
        5.3.2 Single-Phase/Three-Phase SSSC Models with Separate Single Phase
        Transformers ............................................................................................. 159
        5.3.3 Numerical Examples ........................................................................ 162
    5.4 UPFC Modeling in Three-Phase Newton Power Flow in Polar Coordinates
    ............................................................................................................... 166
        5.4.1 Operation Principles of the Three-Phase UPFC ............................... 166
        5.4.2 Three-Phase Converter Transformer Models ................................... 167
        5.4.3 Power Flow Constraints of the Three-Phase UPFC ......................... 169
        5.4.4 Symmetrical Components Control Model for Three-Phase UPFC... 172
        5.4.5 General Three-Phase Control Model for Three-Phase UPFC .......... 175
        5.4.6 Hybrid Control Model for Three-Phase UPFC................................. 176
        5.4.7 Numerical Examples ........................................................................ 178
    5.5 Three-Phase Newton OPF in Polar Coordinates ...................................... 183
    5.6 Appendix A - Definition of $Yg_i$............................................................. 185
    5.7 Appendix B - 5-Bus Test System............................................................. 185
    References............................................................................................ 186

## 6 Steady State Power System Voltage Stability Analysis and Control with FACTS.......................................................................................................... 189
    6.1 Continuation Power Flow Methods for Steady State Voltage Stability
    Analysis ........................................................................................................ 189
        6.1.1 Formulation of Continuation Power Flow........................................ 189
        6.1.2 Modeling of Operating Limits of Synchronous Machines ............... 191
        6.1.3 Solution Procedure of Continuation Power Flow ............................ 192
        6.1.4 Modeling of FACTS-Control in Continuation Power Flow ............. 193
        6.1.5 Numerical Results ............................................................................ 193
    6.2 Optimization Methods for Steady State Voltage Stability Analysis ........ 198

6.2.1 Optimization Method for Voltage Stability Limit Determination .... 198
6.2.2 Optimization Method for Voltage Security Limit Determination .... 199
6.2.3 Optimization Method for Operating Security Limit Determination . 200
6.2.4 Optimization Method for Power Flow Unsolvability ...................... 200
6.2.5 Numerical Examples ........................................................................ 202
6.3 Security Constrained Optimal Power Flow for Transfer Capability Calculations ................................................................................................. 204
6.3.1 Unified Transfer Capability Computation Method with Security Constraints .................................................................................................. 205
6.3.2 Solution of Unified Security Constrained Transfer Capability Problem by Nonlinear Interior Point Method ........................................................... 206
6.3.3 Solution Procedure of the Security Constrained Transfer Capability Problem ........................................................................................................ 211
6.3.4 Numerical Results ............................................................................ 211
References .................................................................................................. 214

# 7 Steady State Voltage Stability of Unbalanced Three-Phase Power Systems ........................................................................................................... 217

7.1 Steady State Unbalanced Three-Phase Power System Voltage Stability . 217
7.2 Continuation Three-Phase Power Flow Approach ................................. 218
7.2.1 Modeling of Synchronous Machines with Operating Limits ........... 218
7.2.2 Three-Phase Power Flow in Polar Coordinates ................................ 219
7.2.3 Formulation of Continuation Three-Phase Power Flow ................... 220
7.2.4 Solution of the Continuation Three-Phase Power Flow ................... 222
7.2.5 Implementation Issues of Continuation Three-Phase Power Flow ... 223
7.2.6 Numerical Results ............................................................................ 224
7.3 Steady State Unbalanced Three-Phase Voltage Stability with FACTS ... 232
7.3.1 STATCOM ....................................................................................... 232
7.3.2 SSSC ................................................................................................ 234
7.3.3 UPFC ................................................................................................ 235
References .................................................................................................. 236

# 8 Congestion Management and Loss Optimization with FACTS ................. 239

8.1 Fast Power Flow Control in Energy Markets .......................................... 239
8.1.1 Operation Strategy ............................................................................ 239
8.1.2 Control Scheme ................................................................................ 241
8.2 Placement of Power Flow Controllers ..................................................... 242
8.3 Economic Evaluation Method ................................................................. 245
8.3.1 Modelling of LFC for Cross-Border Congestion Management ........ 245
8.3.2 Determination of Cross-Border Transmission Capacity ................... 247
8.3.3 Estimation of Economic Welfare Gain through LFC ....................... 248
8.4 Quantified Benefits of Power Flow Controllers ...................................... 252
8.4.1 Transmission Capacity Increase ....................................................... 252
8.4.2 Loss Reduction ................................................................................. 254
References .................................................................................................. 257

## 9 Non-Intrusive System Control of FACTS ..... 259
9.1 Requirement Specification ..... 259
9.1.1 Modularized Network Controllers ..... 260
9.1.2 Controller Specification ..... 261
9.2 Architecture ..... 262
9.2.1 NISC-Approach for Regular Operation ..... 264
9.2.2 NISC-Approach for Contingency Operation ..... 265
References ..... 267

## 10 Autonomous Systems for Emergency and Stability Control of FACTS .. 269
10.1 Autonomous System Structure ..... 269
10.2 Autonomous Security and Emergency Control ..... 271
10.2.1 Model and Control Structure ..... 271
10.2.2 Generic Rules for Coordination ..... 271
10.2.3 Synthesis of the Autonomous Control System ..... 274
10.3 Adaptive Small Signal Stability Control ..... 281
10.3.1 Autonomous Components for Damping Control ..... 281
10.4 Verification ..... 282
10.4.1 Failure of a Transmission Line ..... 284
10.4.2 Increase of Load ..... 286
References ..... 288

## 11 Wide Area Control of FACTS ..... 289
11.1 Wide Area Monitoring and Control System ..... 289
11.2 Wide Area Monitoring Applications ..... 292
11.2.1 Corridor Voltage Stability Monitoring ..... 292
11.2.2 Thermal Limit Monitoring ..... 296
11.2.3 Oscillatory Stability Monitoring ..... 296
11.2.4 Topology Detection and State Calculation ..... 301
11.2.5 Loadability Calculation based on OPF Techniques ..... 303
11.2.6 Voltage Stability Prediction ..... 304
11.3 Wide Area Control Applications ..... 307
11.3.1 Predictive Control with Setpoint Optimization ..... 307
11.3.2 Remote Feedback Control ..... 310
References ..... 317

## 12 Modeling of Power Systems for Small Signal Stability Analysis with FACTS ..... 319
12.1 Small Signal Modeling ..... 320
12.1.1 Synchronous Generators ..... 320
12.1.2 Excitation Systems ..... 322
12.1.3 Turbine and Governor Model ..... 324
12.1.5 Network and Power Flow Model ..... 326
12.1.6 FACTS-Models ..... 327
12.1.7 Study System ..... 333
12.2 Eigenvalue Analysis ..... 334

12.2.1 Small Signal Stability Results of Study System ............................. 334
12.2.2 Eigenvector, Mode Shape and Participation Factor ...................... 340
12.3 Modal Controllability, Observability and Residue ................................ 343
References ........................................................................................................ 346

**13 Linear Control Design and Simulation of Power System Stability with FACTS ............................................................................................................ 347**
13.1 H-Infinity Mixed-Sensitivity Formulation ............................................. 348
13.2 Generalized H-Infinity Problem with Pole Placement .......................... 349
13.3 Matrix Inequality Formulation ............................................................... 351
13.4 Linearization of Matrix Inequalities ...................................................... 352
13.5 Case Study ............................................................................................. 354
13.5.1 Weight Selection ......................................................................... 354
13.5.2 Control Design ............................................................................ 355
13.5.3 Performance Evaluation ............................................................. 357
13.5.4 Simulation Results ...................................................................... 358
13.6 Case Study on Sequential Design ........................................................ 361
13.6.1 Test System ................................................................................. 361
13.7.2 Control Design ............................................................................ 362
13.6.3 Performance evaluation .............................................................. 362
13.6.4 Simulation Results ...................................................................... 363
13.7 H-Infinity Control for Time Delayed Systems ...................................... 366
13.8 Smith Predictor for Time-Delayed Systems ......................................... 367
13.9 Problem Formulation using Unified Smith Predictor ........................... 370
13.10 Case Study ........................................................................................... 372
13.10.1 Control Design .......................................................................... 373
13.10.2 Performance Evaluation ........................................................... 375
13.10.3 Simulation Results .................................................................... 375
References ........................................................................................................ 379

**Index ................................................................................................................ 381**

# 1 FACTS-Devices and Applications

Flexible AC Transmission Systems, called FACTS, got in the recent years a well-known term for higher controllability in power systems by means of power electronic devices. Several FACTS-devices have been introduced for various applications worldwide. A number of new types of devices are in the stage of being introduced in practice. Even more concepts of configurations of FACTS-devices are discussed in research and literature.

In most of the applications the controllability is used to avoid cost intensive or landscape requiring extensions of power systems, for instance like upgrades or additions of substations and power lines. FACTS-devices provide a better adaptation to varying operational conditions and improve the usage of existing installations. The basic applications of FACTS-devices are:

- power flow control,
- increase of transmission capability,
- voltage control,
- reactive power compensation,
- stability improvement,
- power quality improvement,
- power conditioning,
- flicker mitigation,
- interconnection of renewable and distributed generation and storages.

In all applications the practical requirements, needs and benefits have to be considered carefully to justify the investment into a complex new device. Figure 1.1 shows the basic idea of FACTS for transmission systems. The usage of lines for active power transmission should be ideally up to the thermal limits. Voltage and stability limits shall be shifted with the means of the several different FACTS-devices. It can be seen that with growing line length, the opportunity for FACTS-devices gets more and more important.

The influence of FACTS-devices is achieved through switched or controlled shunt compensation, series compensation or phase shift control. The devices work electrically as fast current, voltage or impedance controllers. The power electronic allows very short reaction times down to far below one second.

In the following a structured overview on FACTS-devices is given. These devices are mapped to their different fields of applications. Detailed introductions in FACTS-devices can also be found in the literature [1]-[5] with the main focus on basic technology, modeling and control.

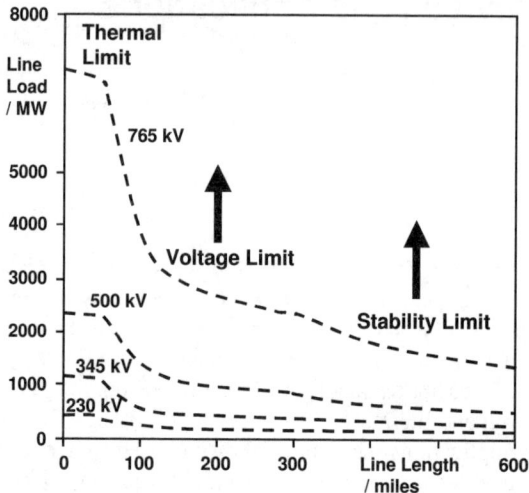

**Fig. 1.1.** Operational limits of transmission lines for different voltage levels

## 1.1 Overview

The development of FACTS-devices has started with the growing capabilities of power electronic components. Devices for high power levels have been made available in converters for high and even highest voltage levels. The overall starting points are network elements influencing the reactive power or the impedance of a part of the power system. Figure 1.2 shows a number of basic devices separated into the conventional ones and the FACTS-devices.

For the FACTS side the taxonomy in terms of 'dynamic' and 'static' needs some explanation. The term 'dynamic' is used to express the fast controllability of FACTS-devices provided by the power electronics. This is one of the main differentiation factors from the conventional devices. The term 'static' means that the devices have no moving parts like mechanical switches to perform the dynamic controllability. Therefore most of the FACTS-devices can equally be static and dynamic.

The left column in Figure 1.2 contains the conventional devices build out of fixed or mechanically switchable components like resistance, inductance or capacitance together with transformers. The FACTS-devices contain these elements as well but use additional power electronic valves or converters to switch the elements in smaller steps or with switching patterns within a cycle of the alternating current. The left column of FACTS-devices uses Thyristor valves or converters. These valves or converters are well known since several years. They have low losses because of their low switching frequency of once a cycle in the converters or the usage of the Thyristors to simply bridge impedances in the valves.

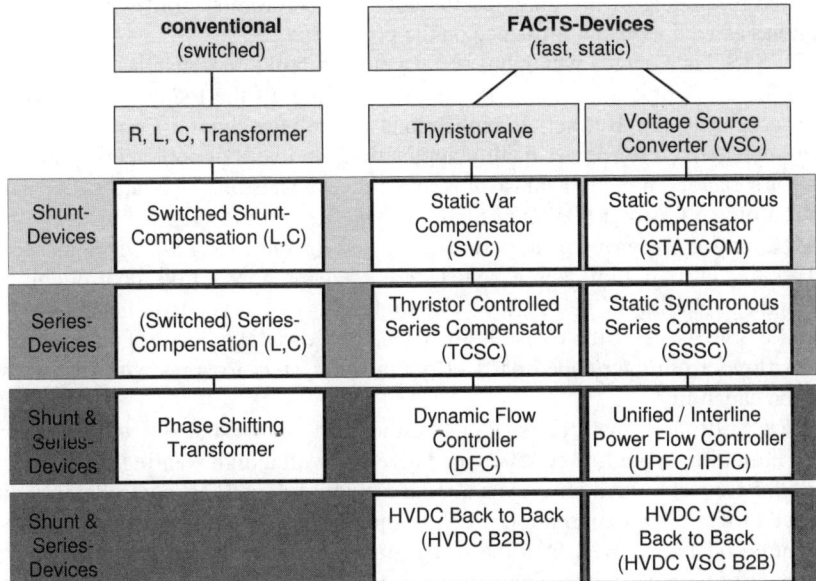

**Fig. 1.2.** Overview of major FACTS-Devices

The right column of FACTS-devices contains more advanced technology of voltage source converters based today mainly on Insulated Gate Bipolar Transistors (IGBT) or Insulated Gate Commutated Thyristors (IGCT). Voltage Source Converters provide a free controllable voltage in magnitude and phase due to a pulse width modulation of the IGBTs or IGCTs. High modulation frequencies allow to get low harmonics in the output signal and even to compensate disturbances coming from the network. The disadvantage is that with an increasing switching frequency, the losses are increasing as well. Therefore special designs of the converters are required to compensate this.

In each column the elements can be structured according to their connection to the power system. The shunt devices are primarily for reactive power compensation and therefore voltage control. The SVC provides in comparison to the mechanically switched compensation a smoother and more precise control. It improves the stability of the network and it can be adapted instantaneously to new situations. The STATCOM goes one step further and is capable of improving the power quality against even dips and flickers.

The series devices are compensating reactive power. With their influence on the effective impedance on the line they have an influence on stability and power flow. These devices are installed on platforms in series to the line. Most manufacturers count Series Compensation, which is usually used in a fixed configuration, as a FACTS-device. The reason is, that most parts and the system setup require the same knowledge as for the other FACTS-devices. In some cases the Series Compensator is protected with a Thyristor-bridge. The application of the TCSC is primarily for damping of inter-area oscillations and therefore stability improvement, but it has as well a certain influence on the power flow.

The SSSC is a device which has so far not been build on transmission level because Series Compensation and TCSC are fulfilling all the today's requirements more cost efficient. But series applications of Voltage Source Converters have been implemented for power quality applications on distribution level for instance to secure factory infeeds against dips and flicker. These devices are called Dynamic Voltage Restorer (DVR) or Static Voltage Restorer (SVR).

More and more growing importance are getting the FACTS-devices in shunt and series configuration. These devices are used for power flow controllability. The higher volatility of power flows due to the energy market activities requires a more flexible usage of the transmission capacity. Power flow control devices shift power flows from overloaded parts of the power system to areas with free transmission capability.

Phase Shifting Transformers (PST) are the most common device in this sector. Their limitation is the low control speed together with a high wearing and maintenance for frequent operation. As an alternative with full and fast controllability the Unified Power Flow Controller (UPFC) is known since several years mainly in the literature and but as well in some test installations. The UPFC provides power flow control together with independent voltage control. The main disadvantage of this device is the high cost level due to the complex system setup. The relevance of this device is given especially for studies and research to figure out the requirements and benefits for a new FACTS-installation. All simpler devices can be derived from the UPFC if their capability is sufficient for a given situation. Derived from the UPFC there are even more complex devices called Interline Power Flow Controller (IPFC) and Generalized Unified Power Flow Controller (GUPFC) which provide power flow controllability in more than one line starting from the same substation.

Between the UPFC and the PST there was a gap for a device with dynamic power flow capability but with a simpler setup than the UPFC. The Dynamic Power Flow Controller (DFC) was introduced recently to fill this gap. The combination of a small PST with Thyristor switched capacitors and inductances provide the dynamic controllability over parts of the control range. The practical requirements are fulfilled good enough to shift power flows in market situations and as well during contingencies.

The last line of HVDC is added to this overview, because such installations are fulfilling all criteria to be a FACTS-device, which is mainly the full dynamic controllability. HVDC Back-to-Back systems allow power flow controllability while additionally decoupling the frequency of both sides. While the HVDC Back-to-Back with Thyristors only controls the active power, the version with Voltage Source Converters allows additionally a full independent controllability of reactive power on both sides. Such a device ideally improves voltage control and stability together with the dynamic power flow control. For sure HVDC with Thyristor or Voltage Source Converters together with lines or cables provide the same functionality and can be seen as very long FACTS-devices.

FACTS-devices are usually perceived as new technology, but hundreds of installations worldwide, especially of SVC since early 1970s with a total installed power of 90.000 MVAr, show the acceptance of this kind of technology. Table 1.1 shows the estimated number of worldwide installed FACTS devices and the estimated total installed power. Even the newer developments like STATCOM or TCSC show a quick growth rate in their specific application areas.

**Table 1.1.** Estimated number of worldwide installed FACTS-devices and their estimated total installed power

| Type | Number | Total Installed Power in MVA |
|---|---|---|
| SVC | 600 | 90.000 |
| STATCOM | 15 | 1.200 |
| Series Compensation | 700 | 350.000 |
| TCSC | 10 | 2.000 |
| HVDC B2B | 41 | 14.000 |
| HVDC VSC B2B | 1 + (7 with cable) | 900 |
| UPFC | 2-3 | 250 |

## 1.2 Power Electronics

Power electronics have a widely spread range of applications from electrical machine drives to excitation systems, industrial high current rectifiers for metal smelters, frequency controllers or electrical trains. FACTS-devices are just one application beside others, but use the same technology trends. It has started with the first Thyristor rectifiers in 1965 and goes to the nowadays modularized IGBT or IGCT voltage source converters.

Without repeating lectures in Semiconductors or Converters, the following sections provide some basic information.

### 1.2.1 Semiconductors

Since the first development of a Thyristor by General Electric in 1957, the targets for power semiconductors are low switching losses for high switching rates and minimal conduction losses. The innovation in the FACTS area is mainly driven by these developments. Today, there are Thyristor and Transistor technologies available. Figure 1.3 shows the ranges of power and voltage for the applications of the specific semiconductors.

The Thyristor is a device, which can be triggered with a pulse at the gate and remains in the on-stage until the next current zero crossing. Therefore only one switching per half-cycle is possible, which limits the controllability.

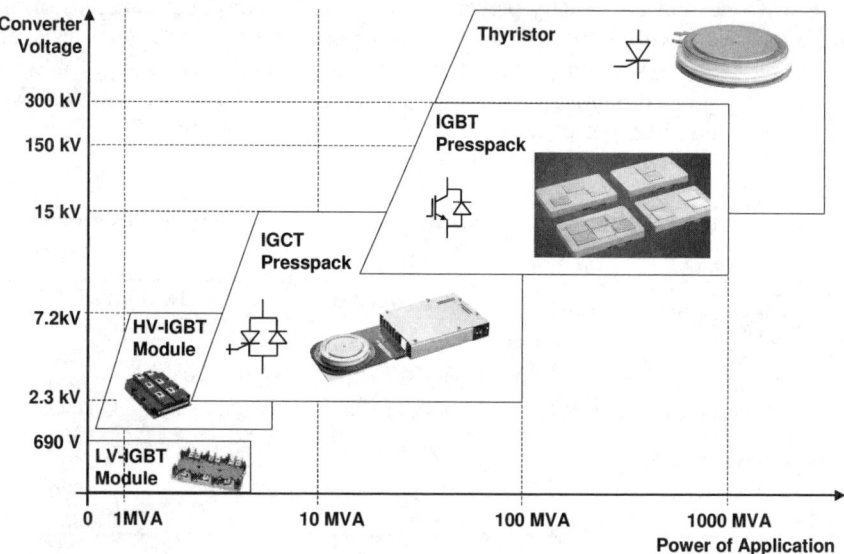

**Fig. 1.3.** Ranges of converter voltages and power of applications for power semiconductors

Thyristors have the highest current and blocking voltage. This means that fewer semiconductors need to be used for an application. Thyristors are used as switches for capacities or inductances, in converters for reactive power compensators or as protection switches for less robust power converters.

The Thyristors are still the devices for applications with the highest voltage and power levels. They are part of the mostly used FACTS-devices up to the biggest HVDC-Transmissions with a voltage level above 500 kV and power above 3000 MVA.

To increase the controllability, GTO-Thyristors have been developed, which can be switched off with a voltage peak at the gate. These devices are nowadays replaced by Insulated Gate Commutated Thyristors (IGCT), which combine the advantage of the Thyristor, the low on stage losses, with low switching losses. These semiconductors are used in smaller FACTS-devices and drive applications.

The Insulated Gate Bipolar Transistor (IGBT) is getting more and more importance in the FACTS area. An IGBT can be switched on with a positive voltage and switched off with a zero voltage. This allows a very simple gate drive unit to control the IGBT. The voltage and power level of the applications is on the way to grow up to 300 kV and 1000 MVA for HVDC with Voltage Source Converters. The IGBT capability covers nowadays the whole range of power system applications.

An important issue for power semiconductors is the packaging to ensure a reliable connection to the gate drive unit. This electronic circuit ensures beside the control of the semiconductor as well its supervision and protection. A development in the Thyristor area tries to trigger the Thyristor with a light signal through an optical fiber. This allows the decoupling of the Semiconductor and the gate drive unit. The advantage is that the electronic circuit can be taken out of the high electromagnetic field close to the Thyristor. The disadvantage is, that the protection of the Thyristor has to be implemented in the Thyristor itself, which leads to an extremely complex component. A supervision of the Thyristor by the gate drive unit is as well impossible in this case, which leads to disadvantages for the entire converter.

A second issue for the packing is the stacking of the semiconductor devices. A number of devices need to be stacked to achieve the required voltage level for the power system application. A mechanically stable packaging needs to ensure an equal current distribution in the semiconductor. Figure 1.4 shows three examples of stacked IGCTs, Thyristors and IGBTs.

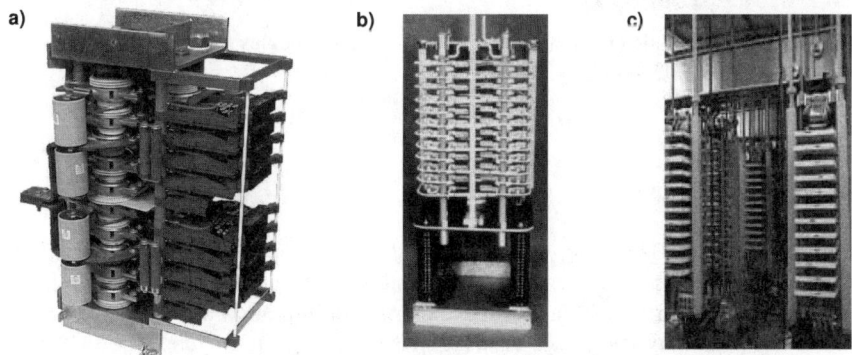

**Fig. 1.4.** Semiconductor stacks, a) Medium Voltage IGCT 3-level topology, 9 MVA power stack, b) SVC Thyristor Valve c) High Voltage IGBT stack for STATCOM (Source: ABB)

As an example the IGBT packaging shall be explained in detail. In Figure 1.5 an IGBT Presspack is shown. Each sub-module contains nine pins of which six are IGBT chips and three are Diode chips. Between two and six sub-modules can be integrated in one frame. The pins are designed to press the chip with a spring on an Aluminum plate. If the entire module is stacked, the sub-modules with the pins are pressed into the frame until the frames are laying tight on each other. With this a well-defined pressure is equally distributed throughout all chips.

Due to the enormous number of chips in power system converter, a single chip failure shall not lead to a disturbance of the entire FACTS-device. In the case of a short circuit of a chip it is melting together with the Aluminum plate providing a long-term stable short circuit of the module. The converter is designed in a way that more modules are stacked than necessary, so that between maintenance intervals a defined number can fail.

All these developments in the power semiconductor and its packaging area lead to reliable system setups today.

8    1 FACTS-Devices and Applications

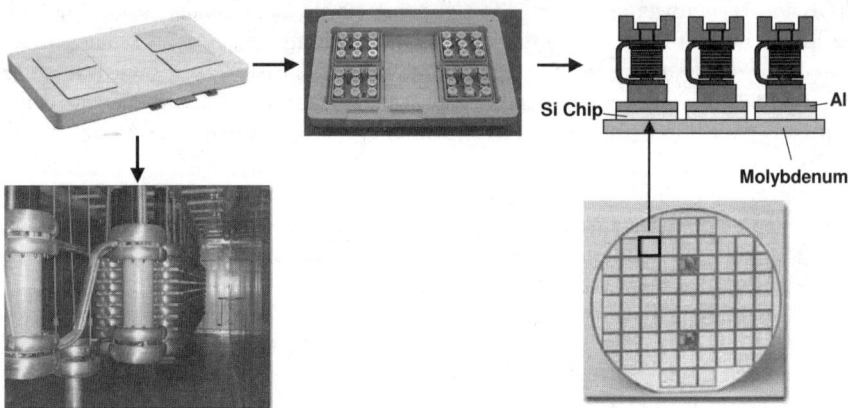

**Fig. 1.5.** IGBT-Module (1 kA$_{rms}$, 2.5 kV) with four sub-modules for Voltage Source Converter (+/-150kV$_{DC}$, 300 MVA) (Source: ABB)

### 1.2.2 Power Converters

Starting with the Thyristor, it can be used most simply as a switch. Thyristor switched capacities or inductances are possible applications. The next step is the Thyristor converter as shown in a most simple configuration in Figure 1.6. In this half-bridge the Thyristors can be triggered once in a half-cycle. The next zero crossing will block the Thyristor. In an ideal case, where the feeding inductance on the DC side is infinity, the output AC current is rectangular, which means it has a high harmonic content. But due the small number of switchings, the switching losses are low. The operational diagram is a half cycle, which means, that the active power flow can be controlled, but the reactive power is fixed with a certain ratio.

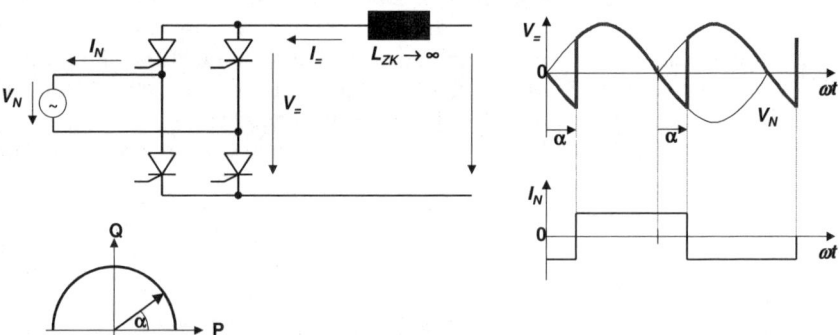

**Fig. 1.6.** Thyristor half-bridge converter and operational diagram

To overcome these disadvantages for FACTS-applications, where the controllability as well of reactive power is a prime target, on and off switchable devices must be used. Figure 1.7 shows on the left a half bridge with IGBTs. The same setup is valid as well for GTO-Thyristors or IGCT's.

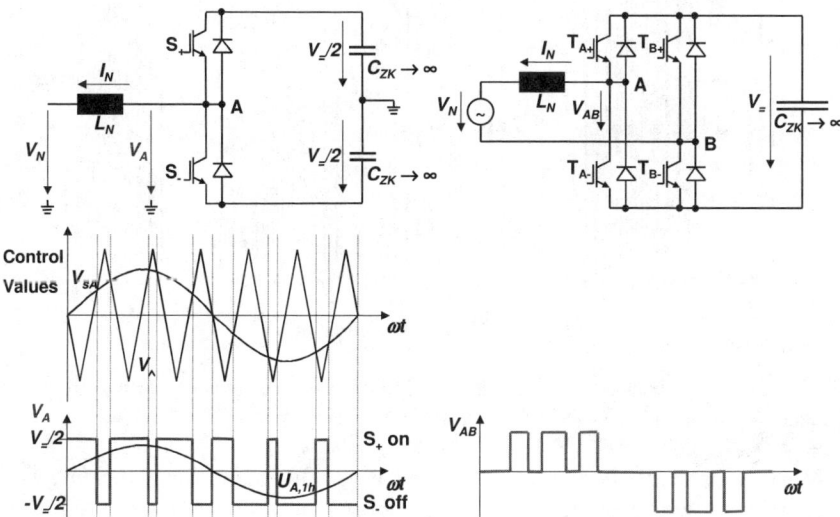

**Fig. 1.7.** 2-Level voltage source converter with pulse width modulation, left: Half-bridge, right: TWIN-circuit

A suitable switching patern must be defined for the switch-on-and-off capability. The simplest solution is the combination of a triangular voltage with a reference voltage as control values. The changing sign of the difference of both signals triggers the IGBTs alternately. The output voltage is jumping between both maximums. With an increasing number of switchings the harmonic content is decreasing.

On the right hand side a TWIN converter uses two IGBT bridges. The output is the voltage between the midpoints. Three stages, plus, minus and zero, are now possible and reducing the harmonics further. This pattern can be achieved as well with a three level converter, where four IGBT and six Diodes are used in the simple bridge.

While the increasing number of switching reduces the harmonics, the switching losses are increasing. For practical applications a compromise between harmonics, which means output filtering, and losses must be found. For HVDC converters, the losses of one converter station are around 1% for Thyristor converters and a little above 2% for IGBT Voltage Source Converters. A switching pattern of an IGBT 2-level converter is shown in Figure 1.8. A special switching scheme, called harmonic cancellation, is applied here. During some time intervals the switching is interrupted to reduce harmonics.

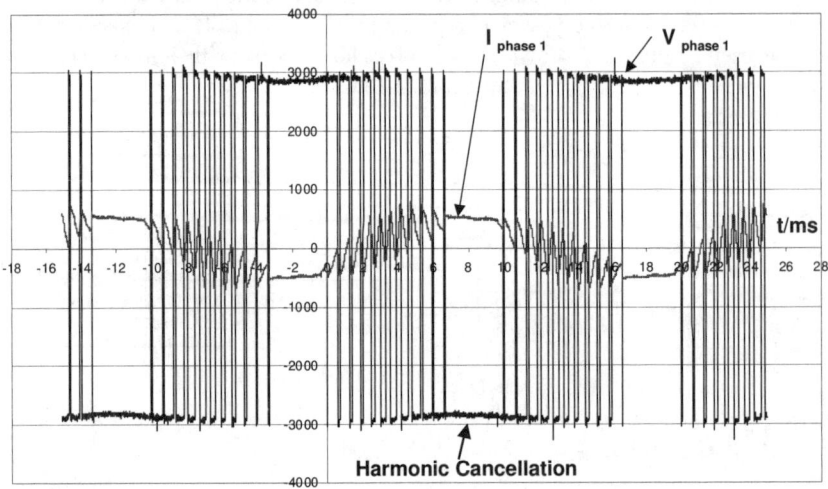

**Fig. 1.8.** Output current and voltage of 2-level voltage source converter with pulse width modulation and harmonic cancellation, modulation frequency 21 f ($V_N$)

More complex converters are proposed in the literature, but the number of semiconductor elements increases the cost more than loss or harmonic reduction would justify.

## 1.3 Configurations of FACTS-Devices

### 1.3.1 Shunt Devices

The most used FACTS-device is the SVC or the version with Voltage Source Converter called STATCOM. These shunt devices are operating as reactive power compensators. The main applications in transmission, distribution and industrial networks are:

- reduction of unwanted reactive power flows and therefore reduced network losses,
- keeping of contractual power exchanges with balanced reactive power,
- compensation of consumers and improvement of power quality especially with huge demand fluctuations like industrial machines, metal melting plants, railway or underground train systems,
- compensation of Thyristor converters e.g. in conventional HVDC lines,
- improvement of static or transient stability.

Almost half of the SVC and more than half of the STATCOMs are used for industrial applications. Industry as well as commercial and domestic groups of users require power quality. Flickering lamps are no longer accepted, nor are interruptions of industrial processes due to insufficient power quality. For example demands for increased steel production and rules for network disturbances have, together with increasing cost of energy, made reactive power compensation a requirement in the steel industry. A special attention is given to weak network connections with severe voltage support problems.

A steel melting process demands a stable and steady voltage support for the electric arc furnace. With dynamic reactive power compensation, the random voltage variations characterized by an arc furnace are minimized. The minimized voltage variations are achieved by continuously compensating the reactive power consumption from the arc furnace. The result is an overall improvement of the furnace operation, which leads to better process and production economy.

Railway or underground systems with huge load variations require SVCs or STATCOMs similar to the application above. SVC or STATCOM for even stricter requirements on power quality are used in other kinds of critical factory processes, like electronic or semiconductor productions.

A growing area of application is the renewable or distributed energy sector. Especially offshore wind farms with its production fluctuation have to provide a balanced reactive power level and keep the voltage limitations within the wind farm, but as well on the interconnection point with the main grid. A lot distributed generation devices are interconnected with the grid through a voltage source converter similar to the STATCOM fulfilling all requirements on a stable network operation.

### *1.3.1.1 SVC*

Electrical loads both generate and absorb reactive power. Since the transmitted load varies considerably from one hour to another, the reactive power balance in a grid varies as well. The result can be unacceptable voltage amplitude variations or even a voltage depression, at the extreme a voltage collapse. A rapidly operating Static Var Compensator (SVC) can continuously provide the reactive power required to control dynamic voltage oscillations under various system conditions and thereby improve the power system transmission and distribution stability. Installing an SVC at one or more suitable points in the network can increase transfer capability and reduce losses while maintaining a smooth voltage profile under different network conditions. In addition an SVC can mitigate active power oscillations through voltage amplitude modulation.

SVC installations consist of a number of building blocks. The most important is the Thyristor valve, i.e. stack assemblies of series connected anti-parallel Thyristors to provide controllability. Air core reactors and high voltage AC capacitors are the reactive power elements used together with the Thyristor valves. The step-up connection of this equipment to the transmission voltage is achieved through a power transformer. The Thyristor valves together with auxiliary systems are located indoors in an SVC building, while the air core reactors and capacitors, together with the power transformer are located outdoors.

12    1 FACTS-Devices and Applications

In principle the SVC consists of Thyristor Switched Capacitors (TSC) and Thyristor Switched or Controlled Reactors (TSR / TCR). The coordinated control of a combination of these branches varies the reactive power as shown in Figure 1.9.

The first commercial SVC was installed in 1972 for an electric arc furnace. On transmission level the first SVC was used in 1979. Since then it is widely used and the most accepted FACTS-device. A recent installation is shown in Figure 1.10.

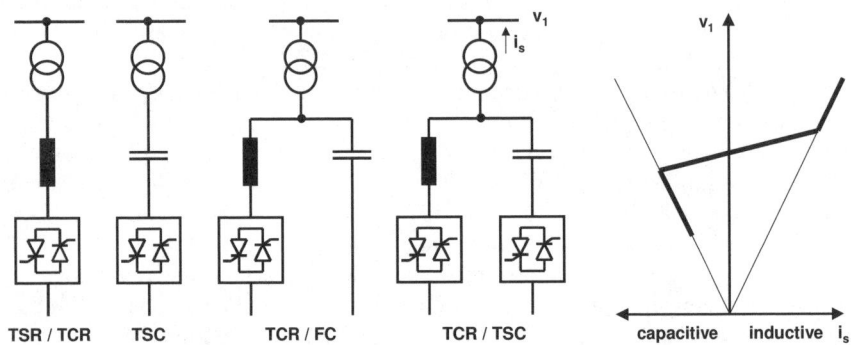

**Fig. 1.9.** SVC building blocks and voltage / current characteristic

**Fig. 1.10.** SVC (Source: ABB)

### 1.3.1.2 STATCOM

In 1999 the first SVC with Voltage Source Converter called STATCOM (STATic COMpensator) went into operation. The STATCOM has a characteristic similar to the synchronous condenser, but as an electronic device it has no inertia and is superior to the synchronous condenser in several ways, such as better dynamics, a lower investment cost and lower operating and maintenance costs.

A STATCOM is build with Thyristors with turn-off capability like GTO or today IGCT or with more and more IGBTs. The structure and operational characteristic is shown in Figure 1.11. The static line between the current limitations has a certain steepness determining the control characteristic for the voltage. The advantage of a STATCOM is that the reactive power provision is independent from the actual voltage on the connection point. This can be seen in the diagram for the maximum currents being independent of the voltage in comparison to the SVC in Figure 1.9. This means, that even during most severe contingencies, the STATCOM keeps its full capability.

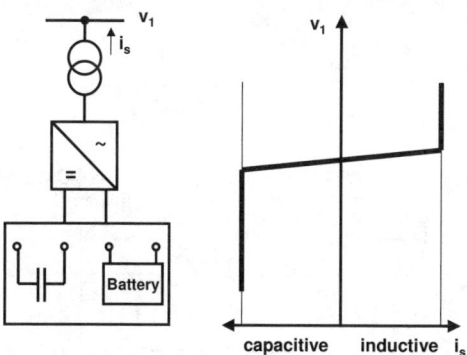

**Fig. 1.11.** STATCOM structure and voltage / current characteristic

In the distributed energy sector the usage of Voltage Source Converters for grid interconnection is common practice today. The next step in STATCOM development is the combination with energy storages on the DC-side. The performance for power quality and balanced network operation can be improved much more with the combination of active and reactive power

Figure 1.12 to Figure 1.14 show a typical STATCOM layout on transmission level as part of a substation.

14    1 FACTS-Devices and Applications

**Fig. 1.12.** Substation with a STATCOM (Source: ABB)

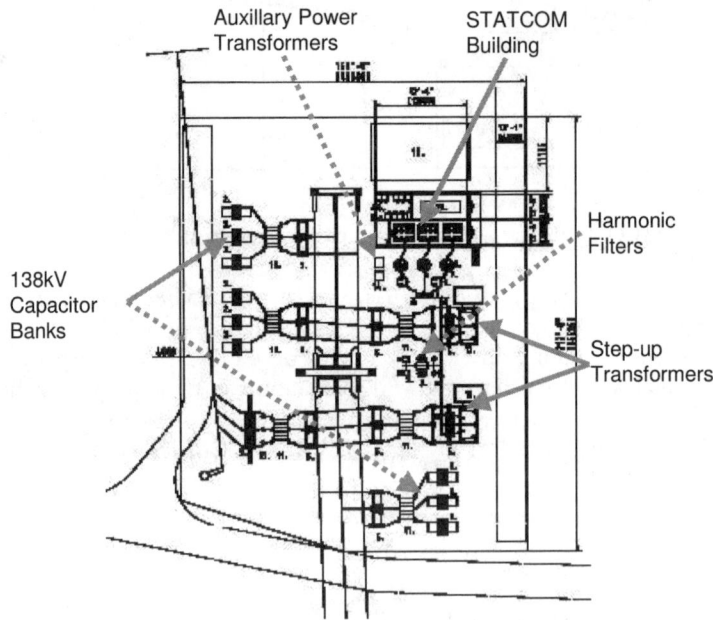

**Fig. 1.13.** Typical substation layout with STATCOM (Source: ABB)

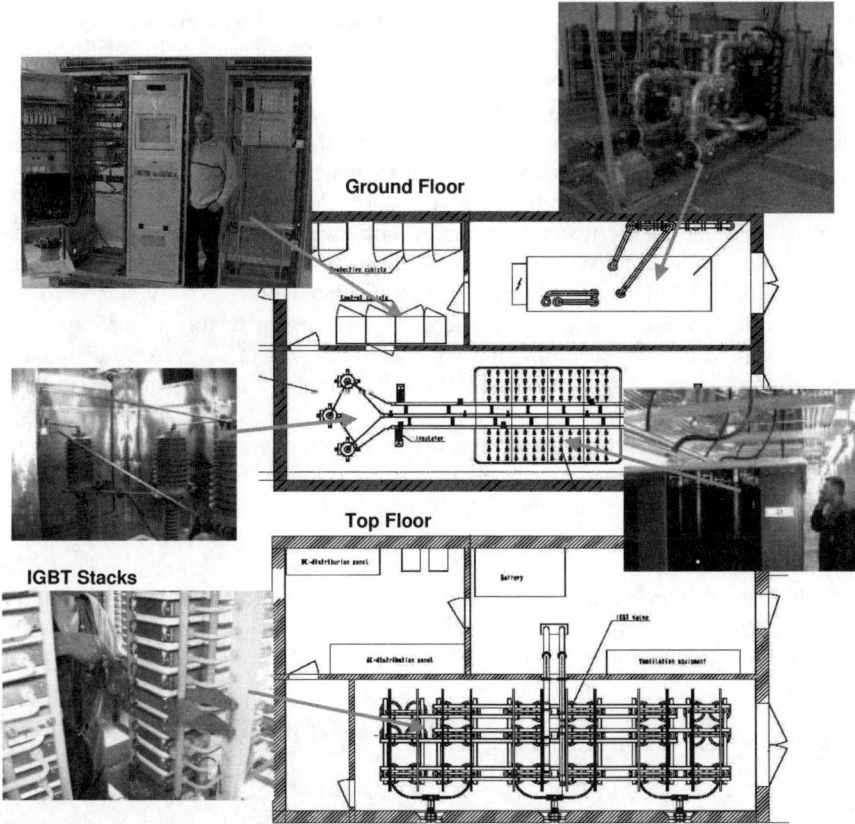

**Fig. 1.14.** Typical layout of a STATCOM-building (Source: ABB)

### 1.3.2 Series Devices

Series devices have been further developed from fixed or mechanically switched compensations to the Thyristor Controlled Series Compensation (TCSC) or even Voltage Source Converter based devices. The main applications are:

- reduction of series voltage decline in magnitude and angle over a power line,
- reduction of voltage fluctuations within defined limits during changing power transmissions,
- improvement of system damping resp. damping of oscillations,
- limitation of short circuit currents in networks or substations,
- avoidance of loop flows resp. power flow adjustments.

### 1.3.2.1 Series Compensation

The world's first Series Compensation on transmission level, counted nowadays by the manufacturers as a FACTS-device, went into operation in 1950. Series Compensation is used in order to decrease the transfer reactance of a power line at rated frequency. A series capacitor installation generates reactive power that in a self-regulating manner balances a fraction of the line's transfer reactance. The result is that the line is electrically shortened, which improves angular stability, voltage stability and power sharing between parallel lines.

Series Capacitors are installed in series with a transmission line, which means that all the equipment has to be installed on a fully insulated platform. On this steel platform the main capacitor is located together with the overvoltage protection circuits. The overvoltage protection is a key design factor, as the capacitor bank has to withstand the throughput fault current, even at a severe nearby fault. The primary overvoltage protection typically involves non-linear varistors of metal-oxide type, a spark gap and a fast bypass switch. Secondary protection is achieved with ground mounted electronics acting on signals from optical current transducers in the high voltage circuit.

Even if the device is known since several years, improvements are ongoing. One recent achievement is the usage of dry capacitors with a higher energy density and higher environmental friendliness. As a primary protection Thyristor switchs can be used, but cheaper alternatives with almost the same capability based on triggered spark gaps and special breakers without power electronics have recently been developed.

**Fig. 1.15.** Series Compensation (Series Capacitor) (Source: ABB)

A special application of Series Compensation can be achieved by combining it with a series reactance to get a fault current limiter. Both components are neutralizing each other in normal operation. In the case of a fault, die Series Compensation is bridged with a fast protection device or a Thyristor bridge. The remaining reactance is limiting the fault current. Pilot installations of such a system configuration are already in use.

### 1.3.2.2 TCSC

Thyristor Controlled Series Capacitors (TCSC) address specific dynamical problems in transmission systems. Firstly it increases damping when large electrical systems are interconnected. Secondly it can overcome the problem of Sub-Synchronous Resonance (SSR), a phenomenon that involves an interaction between large thermal generating units and series compensated transmission systems. The TCSC's high speed switching capability provides a mechanism for controlling line power flow, which permits increased loading of existing transmission lines, and allows for rapid readjustment of line power flow in response to various contingencies. The TCSC also can regulate steady-state power flow within its rating limits.

From a principal technology point of view, the TCSC resembles the conventional series capacitor. All the power equipment is located on an isolated steel platform, including the Thyristor valve that is used to control the behavior of the main capacitor bank. Likewise the control and protection is located on ground potential together with other auxiliary systems. Figure 1.16 shows the principle setup of a TCSC and its operational diagram. The firing angle and the thermal limits of the Thyristors determine the boundaries of the operational diagram.

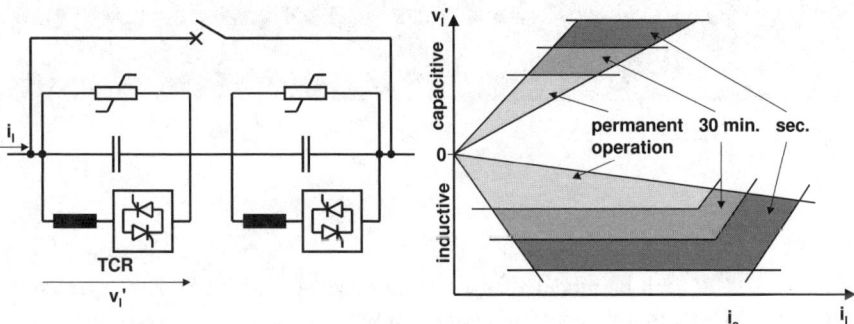

**Fig. 1.16.** Principle setup and operational diagram of a Thyristor Controlled Series Compensation (TCSC)

The main principles of the TCSC concept are two; firstly, to provide electromechanical damping between large electrical systems by changing the reactance of a specific interconnecting power line, i.e. the TCSC will provide a variable capacitive reactance. Secondly, the TCSC shall change its apparent impedance (as seen by the line current) for sub-synchronous frequencies, such that a prospective sub-synchronous resonance is avoided. Both objectives are achieved with the TCSC, using control algorithms that work concurrently. The controls will function on the Thyristor circuit in parallel to the main capacitor bank such that controlled charges are added to the main capacitor, making it a variable capacitor at fundamental frequency but a "virtual inductor" at sub-synchronous frequencies. Figure 1.17 shows a TCSC on transmission level. The first TCSC was commissioned in 1996.

**Fig. 1.17.** TCSC (Source: ABB)

### 1.3.2.3 SSSC

While the TCSC can be modeled as a series impedance, the SSSC is a series voltage source. The principle configuration is shown in Figure 1.18, which looks basically the same as the STATCOM. But in reality this device is more complicated because of the platform mounting and the protection. A Thyristor protection is absolutely necessary, because of the low overload capacity of the semiconductors, especially when IGBTs are used.

The voltage source converter plus the Thyristor protection makes the device much more costly, while the better performance cannot be used on transmission level. The picture is quite different if we look into power quality applications. This device is then called Dynamic Voltage Restorer (DVR). The DVR is used to keep the voltage level constant, for example in a factory infeed. Voltage dips and flicker can be mitigated. The duration of the action is limited by the energy stored in the DC capacitor. With a charging mechanism or battery on the DC side, the device could work as an uninterruptible power supply. A picture of a modularized installation with 22 MVA is shown on the right in Figure 1.18.

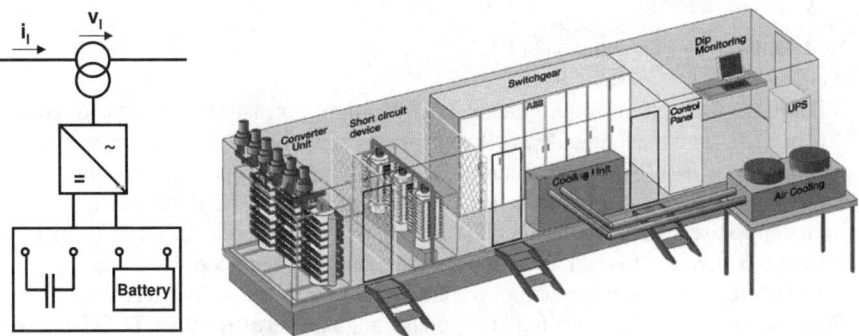

**Fig. 1.18.** Principle setup of SSSC and implementation as DVR for power quality applications (Source: ABB)

### 1.3.3 Shunt and Series Devices

Power flow capability is getting more and more importance with the growing restrictions for new power lines and the more volatile power flow due to the energy market activities.

#### *1.3.3.1 Dynamic Flow Controller*

A new device in the area of power flow control is the Dynamic Power Flow Controller (DFC). The DFC is a hybrid device between a Phase Shifting Transformer (PST) and switched series compensation.

A functional single line diagram of the Dynamic Flow Controller is shown in Figure 1.19. The Dynamic Flow Controller consists of the following components:

- a standard phase shifting transformer with tap-changer (PST)
- series-connected Thyristor Switched Capacitors and Reactors (TSC / TSR)
- A mechanically switched shunt capacitor (MSC). (This is optional depending on the system reactive power requirements)

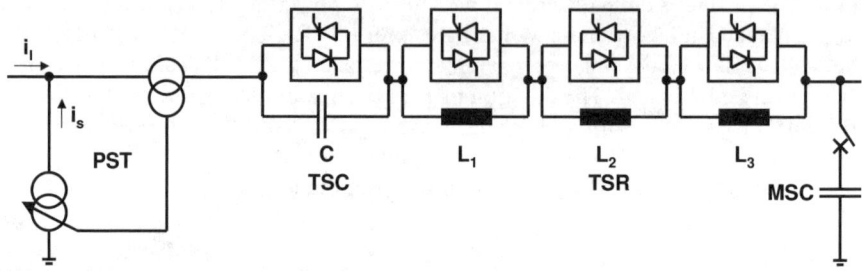

**Fig. 1.19.** Principle configuration of DFC

Based on the system requirements, a DFC might consist of a number of series TSC or TSR. The mechanically switched shunt capacitor (MSC) will provide voltage support in case of overload and other conditions. Normally the reactances of reactors and the capacitors are selected based on a binary basis to result in a desired stepped reactance variation. If a higher power flow resolution is needed, a reactance equivalent to the half of the smallest one can be added.

The switching of series reactors occurs at zero current to avoid any harmonics. However, in general, the principle of phase-angle control used in TCSC can be applied for a continuous control as well. The operation of a DFC is based on the following rules:

- TSC / TSR are switched when a fast response is required.
- The relieve of overload and work in stressed situations is handled by the TSC / TSR.
- The switching of the PST tap-changer should be minimized particularly for the currents higher than normal loading.
- The total reactive power consumption of the device can be optimized by the operation of the MSC, tap changer and the switched capacities and reactors.

In order to visualize the steady state operating range of the DFC, we assume an inductance in parallel representing parallel transmission paths. The overall control objective in steady state would be to control the distribution of power flow between the branch with the DFC and the parallel path. This control is accomplished by control of the injected series voltage.

The PST (assuming a quadrature booster) will inject a voltage in quadrature with the node voltage. The controllable reactance will inject a voltage in quadrature with the throughput current. Assuming that the power flow has a load factor close to one, the two parts of the series voltage will be close to collinear. However, in terms of speed of control, influence on reactive power balance and effectiveness at high/low loading the two parts of the series voltage has quite different characteristics. The steady state control range for loadings up to rated current is illustrated in Figure 1.20, where the x-axis corresponds to the throughput current and the y-axis corresponds to the injected series voltage.

## 1.3 Configurations of FACTS-Devices

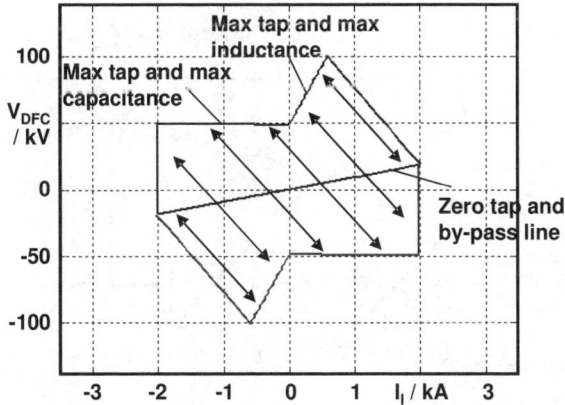

**Fig. 1.20.** Operational diagram of a DFC

Operation in the first and third quadrants corresponds to reduction of power through the DFC, whereas operation in the second and fourth quadrants corresponds to increasing the power flow through the DFC. The slope of the line passing through the origin (at which the tap is at zero and TSC / TSR are bypassed) depends on the short circuit reactance of the PST.

Starting at rated current (2 kA) the short circuit reactance by itself provides an injected voltage (approximately 20 kV in this case). If more inductance is switched in and/or the tap is increased, the series voltage increases and the current through the DFC decreases (and the flow on parallel branches increases). The operating point moves along lines parallel to the arrows in the figure. The slope of these arrows depends on the size of the parallel reactance. The maximum series voltage in the first quadrant is obtained when all inductive steps are switched in and the tap is at its maximum.

Now, assuming maximum tap and inductance, if the throughput current decreases (due e.g. to changing loading of the system) the series voltage will decrease. At zero current, it will not matter whether the TSC / TSR steps are in or out, they will not contribute to the series voltage. Consequently, the series voltage at zero current corresponds to rated PST series voltage. Next, moving into the second quadrant, the operating range will be limited by the line corresponding to maximum tap and the capacitive step being switched in (and the inductive steps by-passed). In this case, the capacitive step is approximately as large as the short circuit reactance of the PST, giving an almost constant maximum voltage in the second quadrant.

### *1.3.3.2 Unified Power Flow Controller*

The UPFC is a combination of a static compensator and static series compensation. It acts as a shunt compensating and a phase shifting device simultaneously.

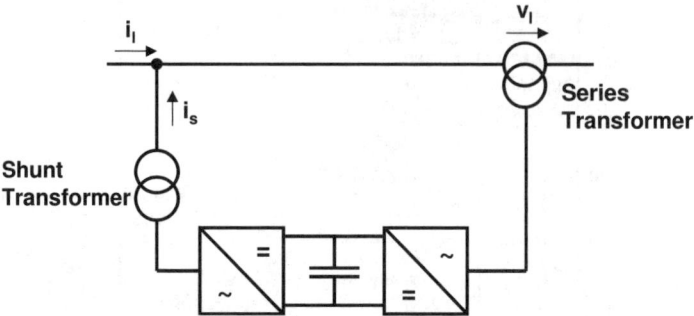

**Fig. 1.21.** Principle configuration of an UPFC

The UPFC consists of a shunt and a series transformer, which are connected via two voltage source converters with a common DC-capacitor. The DC-circuit allows the active power exchange between shunt and series transformer to control the phase shift of the series voltage. This setup, as shown in Figure 1.21, provides the full controllability for voltage and power flow. The series converter needs to be protected with a Thyristor bridge. Due to the high efforts for the Voltage Source Converters and the protection, an UPFC is getting quite expensive, which limits the practical applications where the voltage and power flow control is required simultaneously.

### 1.3.3.3 Interline Power Flow Controller

One of the latest FACTS-devices is named convertible static compensator (CSC) and was recently installed as a pilot by the New York Power Authority (NYPA) [6][7]. The CSC-project shall increase power transfer capability and maximise the use of the existing transmission network. Within the general conceptual framework of the CSC, two multi-converter FACTS-devices, the Interline Power Flow Controller (IPFC) [8] and the Generalized Unified Power Flow Controller (GUPFC) [9] (see section 1.3.5), are among many possible configurations. The target is to control power flows of multi-lines or a subnetwork rather than control the power flow of a single line by for instance DFC or UPFC. The IPFC combines two or more series converters and the GUPFC combines one shunt converter and two or more series converters. The current NYPA's CSC installation is a two-converter one and can operate as an IPFC but not as a GUPFC.

When the power flows of two lines starting in one substation need to be controlled, an Interline Power Flow Controller (IPFC) can be used. The IPFC consists of two series VSCs whose DC capacitors are coupled. This allows active power to circulate between the VSCs. Figure 1.22 shows the principle configuration of an IPFC. With this configuration two lines can be controlled simultaneously to optimize the network utilization. In general, due to its complex setup, specific application cases need to be identified justifying the investment.

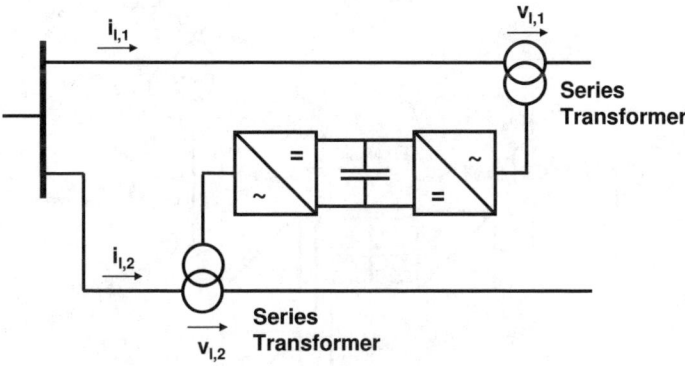

**Fig. 1.22.** Principle configuration of an IPFC

### 1.3.3.4 Generalized Unified Power Flow Controller

The GUPFC combines three or more shunt and series converters [9]. It extends the concept of voltage and power flow control beyond what is achievable with the known two-converter UPFC. The simplest GUPFC consists of three converters, one connected in shunt and the other two in series with two transmission lines in a substation. Figure 1.23 shows the principle configuration. The basic GUPFC can control total five power system quantities such as a bus voltage and independent active and reactive power flows of two lines.

The concept of GUPFC can be extended for more lines if necessary. The device may be installed in some central substations to manage power flows of multi-lines or a group of lines and provide voltage support as well. By using GUPFC-devices, the transfer capability of transmission lines can be increased significantly. Further more, by using the multi-line management capability of the GUPFC, active power flows on lines can not only be increased, but also be decreased with respect to operating and market transaction requirements. In general the GUPFC can be used to increase transfer capability and relieve congestions in a flexible way.

The complexity of its configuration and control scheme needs specific applications cases.

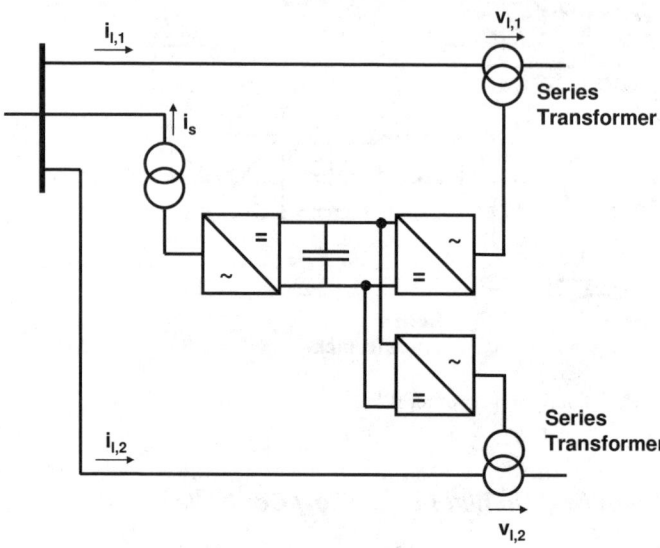

**Fig. 1.23.** Principle configuration of a GUPFC

### 1.3.4 Back-to-Back Devices

The Back-to-Back devices provide in general a full power flow controllability and power flow limitation. An overload of these devices is therefore impossible. They can resist cascading outages, which might occur due to line outages when one line after the other is overloaded. This gives a great benefit even if the frequency decoupling characteristic is not needed.

Conventional HVDC Back-to-Back systems with Thyristor converters need space consuming filters to reduce the harmonic distortion. The reactive power is not controllable. These devices are mainly used when two asynchronous networks need to be coupled or in the usual application as power transmission line over long distances.

The HVDC with Voltage Source Converters instead provides benefits as well within synchronous operated networks. It has a much smaller footprint and provides the full voltage controllability to the network on both ends. Therefore it can be operated in addition to the power flow control as two STATCOMS. On both ends a full four quadrant circular operational diagram is provided. This reactive power provision can be used to increase the transmission capability of surrounding transmission lines in addition to balancing the power flow.

Figure 1.24 shows the principle configuration of a HVDC Back-to-Back with Voltage Source Converters. A practical implementation is shown in Figure 1.25, which is based on the design of two STATCOM converters with IGBTs.

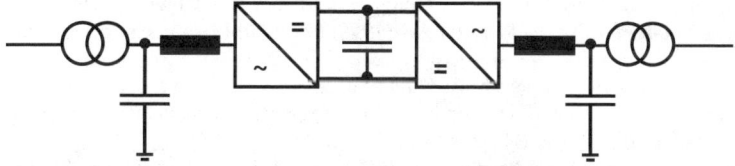

**Fig. 1.24.** Schematic configuration of a HVDC Back-to-Back with Voltage Source Converters

**Fig. 1.25.** HVDC Back-to-Back with Voltage Source Converters, 2 x 36 MVA (Source: ABB)

## References

[1] Sood VK (2004) HVDC and FACTS Controllers: Applications of Static Converters in Power Systems. Kluwer Academic Publishers
[2] Acha E, Fuerte-Esquivel C, Ambiz-Perez H (2004) FACTS Modelling and Simulation in Power Networks. John Wiley & Sons
[3] Mathur RM, Varma RK (2002) Thyristor Based FACTS Controllers for Electrical Transmission Systems. IEEE Computer Society Press
[4] Hingorani NG, Gyugyi L (1999) Understanding FACTS: Concepts and Technology of Flexible AC Transmission Systems. IEEE Computer Society Press
[5] Song YH, Johns T (1999) Flexible Ac Transmission Systems (Facts). IEE Power Series 30

[6] Fardanesh B, Henderson M, Shperling B, Zelingher S, Gyugyi L, Schauder C, Lam B, Moundford J, Adapa R, Edris AA (1998) Convertible static compensator: application to the New York transmission system. CIGRE 14-103, Paris, France
[7] Wei X, Chow JH, Fardanesh, B, Edris AA (2004) A common modeling framework of voltage sourced converters for load flow, sensitivity, and dispatch analysis. IEEE Transactions on Power Systems, vol 19, no 2
[8] Gyugyi L, Sen KK, Schauder CD (1999) The Interline Power Flow Controller: A New Approach to Power Flow Management in Transmission Systems. IEEE Transaction on Power Delivery, vol 14, no 3, pp 1115–1123
[9] Fardanesh B, Shperling B, Uzunovic E, Zelingher S (2000) Multi-Converter FACTS Devices: the Generalized Unified Power Flow Controller (GUPFC). Proc. IEEE PES Summer Meeting, Seattle, USA

# 2 Modeling of Multi-Functional Single Converter FACTS in Power Flow Analysis

This chapter discusses the recent developments in modeling of multi-functional single converter FACTS-devices in power flow analysis. The objectives of this chapter are:

1. to model the well-recognized FACTS devices such as STATOM, SVC, SSSC and TCSC in power flow analysis,
2. to establish multi-control functional models of these FACTS-devices,
3. to handle various internal and external operating constraints of FACTS-devices.

## 2.1 Power Flow Calculations

### 2.1.1 Power Flow Methods

It is well known that power flow calculations are the most frequently performed routine power network calculations, which can be used in power system planning, operational planning, and operation/control. It is also considered as the fundamental of power system network calculations. The calculations are required for the analysis of steady-state as well as dynamic performance of power systems.

In the past, various power flow solution methods such as impedance matrix methods, Newton-Raphson methods, decoupled Newton power flow methods, etc have been proposed [1]-[13]. Among the power flow methods proposed, the Newton's methods using sparse matrix elimination techniques [14] have been considered as the most efficient power flow solution techniques for large-scale power system analysis. A detailed review of power flow methods can be found in [1]. In this chapter, FACTS models for power flow analysis as well as the implementation of these models in Newton power flow will be discussed in detail.

### 2.1.2 Classification of Buses

In power flow analysis, all buses can be classified into the following categories:

**Slack bus.** At a slack bus, the voltage angle and magnitude are specified while the active and reactive power injections are unknown. The voltage angle of the slack bus is taken as the reference for the angles of all other buses. Usually there is only one slack bus in a system. However, in some production grade programs, it may be possible to include more than one bus as distributed slack buses.

**PV Buses.** At a PV bus, the active power injection and voltage magnitude are specified while the voltage angle and reactive power injection are unknown. Usually buses of generators, synchronous condensers are considered as PV buses.

**PQ Buses.** At a PQ bus, the active and reactive power injections are specified while the voltage magnitude and angle at the bus are unknown. Usually a load bus is considered as a PQ bus.

### 2.1.3 Newton-Raphson Power Flow in Polar Coordinates

The Newton-Raphson Power Flow can be formulated either in polar coordinates or in rectangular coordinates. In this chapter, the implementation of FACTS models in the Newton-Raphson power flow will be discussed since the popularity of the methods. Basically, the Newton-Raphson power flow equations in polar coordinates may be given by [1]:

$$\begin{bmatrix} \dfrac{\partial \Delta P}{\partial \theta} & \dfrac{\partial \Delta P}{\partial V} \\ \dfrac{\partial \Delta Q}{\partial \theta} & \dfrac{\partial \Delta Q}{\partial V} \end{bmatrix} \begin{bmatrix} \Delta \theta \\ \Delta V \end{bmatrix} = \begin{bmatrix} -\Delta P \\ -\Delta Q \end{bmatrix} \tag{2.1}$$

where $\Delta P$ and $\Delta Q$ are bus active and reactive power mismatches while $\theta$ and $V$ are bus magnitude and angle, respectively.

## 2.2 Modeling of Multi-Functional STATCOM

In recent years, energy, environment, deregulation of power utilities have delayed the construction of both generation facilities and new transmission lines. These problems have necessitated a change in the traditional concepts and practices of power systems. There are emerging technologies available, which can help electric companies to deal with above problems. One of such technologies is Flexible AC Transmission System (FACTS) [15][16]. As discussed in Chapter 1, within the family of the converter based FACTS, there are a number of FACTS devices available, including the Static Synchronous Compensator (STATCOM) [17], the Static Synchronous Series Compensator (SSSC) [18][19], the Unified Power Flow Controller (UPFC) [20][21], and the latest FACTS devices [22]-[32], etc.

Among the converter based FACTS-devices, STATCOM may be one of the popular FACTS-devices, which has many installations in electric utilities worldwide. Considering the practical applications of the STATCOM in power systems, it is of importance and interest to investigate the possible multi-control functions of the STATCOM as well as model these functions in power system steady state operation and control, such that the various control capabilities can be fully employed, and the benefits of applications of the STATCOM may be fully realized. Nine multi-control functions of the STATCOM will be presented:

- There are two solutions associated with the current magnitude control function, which are discussed. Alternative formulations of the control function to avoid the multiple solutions of the current magnitude control are proposed. Two reactive power control functions are proposed, which are interesting and attractive, and they can be used in either normal control or security control of deregulated electric power systems.
- Full consideration of the current and voltage operating constraints associated with the STATCOM and their detailed implementation in Newton power flow will be described. Effort is particularly made on the enforcement of simultaneous multiple violated internal and external constraints associated with the STATCOM. A strategy will be presented to deal with the multiple constraints enforcement problem.

### 2.2.1 Multi-Control Functional Model of STATCOM for Power Flow Analysis

#### 2.2.1.1 Operation Principles of the STATCOM

A STATCOM is usually used to control transmission voltage by reactive power shunt compensation. Typically, a STATCOM consists of a coupling transformer, an inverter and a DC capacitor, which is shown in Fig. 1.11. For such an arrangement, in ideal steady state analysis, it can be assumed that the active power exchange between the AC system and the STATCOM can be neglected, and only the reactive power can be exchanged between them.

#### 2.2.1.2 Power Flow Constraints of the STATCOM

Based on the operating principle of the STATCOM, the equivalent circuit can be derived, which is given in Fig. 2.1. In the derivation, it is assumed that (a) harmonics generated by the STATCOM are neglected; (b) the system as well as the STATCOM are three phase balanced.

Then the STATCOM can be equivalently represented by a controllable fundamental frequency positive sequence voltage source $V_{sh}$. In principle, the STATCOM output voltage can be regulated such that the reactive power of the STATCOM can be changed.

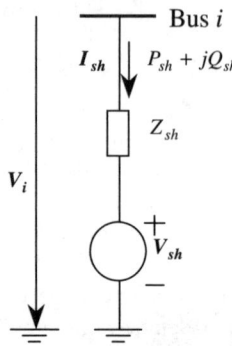

**Fig. 2.1.** STATCOM equivalent circuit

According to the equivalent circuit of the STATCOM shown in Fig. 2.1, suppose $V_{sh} = V_{sh} \angle \theta_{sh}$, $V_i = V_i \angle \theta_i$, then the power flow constraints of the STATCOM are:

$$P_{sh} = V_i^2 g_{sh} - V_i V_{sh}(g_{sh} \cos(\theta_i - \theta_{sh}) + b_{sh} \sin(\theta_i - \theta_{sh})) \quad (2.2)$$

$$Q_{sh} = -V_i^2 b_{sh} - V_i V_{sh}(g_{sh} \sin(\theta_i - \theta_{sh}) - b_{sh} \cos(\theta_i - \theta_{sh})) \quad (2.3)$$

where $g_{sh} + jb_{sh} = 1/Z_{sh}$.

The operating constraint of the STATCOM is the active power exchange via the DC-link as described by:

$$PE = \text{Re}(V_{sh} I_{sh}^*) = 0 \quad (2.4)$$

where $\text{Re}(V_{sh} I_{sh}^*) = V_{sh}^2 g_{sh} - V_i V_{sh}(g_{sh} \cos(\theta_i - \theta_{sh}) - b_{sh} \sin(\theta_i - \theta_{sh}))$.

### 2.2.1.3 Multi-Control Functions of the STATCOM

In the practical applications of a STATCOM, it may be used for controlling one of the following parameters [34]:

1. voltage magnitude of the local bus, to which the STATCOM is connected;
2. reactive power injection to the local bus, to which the STATCOM is connected;
3. impedance of the STATCOM;
4. current magnitude of the STATCOM while the current $I_{sh}$ leads the voltage injection $V_{sh}$ by $90°$;
5. current magnitude of the STATCOM, while the current $I_{sh}$ lags the voltage injection $V_{sh}$ by $90°$;
6. voltage injection;
7. voltage magnitude at a remote bus;

8. reactive power flow;
9. apparent power or current control of a local or remote transmission line.

Among these control options, control of the voltage of the local bus, which the STATCOM is connected to, is the most-recognized control function. The other control possibilities have not fully been investigated in power flow analysis. The mathematical descriptions of the control functions are presented as follows.

*Control mode 1: Bus voltage control*

The bus control constraint is as follows:

$$V_i - V_i^{Spec} = 0 \tag{2.5}$$

where $V_i^{Spec}$ is the bus voltage control reference.

*Control mode 2: Reactive power control*

In this control mode, the reactive power generated by the STATCOM is controlled to a reactive power injection reference. Mathematically, such a control constraint is described as follows:

$$Q_{sh} - Q_{sh}^{Spec} = 0 \tag{2.6}$$

where $Q_{sh}^{Spec}$ is the specified reactive power injection control reference. $Q_{sh}$, which is given by (2.3), is the actual reactive power generated by the STATCOM.

*Control mode 3: Control of equivalent impedance*

In principle, a STATCOM compensation can be equivalently represented by an imaginary impedance or reactance. In this control mode, $V_{sh}$ is regulated to control the equivalent reactance of the STATCOM to a specified reactance reference:

$$X_{shunt} - X_{shunt}^{Spec} = 0 \tag{2.7}$$

where $X_{shunt}^{Spec}$ is the specified reactance control reference of the STATCOM. $X_{shunt}$ is the equivalent reactance of the STATCOM. $X_{shunt}$, which is a function of the state variables $V_i$ and $V_{sh}$, is defined as:

$$X_{shunt} = \text{Im}(V_{sh}/I_{sh}) = \text{Im}[V_{sh}Z_{sh}/(V_i - V_{sh})] \tag{2.8}$$

*Control mode 4: Control of current magnitude - Capacitive compensation*

In this control mode, a STATCOM is used to control the magnitude of the current $I_{sh}$ of the STATCOM to a specified current magnitude control reference. The control constraint may be represented by $I_{sh} - I_{sh}^{Spec} = 0$. However it is found that there are two solutions corresponding to this control constraint. Due to the problem incurred, the power flow solution with such a constraint may arbitrarily converge to one of the two solutions. In section 2.2.4, further analysis is given to show the two solutions associated with this current magnitude control constraint.

In order to avoid the above non-unique solution problem, an alternative formulation of the current magnitude control is introduced here. Since $I_{sh} = I_{sh}^{Spec}$, if further assume $I_{sh}$ leads $V_{sh}$ by $90°$, then $I_{sh} = I_{sh}^{Spec} \angle(\theta_{sh} + 90°)$. $I_{sh}$ can also be defined by $I_{sh} = \dfrac{V_i - V_{sh}}{Z_{sh}}$, then we have:

$$I_{sh}^{Spec} \angle(\theta_{sh} + 90°) = (V_i - V_{sh})/Z_{sh} \qquad (2.9)$$

Mathematically, such a control mode can be described by one of the following equations:

$$\begin{aligned} &\text{Re}(I_{sh}^{Spec} \angle(\theta_{sh} + 90°)) = \text{Re}[(V_i - V_{sh})Z_{sh}] \\ \text{or} \quad &\text{Im}(I_{sh}^{Spec} \angle(\theta_{sh} + 90°)) = \text{Im}[(V_i - V_{sh})/Z_{sh}] \end{aligned} \qquad (2.10)$$

The formulation of (2.10) can force the power flow to converge to one of the two solutions. This control mode has a clear physical meaning. Since $I_{sh}$ leads $V_{sh}$ by $90°$, this control mode provides capacitive reactive power compensation while keeping the current magnitude constant.

*Control mode 5: Control of current magnitude - Inductive compensation*

In order to circumvent the same problem mentioned above, new formulation of the current control constraint needs to be introduced. In this control mode, the STATCOM is used to control the magnitude of the current $I_{sh}$ of the STATCOM while $I_{sh}$ lags $V_{sh}$ by $90°$. Mathematically, such a control mode may be described by:

$$\begin{aligned} &\text{Re}(I_{sh}^{Spec} \angle(\theta_{sh} - 90°)) = \text{Re}[(V_i - V_{sh})/Z_{sh}] \\ \text{or} \quad &\text{Im}(I_{sh}^{Spec} \angle(\theta_{sh} - 90°)) = \text{Im}[(V_i - V_{sh})/Z_{sh}] \end{aligned} \qquad (2.11)$$

Similar to that of (2.10), the formulation of (2.11) can force the power flow to converge to the other one of the two possible solutions. This control mode also has a clear physical meaning, that is, it provides inductive reactive power compensation while keeping the current magnitude constant.

*Control mode 6: Control of equivalent injected voltage magnitude $V_{sh}$ of STATCOM*

In this control mode, a STATCOM is used to control the magnitude of the voltage $V_{sh}$ of the STATCOM to a specified voltage magnitude control reference. The control constraint is as follows:

$$V_{sh} - V_{sh}^{Spec} = 0 \qquad (2.12)$$

where $V_{sh}$ is the voltage magnitude of the equivalent injected voltage $V_{sh}$ of the STATCOM. $V_{sh}^{Spec}$ is the voltage control reference.

*Control mode 7: Remote voltage magnitude control*

In this control mode, the STATCOM is used to control a remote voltage magnitude at bus $j$ to a specified voltage control reference. Mathematically, such a control constraint is described as follows:

$$V_j - V_j^{Spec} = 0 \tag{2.13}$$

where $V_j$ is the voltage magnitude of a remote bus, and $V_j^{Spec}$ is the specified remote bus voltage control reference.

*Control mode 8: Local or remote reactive power flow control*

In this control mode, the STATCOM is used to control either the local reactive power flow of a transmission line connected to the local bus or the reactive power flow of a remote transmission line to a specified reactive power flow control reference. Mathematically, such a control constraint is described as follows:

$$Q_{jk} - Q_{jk}^{Spec} = 0 \tag{2.14}$$

where $Q_{jk}$ is the reactive power flow leaving bus $j$ on the transmission line $j$-$k$. $Q_{jk}^{Spec}$ is the reactive power flow control reference.

*Control mode 9: Local or remote control of (maximum) apparent power*

In this control mode, the STATCOM is used to control either the apparent power of a transmission line connected to the local bus or the apparent power of a remote transmission line to a specified power control reference. Mathematically, such a control constraint is described as follows:

$$S_{jk} - S_{jk}^{Spec} = 0 \tag{2.15}$$

where $S_{jk} = \sqrt{(P_{jk})^2 + (Q_{jk})^2}$ is the apparent power of the transmission line $j$-$k$ while $P_{jk}$ and $Q_{jk}$ are the active and reactive power of the transmission line. $S_{jk}^{Spec}$ is the apparent power control reference, which may be the power rating of the transmission line.

Alternatively, the current magnitude of the transmission line may be controlled. The constraint can be represented by:

$$I_{jk} - I_{jk}^{Spec} = 0 \tag{2.16}$$

where $I_{jk} = \dfrac{\sqrt{(P_{jk})^2 + (Q_{jk})^2}}{V_j}$ is the actual current magnitude on the transmission line j-k. $I_{jk}^{Spec}$ is the current control reference, which may be the current rating of the transmission line.

*Remarks on control modes 8 and 9:*

- It is well recognized that a STATCOM may control a local bus voltage. However, it has not been recognized that a STATCOM may be used to control power flow of a transmission line. In addition to the local voltage control mode, it is important to explore other possible applications, such that the capabilities of the STATCOM can be fully employed.
- The control modes 8 and 9 presented in (2.14) and (2.15) or (2.16) introduce possible innovative applications of the STATCOM in power flow control.
- The reactive power flow control mode 8 can be used to control reactive power flow of an adjacent transmission line.
- The apparent power or current control mode 9 can be used to control the apparent power or current of an adjacent transmission line.
- In an electricity market, transmission congestion management by shunt reactive power control resources like STATCOM may be cheaper than by redispatching of active generating power. In this situation, control modes 8 and 9 may be very attractive. However, the control modes should not be overestimated. The controls may be very effective when there is excessive reactive power flow on a transmission line.
- Both control modes 8 and 9 of the STATCOM may be used in not only normal control when there is excessive reactive flowing on a transmission line but also security control of electric power systems when there is a violation of the thermal constraint of a transmission line.

Equations (2.5) - (2.7), (2.10) - (2.16) can be generally written as:

$$\Delta F(x) = F(x, f^{Spec}) = 0 \qquad (2.17)$$

where $x = [\theta_i, V_i, \theta_j, V_j, \theta_k, V_k, \theta_{sh}, V_{sh}]^t$. $f^{Spec}$ is the control reference.

### 2.2.1.4 Voltage and Thermal Constraints of the STATCOM

The equivalent voltage injection $V_{sh}$ bound constraints:

$$V_{sh}^{\min} \le V_{sh} \le V_{sh}^{\max} \qquad (2.18)$$

$$-\pi \le \theta_{sh} \le \pi \qquad (2.19)$$

where $V_{sh}^{max}$ is the voltage rating of the STATCOM, while $V_{sh}^{min}$ is the minimal voltage limit of the STATCOM.

The current flowing through a STATCOM should be less than its current rating:

$$I_{sh} \leq I_{sh}^{max} \tag{2.20}$$

where $I_{sh}^{max}$ is current rating of the STATCOM converter while $I_{sh}$ is the magnitude of the current through the STATCOM and given by:

$$I_{sh} = |(V_i - V_{sh})/Z_{sh}| = \sqrt{V_i^2 + V_{sh}^2 - 2V_iV_{sh}\cos(\theta_i - \theta_{sh})}/|Z_{sh}| \tag{2.21}$$

The constraints (2.18) and (2.20) are the internal constraints of the STATCOM.

### 2.2.1.5 External Voltage Constraints

In the practical operation of power systems, normally, a bus voltage should be within its operating limits. For all the control modes except the typical control mode 1, the voltage of bus $i$, to which the STATCOM is connected, should be constrained by:

$$V_i^{min} \leq V_i \leq V_i^{max} \tag{2.22}$$

For all the control modes except the control mode 7, the voltage of the remote bus $j$ may be monitored. The operating constraints of the voltage may be described by:

$$V_j^{min} \leq V_j \leq V_j^{max} \tag{2.23}$$

where $V_j^{max}$ and $V_j^{min}$ are the specified maximal and minimal voltage limits, respectively, at the remote bus $j$.

It should be pointed out, that other types of external limits other than (2.22), (2.23) may also be included.

## 2.2.2 Implementation of Multi-Control Functional Model of STATCOM in Newton Power Flow

### 2.2.2.1 Multi-Control Functional Model of STATCOM in Newton Power Flow

A STATCOM has only one degree of freedom for control since the active power exchange with the DC link should be zero at any time. The STATCOM may be used to control one of the nine parameters. The Newton power flow equation including power mismatch constraints of buses $i$, $j$, $k$, and the STATCOM control constraints may be represented by:

$$\begin{bmatrix} \frac{\partial PE}{\partial \theta_{sh}} & \frac{\partial PE}{\partial V_{sh}} & \frac{\partial PE}{\partial \theta_i} & \frac{\partial PE}{\partial V_i} & 0 & 0 & 0 & 0 \\ \frac{\partial F}{\partial \theta_{sh}} & \frac{\partial F}{\partial V_{sh}} & \frac{\partial F}{\partial \theta_i} & \frac{\partial F}{\partial V_i} & \frac{\partial F}{\partial \theta_j} & \frac{\partial F}{\partial V_j} & \frac{\partial F}{\partial \theta_k} & \frac{\partial F}{\partial V_k} \\ \frac{\partial P_i}{\partial \theta_{sh}} & \frac{\partial P_i}{\partial V_{sh}} & \frac{\partial P_i}{\partial \theta_i} & \frac{\partial P_i}{\partial V_i} & \frac{\partial P_i}{\partial \theta_j} & \frac{\partial P_i}{\partial V_j} & \frac{\partial P_i}{\partial \theta_k} & \frac{\partial P_i}{\partial V_k} \\ \frac{\partial Q_i}{\partial \theta_{sh}} & \frac{\partial Q_i}{\partial V_{sh}} & \frac{\partial Q_i}{\partial \theta_i} & \frac{\partial Q_i}{\partial V_i} & \frac{\partial Q_i}{\partial \theta_j} & \frac{\partial Q_i}{\partial V_j} & \frac{\partial Q_i}{\partial \theta_k} & \frac{\partial Q_i}{\partial V_k} \\ 0 & 0 & \frac{\partial P_j}{\partial \theta_i} & \frac{\partial P_j}{\partial V_i} & \frac{\partial P_j}{\partial \theta_j} & \frac{\partial P_j}{\partial V_j} & \frac{\partial P_j}{\partial \theta_k} & \frac{\partial P_j}{\partial V_k} \\ 0 & 0 & \frac{\partial Q_j}{\partial \theta_i} & \frac{\partial Q_j}{\partial V_i} & \frac{\partial Q_j}{\partial \theta_j} & \frac{\partial Q_j}{\partial V_j} & \frac{\partial Q_j}{\partial \theta_k} & \frac{\partial Q_j}{\partial V_k} \\ 0 & 0 & \frac{\partial P_k}{\partial \theta_i} & \frac{\partial P_k}{\partial V_i} & \frac{\partial P_k}{\partial \theta_j} & \frac{\partial P_k}{\partial V_j} & \frac{\partial P_{ki}}{\partial \theta_k} & \frac{\partial P_k}{\partial V_k} \\ 0 & 0 & \frac{\partial Q_k}{\partial \theta_i} & \frac{\partial Q_k}{\partial V_i} & \frac{\partial Q_k}{\partial \theta_j} & \frac{\partial Q_k}{\partial V_j} & \frac{\partial Q_k}{\partial \theta_k} & \frac{\partial Q_k}{\partial V_k} \end{bmatrix} \begin{bmatrix} \Delta \theta_{sh} \\ \Delta V_{sh} \\ \Delta \theta_i \\ \Delta V_i \\ \Delta \theta_j \\ \Delta V_j \\ \Delta \theta_k \\ \Delta V_k \end{bmatrix} = \begin{bmatrix} -PE \\ -\Delta F \\ -\Delta P_i \\ -\Delta Q_i \\ -\Delta P_j \\ -\Delta Q_j \\ -\Delta P_k \\ -\Delta Q_k \end{bmatrix} \quad (2.24)$$

where $\Delta P_l$ and $\Delta Q_l$ ($l = i, j, k$) are, respectively, the real and reactive power mismatches at bus $l$.

The STATCOM has two state variables $\theta_{sh}$ and $V_{sh}$, and two equalities. The two equalities formulate the first two rows of the above Newton equation. The first equality is the active power balance equation described by (2.4), while the second equality is the control constraint of the STATCOM, which is generally described by (2.17).

### 2.2.2.2 Modeling of Constraint Enforcement in Newton Power Flow

If the injected voltage $V_{sh}$ violates its voltage limit either $V_{sh}^{max}$ or $V_{sh}^{min}$, $V_{sh}$ is simply kept at the limit. Mathematically, the following equality should hold:

$$\begin{aligned} V_{sh} - V_{sh}^{max} = 0, & \quad \text{if } V_{sh} \geq V_{sh}^{max} \\ V_{sh} - V_{sh}^{min} = 0, & \quad \text{if } V_{sh} \leq V_{sh}^{min} \end{aligned} \quad (2.25)$$

In the meantime, the control constraint of (2.17) should be released. Similarly, the violations of the other constraints such as (2.20), (2.22), and (2.23) can be handled. The principle here is that when an inequality constraint is violated, it becomes an equality being kept at its limit while releasing the control constraint (2.17).

Due to the fact that the STATCOM has only one control degree of freedom, it is assumed that each time only one inequality constraint is violated. Similar to (2.25), the general constraint enforcement equation of (2.18), (2.20), (2.22), and (2.23) may be written as:

$$\Delta G(x) = G(x) - G^{Spec} = 0 \qquad (2.26)$$

where $x = [\theta_{sh}, V_{sh}, \theta_i, V_i, \theta_j, V_j, \theta_k, V_k]^t$. $G^{Spec}$ is the limit of the internal voltage or current of the STATCOM or the limit of the external bus voltage.

When any of the inequalities in (2.18), (2.20), (2.22), and (2.23) is violated and enforced, in the Newton power flow equation of (2.24), the control constraint (2.17) is replaced by (2.26).

The constraint enforcement only affects the second row of the Newton equation while other elements are unchanged. However, if two or more inequality constraints associated with a STATCOM are violated simultaneously, the constraint enforcement will become very complex. A strategy will be presented in section 2.2.3. In power flow calculations, a special initialization of the STATCOM is not needed.

### 2.2.3 Multi-Violated Constraints Enforcement

#### 2.2.3.1 Problem of Multi-Violated Constraints Enforcement

Basically, there are internal and external inequality constraints that may need to be considered for the operation of a STATCOM. The practical operation of a STATCOM is primarily constrained by its two internal operation inequalities, i.e., its voltage and thermal constraints given by (2.18) and (2.20), respectively. In the meantime, a STATCOM should also be able to monitor and control the local voltage at bus $i$, and the remote voltage at bus $j$ within their limits. In other words, the two external voltage constraints given by (2.22), (2.23) should be satisfied for some operating modes while a STATCOM's internal constraints are not violated.

When any one of the inequality constraints is violated, it should be enforced while the associated control constraint described by (2.17) needs to be released. As pointed out in the previous section, in principle, the STATCOM is only able to enforce one of the inequalities each time since it has only one control degree of freedom. Thus, difficulty will appear if two or more internal or external constraints of a STATCOM are violated at the same time.

#### 2.2.3.2 Concepts of Dominant Constraint and Dependent Constraint

Suppose there are two constraints, say constraint A and B, associated with a STATCOM. Assume the two constraints are violated simultaneously, if after the constraint A is enforced, the violation of the constraint B is automatically resolved. In this case, constraint A is called a dominant constraint, and constraint B is called a dependent constraint. The concept of dominant and dependent constraints is applicable to situations when there are more than two violated constraints.

In the following, a strategy for enforcement of two or more simultaneous violated constraints will be discussed based on the concepts of the dominant constraint and dependent constraint.

### 2.2.3.3 Strategy for Multi-Violated Constraints Enforcement

Generally, an internal constraint has priority to be enforced if both the internal and external constraints are violated simultaneously. When multiple constraints associated with the STATCOM are violated, a strategy is proposed as follows:

*Step 0: Formulate violated constraints set of STATCOM*
- (a) after the $K^{th}$ power flow iterations, formulate the violated constraints set of the STATCOM.
- (b) find the priority order of the violated cosntraints.
- (c) suppose that there are $n$ violated constraints, and they are in the priority order of $A_1, A_2, \ldots A_n$.

*Step 1: Identification of dominant constraint*

　　KK = K

　　Loop i = 1, n
- (a) At $KK^{th}$ power flow iteration, constraint $A_i$ is enforced
- (b) After the iteration, check whether the other violated constraints are within their limits. If yes, constraint $A_i$ is the domoniant constraint, go to Step 3.
- (c) Set power flow iteration count KK = KK+1

　　End loop

*Step 2: Dominant constraint not found*
- (a) If there are violated internal constrains, choose one of them to be enforced until the power flow converges. For the violated internal constraint chosen to be enforced, it may be the dominant constraint of all the violated internal constraint.
- (b) Otherwise choose one of the external constraints to be enforced until the power flow converges.

*Step 3: Dominant constraint found*

　　The dominant constraint out of all the violated constraints is found, the constraint is to be continuously enforced until the power flow converges.

The above constraint enforcement algorithm embedded in the Newton power flow calculations may be introduced after the power flow moderately converges, say after one or two Newton power flow iterations.

## 2.2.4 Multiple Solutions of STATCOM with Current Magnitude Control

For the STATCOM shown in Fig. 2.1, the following active power constraint, which represents the active power exchange between the AC system and the STATCOM, should be held at any instant:

$$\text{Re}(V_{sh}I_{sh}^*) = V_{sh}^2 g_{sh} - V_i V_{sh}(g_{sh}\cos(\theta_i - \theta_{sh}) - b_{sh}\sin(\theta_i - \theta_{sh})) = 0 \quad (2.27)$$

In order to simplify the derivation, assume the resistance of the STATCOM coupling transformer is neglected, the above equation becomes:

$$\text{Re}(V_{sh}I_{sh}^*) = V_i V_{sh} b_{sh} \sin(\theta_i - \theta_{sh}) = 0 \quad (2.28)$$

Since $V_i \neq 0$ and $V_{sh} \neq 0$, from the above equation, we have:

$$\theta_i - \theta_{sh} = 0 \text{ or } \theta_i - \theta_{sh} = \pi \quad (2.29)$$

The conventional current magnitude control is:

$$I_{sh} - I_{sh}^{Spec} = 0 \quad (2.30)$$

Substitute (2.21) into the above equation, we have:

$$I_{sh}^{Spec}|Z_{sh}| = \sqrt{V_i^2 + V_{sh}^2 - 2V_i V_{sh}\cos(\theta_i - \theta_{sh})} \quad (2.31)$$

If $\theta_i - \theta_{sh} = 0$, the above equation becomes:

$$I_{sh}^{Spec}|Z_{sh}| = \sqrt{V_i^2 + V_{sh}^2 - 2V_i V_{sh}} = \pm(V_i - V_{sh}) \quad (2.32)$$

The above equation can be written as (2.33) showing the two solutions to the current magnitude control given by (2.30):

$$V_{sh} = V_i \pm I_{sh}^{Spec}|Z_{sh}| \quad (2.33)$$

If $\theta_i - \theta_{sh} = \pi$, we have:

$$I_{sh}^{Spec}|Z_{sh}| = \sqrt{V_i^2 + V_{sh}^2 + 2V_i V_{sh}} = V_i + V_{sh} \quad (2.34)$$

Then we have:

$$V_{sh} = I_{sh}^{Spec}|Z_{sh}| - V_i \quad (2.35)$$

Note that $I_{sh}^{Spec}|Z_{sh}| \ll V_i$, thus $V_{sh} = I_{sh}^{Spec}|Z_{sh}| - V_i < 0$. This is not a feasible solution of (2.30). Therefore there are only two solutions to (2.30), which are given by (2.33).

## 2.2.5 Numerical Examples

### 2.2.5.1 Multi-Control Capabilities of STATCOM

To verify the STATCOM model and explore the multi-control capabilities of the STATCOM, numerical studies have been carried out on the IEEE 30-bus system, IEEE 118-bus system and IEEE 300-bus system. In the tests, a convergence tolerance of 1.0e-12 p.u. (or 1.0e-10 MW/MVAr) is used for maximal absolute bus power mismatches and power flow control mismatches. The single-line circuit diagram of the IEEE 30-bus system is shown in Fig. 2.2.

In order to show the multi-control capabilities of the STATCOM in power flow studies, cases 1-10 on the IEEE 30-bus system have been carried out. Case 1 is the base case without STATCOM. In cases 2-10, a STATCOM is installed at bus 12. In cases 2-10, nine different control modes of the STATCOM have been simulated. Control references for each control mode and corresponding number of iterations are shown in column 3 and column 4, respectively, in Table 2.1.

Power flow solutions of the STATCOM state variables for case 5 and case 6 are shown in Table 2.2. The two cases are corresponding to two constant current control modes of the STATCOM, respectively. However, if the current magnitude control in (2.30) is applied, the STATCOM solution may converge arbitrarily to one of the above two solutions of case 5 and case 6.

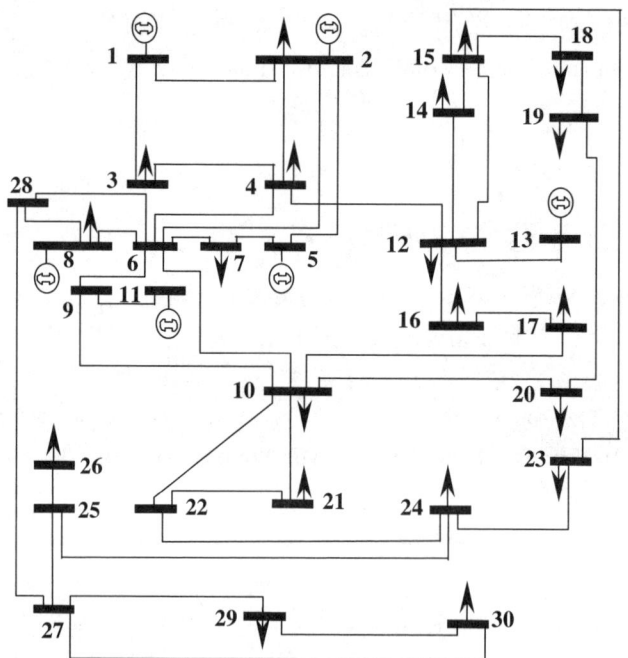

**Fig. 2.2.** Single-line circuit diagram of the IEEE 30-bus system.

## 2.2 Modeling of Multi-Functional STATCOM

**Table 2.1.** Results of STATCOM multi-control on the IEEE 30-bus system

| Case No. | Control mode | Control reference in p.u. | Iterations |
|---|---|---|---|
| 1 | None | None | 4 |
| 2 | 1 | $V_{12}^{Spec} = 1.0$ | 4 |
| 3 | 2 | $Q_{sh}^{Spec} = 1.0$ | 4 |
| 4 | 3 | $X_{shunt}^{Spec} = -10.0$ | 4 |
| 5 | 4 | $I_{sh}^{Spec} = 0.3$ | 4 |
| 6 | 5 | $I_{sh}^{Spec} = 0.3$ | 4 |
| 7 | 6 | $V_{sh}^{Spec} = 1.0$ | 4 |
| 8 | 7 | $V_{17}^{Spec} = 1.0$ | 4 |
| 9 | 8 | $Q_{13,12}^{Spec} = 0.0$ | 4 |
| 10 | 9 | $S_{9,11}^{Spec} = 0.22$ | 5 |

The power flows of line 13-12 and line 9-11 of case 1, case 9 and case 10 are given by Table 2.3. In comparison to the power flows of line 13-12 of case 1 and case 9, the STATCOM of case 9 is able to control the reactive power flow of line 13-12 to the specified control reference 0.0 p.u., while the active power flow is almost unchanged. By driving the reactive power flow on the line to zero using STATCOM, the un-used (available) transmission line capacity can be increased. It can be seen that the base case reactive power of line 13-12 is 0.384 p.u., so the reactive power flow control by the STATCOM is significant.

Comparing the power flows of line 9-11 in case 1 and case 10, it can be found that in case 10, the apparent power of the remote line 9-11 can be controlled to the specified control reference of 0.22 p.u., while the active power is almost unchanged.

**Table 2.2.** Results of case 5 and case 6 for the IEEE 30-bus system

| Case No. | Power flow solution of $V_{sh}$ (in p.u.), $\theta_{sh}$ (in degree) |
|---|---|
| 5 | $V_{sh} = 0.9831$, $\theta_{sh} = -9.3°$, Inductive compensation |
| 6 | $V_{sh} = 1.0449$, $\theta_{sh} = -9.6°$, Capacitive compensation |

**Table 2.3.** Power flows of transmission line of case 1, case 9 and case 10 for the IEEE 30-bus System

| Case No. | Power flows of line 13-12 | Power flow of 9-11 |
|---|---|---|
| 1 | $P_{13,12} = 0.169$, $Q_{13,12} = \mathbf{0.384}$, $S_{13,12} = 0.420$ | $P_{9,11} = -0.179$, $Q_{9,11} = -0.288$, $S_{9,11} = \mathbf{0.420}$ |
| 9 | $P_{13,12} = 0.169$, $Q_{13,12} = \mathbf{0.000}$, $S_{13,12} = 0.169$ | – |
| 10 | – | $P_{9,11} = -0.179$, $Q_{9,11} = -0.127$, $S_{9,11} = \mathbf{0.220}$ |

The control reference is much lower than the apparent power of 0.42 p.u. of line 9-11 in case 1 (base case). The control mode 9 may be used when the thermal limit of a transmission line is violated or the un-used transmission capacity needs to be increased.

Cases 9 and 10 reveal that the STATCOM has very little influence on the active power flow of a transmission line, while it has strong capability of controlling reactive power on a transmission line. In addition, both control modes 8 and 9 of a STATCOM can be used for local control of reactive power flow on a transmission line. The control modes may be attractive when, in electricity market environments, re-dispatching active power becomes much more expensive than controlling reactive power.

Test results on the IEEE 118-bus system and the IEEE-300 bus system can be found in [34].

### 2.2.5.2 Multi-Violated STATCOM Constraints Enforcement

The following case is to show the enforcement of a single constraint violation of a STATCOM, which is as follows:

*Case 12*: This is similar to case 2, but assume that a current limit of $I_{sh}^{max} = 0.9$ p.u. is applied to the STATCOM.

The power flow algorithm converges in 4 iterations. In the tests, it has been found that for single constraint violation of the STATCOM based on the IEEE 30, 118, 300 bus systems, the power flow algorithm can converge in the same number of iterations as that of base case power flow solution. Occasionally, the power flow algorithm needs one or more extra iterations.

The following case is used to illustrate the enforcement of the multiple violated voltage and current constraints associated the STATCOM on the IEEE 30 bus system.

*Case 13*: This is similar to case 3. In this case, it is assumed that the two internal constraints and two external voltage constraints at bus 12 and 17 are violated when the following voltage and current limits are applied:

$$V_{sh}^{max} = 1.065\text{p.u.}, I_{sh}^{max} = 0.46\text{p.u.}, V_{12}^{max} = 1.05\text{p.u.}, V_{17}^{max} = 1.00\text{p.u.}.$$

The detailed enforcement of the four violated constraints is presented in Table 2.4. In the enforcement of the multiple violated constraints, much effort has been made on the identification of the dominant constraint. Case 13 shows the worst scenario, in which the algorithm took four power flow iterations to identify the dominant constraint. However, if another constraint, for instance $V_{sh}$ rather than the external remote voltage constraint were the dominant constraint, the algorithm would complete the identification in one power flow iteration.

**Table 2.4.** Enforcement of multi-violated constraints of a STATCOM in the IEEE 30-bus system

| Iteration count | Enforcement of the multi-violated constraints |
|---|---|
| 0 | None |
| 1 | Form the violated constraint set $\{V_{sh}, I_{sh}, V_{12}, V_{17}\}$ and priority order |
| 2 | $V_{sh}$ is enforced, it is not the dominant constraint |
| 3 | $I_{sh}$ is enforced, it is not the dominant constraint. But it is the dominant constraint between $V_{sh}$ and $I_{sh}$ |
| 4 | $V_{12}$ is enforced, it is not the dominant constraint |
| 5 | $V_{17}$ is enforced, it is the dominant constraint. Identification of dominant and dependent constraints pair complete |
| 6 | $V_{17}$ is enforced |
| 7 | The power flow converges |

The multi-violated STATCOM constraints enforcement algorithm has successfully solved further cases on the IEEE 118-bus system and IEEE 300-bus system [34].

This section has defined a multi-control functional model for STATCOM suitable for power system steady state operational studies. Nine control modes have been incorporated into the STATCOM model. There are two solutions associated with the current magnitude control. Alternative formulations of the control mode are introduced. The reactive power flow and apparent power control modes, which may be used in either normal control or security control of electric power systems, are interesting and attractive when, in electricity market environments, re-dispatching active power becomes expensive, and there is excessive reactive power on a transmission line. Numerical results based on the IEEE 30-bus system, IEEE 118-bus system, and IEEE 300-bus system with the multi-control functional STATCOM have demonstrated the feasibility and effectiveness of the proposed multi-control functional STATCOM model for power system steady state operation and control.

Furthermore, a comprehensive strategy for enforcement of the multi-violated STATCOM internal and external constraints has been described. Numerical results show that the strategy proposed can enforce successfully the multiple violated voltage and current (inequality) constraints associated with a STATCOM on the IEEE systems. The constraint enforcement strategy for multiple violated constraints associated with a STATCOM may be further enhanced by incorporation of advanced expert rules into the algorithm. The key issue is to reduce the effort for identifying the dominant constraint.

## 2.3 Modeling of Multi-Control Functional SSSC

It is found that, in the past, much effort has been paid in the modeling of the UPFC for power flow and optimal power flow analysis, while few work has been published on the modeling of the SSSC, in particular, for power flow analysis. In this section, multi-control modes of the SSSC will be discussed, in particular:

- The multi-control modes of the SSSC will be explored. A multi-control functional model of the SSSC, which can be used for steady state controlling any of the following parameters, (a) the active power flow of the transmission line, (b) the reactive power flow the transmission line, (c) the bus voltage, and (d) the impedance of the transmission line, will be presented.
- Fully consideration of the current and voltage operating constraints of the SSSC and detailed direct implementation of these in Newton power flow will be described.

### 2.3.1 Multi-Control Functional Model of SSSC for Power Flow Analysis

#### 2.3.1.1 Operation Principles of the SSSC

A SSSC [18][19] usually consists of a coupling transformer, an inverter and a capacitor. As shown in Fig. 2.3, the SSSC is series connected with a transmission line through the coupling transformer.

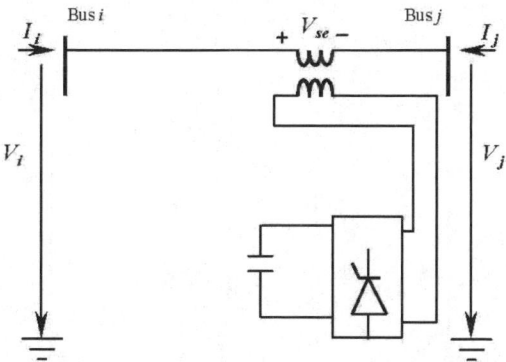

**Fig. 2.3.** SSSC operation principles

It is assumed here that the transmission line is series connected via the SSSC bus $j$. The active and reactive power flows of the SSSC branch $i$-$j$ entering the bus $j$ are equal to the sending end active and reactive power flows of the transmission line, respectively. In principle, the SSSC can generate and insert a series voltage, which can be regulated to change the impedance (more precisely reactance) of the transmission line. In this way, the power flow of the transmission line or the voltage of the bus, which the SSSC is connected with, can be controlled.

### 2.3.1.2 Equivalent Circuit and Power Flow Constraints of SSSC

An equivalent circuit of the SSSC as shown in Fig. 2.4 can be derived based on the operation principle of the SSSC. In the equivalent, the SSSC is represented by a voltage source $V_{se}$ in series with a transformer impedance. In the practical operation of the SSSC, $V_{se}$ can be regulated to control the power flow of line $i$-$j$ or the voltage at bus $i$ or $j$.

In the equivalent circuit, $V_{se} = V_{se} \angle \theta_{se}$, $V_i = V_i \angle \theta_i$, $V_j = V_j \angle \theta_j$, then the bus power flow constraints of the SSSC are:

$$P_{ij} = V_i^2 g_{ii} - V_i V_j (g_{ij} \cos \theta_{ij} + b_{ij} \sin \theta_{ij}) \\ - V_i V_{se} (g_{ij} \cos(\theta_i - \theta_{se}) + b_{ij} \sin(\theta_i - \theta_{se})) \quad (2.36)$$

$$Q_{ij} = -V_i^2 b_{ii} - V_i V_j (g_{ij} \sin \theta_{ij} - b_{ij} \cos \theta_{ij}) \\ - V_i V_{se} (g_{ij} \sin(\theta_i - \theta_{se}) - b_{ij} \cos(\theta_i - \theta_{se})) \quad (2.37)$$

$$P_{ji} = V_j^2 g_{jj} - V_i V_j (g_{ij} \cos \theta_{ji} + b_{ij} \sin \theta_{ji}) \\ + V_j V_{se} (g_{ij} \cos(\theta_j - \theta_{se}) + b_{ij} \sin(\theta_j - \theta_{se})) \quad (2.38)$$

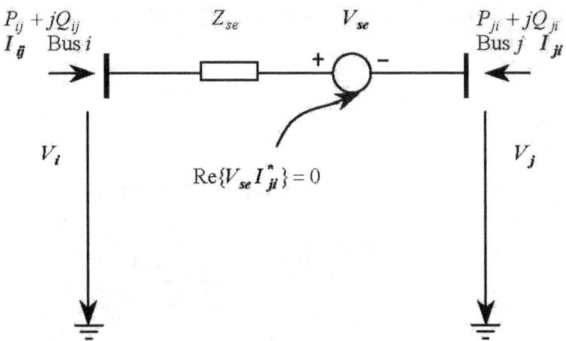

**Fig. 2.4.** SSSC equivalent circuit

$$Q_{ji} = -V_j^2 b_{jj} - V_i V_j (g_{ij} \sin \theta_{ji} - b_{ij} \cos \theta_{ji})$$
$$+ V_j V_{se} (g_{ij} \sin(\theta_j - \theta_{se}) - b_{ij} \cos(\theta_j - \theta_{se})) \quad (2.39)$$

where $g_{ij} + jb_{ij} = 1/Z_{se}$, $g_{ii} = g_{ij}$, $b_{ii} = b_{ij}$, $g_{jj} = g_{ij}$, $b_{jj} = b_{ij}$.

The operating constraint of the SSSC (active power exchange via the DC link) is:

$$PE = \text{Re}(V_{se} I_{ji}^*) = 0 \quad (2.40)$$

where $\text{Re}(V_{se} I_{ji}^*) = -V_i V_{se} (g_{ij} \cos(\theta_i - \theta_{se}) - b_{ij} \sin(\theta_i - \theta_{se}))$
$+ V_j V_{se} (g_{ij} \cos(\theta_j - \theta_{se}) - b_{ij} \sin(\theta_j - \theta_{se}))$

### 2.3.1.3 Multi-Control Functions and Constraints of SSSC

In the practical applications of the SSSC, it may be used for control of any of the following parameters, (a) the active power flow of the transmission line, (b) the reactive power flow of the transmission line, (c) the bus voltage, and (d) the impedance of the transmission line. Therefore, the SSSC may have four control modes. Among the four control modes, the active power flow control mode has been well recognized. The mathematical descriptions of the four control functions of the SSSC are presented as follows.

*Control mode 1: Active power flow control*

The active power flow control constraint is as follows:

$$P_{ji} - P_{ji}^{Spec} = 0 \quad (2.41)$$

where $P_{ji}^{Spec}$ is the specified active power flow control reference.

*Control mode 2: Reactive power flow control*

In this function, the reactive power flow control constraint is as follows:

$$Q_{ji} - Q_{ji}^{Spec} = 0 \qquad (2.42)$$

where $Q_{ji}^{Spec}$ is the specified reactive power flow control reference. As mentioned, $P_{ji}$, $Q_{ji}$ are the SSSC branch active and reactive power flows, respectively, leaving the SSSC bus $j$ while the sending end active and reactive power flows of the transmission line are $-P_{ji}$ and $-Q_{ji}$, respectively.

*Control mode 3: Bus voltage control*
The bus voltage control constraint is given by:

$$V_i - V_i^{Spec} = 0$$
$$\text{or } V_j - V_j^{Spec} = 0 \qquad (2.43)$$

where $V_i^{Spec}$ and $V_j^{Spec}$ are the bus voltage control references.

*Control mode 4: Impedance (reactance) control*
In this function, $V_{se}$ is regulated to control equivalent reactance of the SSSC to a specified reactance reference:

$$X_{comp} - X_{comp}^{Spec} = 0 \qquad (2.44)$$

where $X_{comp}^{Spec}$ is the specified reactance reference. While $X_{comp}$ is a function of the state variables $V_i$, $V_j$ and $V_{se}$.

Theoretically, the control mode 4 is equivalent to replacing the entire SSSC with a fixed reactance. However, the problem of modeling of control mode 4 of the SSSC by a fixed reactance is that it would be very difficult to deal with the voltage and current constraints of the SSSC if not impossible. Subsequently, this would cause the change of the structure and dimension of the Newton Jacobian matrix, and increase the complexity of the code. In contrast, the present reactance control mode of the SSSC in (2.44) has no such limitations. Further advantages of the present formulation of the control modes of the SSSC will be discussed in section 2.3.2.

Equations (2.41)-(2.44) can be generally written as:

$$\Delta F(x) = F(x) - F^{Spec} = 0 \qquad (2.45)$$

where $x = [\theta_i, V_i, \theta_j, V_j, \theta_{se}, V_{se}]^t$.

## 2.3.1.4 Voltage and Current Constraints of the SSSC

The equivalent voltage injection $V_{se}$ bound constraints are as follows:

48  2 Modeling of Multi-Functional Single Converter FACTS in Power Flow Analysis

$$0 \leq V_{se} \leq V_{se}^{\max} \tag{2.46}$$

$$-\pi \leq \theta_{se} \leq \pi \tag{2.47}$$

where $V_{se}^{\max}$ is the voltage rating of $V_{se}$, which may be constant, or may change slightly with changes in the DC bus voltage, depending on the inverter design. In principle, $\theta_{se}$ can be any real. Therefore (2.47) is not a real constraint while (2.46) is a real constraint, which should be hold at any time.

The current through each series converter should be within its current rating:

$$I_{se} \leq I_{se}^{\max} \tag{2.48}$$

where $I_{se}^{\max}$ is the current rating of the series converter.

Note the fact that $I_{se} = I_{se} \angle \theta_{se} = \dfrac{V_i - V_{se} - V_j}{Z_{se}}$, the actual current magnitude through the SSSC can be obtained:

$$\begin{aligned} I_{se} &= I_{ji} \\ &= \sqrt{V_i^2 + V_{se}^2 + V_j^2 - 2V_i V_{se}\cos(\theta_i - \theta_{se}) + 2V_j V_{se}\cos(\theta_j - \theta_{se}) - 2V_i V_j \cos(\theta_i - \theta_j)}/|Z_{se}| \end{aligned} \tag{2.49}$$

### 2.3.2 Implementation of Multi-Control Functional Model of SSSC in Newton Power Flow

#### 2.3.2.1 Multi-Control Functional Model of SSSC in Newton Power Flow

For the SSSC, the power mismatches, at its buses $i$, $j$, respectively, should be held,

$$\Delta P_i = Pg_i - Pd_i - P_i = 0 \tag{2.50}$$

$$\Delta Q_i = Qg_i - Qd_i - Q_i = 0 \tag{2.51}$$

$$\Delta P_j = Pg_j - Pd_j - P_j = 0 \tag{2.52}$$

$$\Delta Q_j = Qg_j - Qd_j - Q_j = 0 \tag{2.53}$$

where $P_k$, $Q_k$ are, respectively, the real and reactive power leaving the bus $k$ ($k=i, j, \ldots$). These are sum of the real and reactive power flows including those given by (2.50)-(2.53), respectively. While $Pg_k$, $Qg_k$ are, respectively, the real and reactive generating power entering the bus $k$, and $Pd_k$, $Qd_k$ are, respectively, the real and reactive load leaving the bus $k$.

## 2.3 Modeling of Multi-Control Functional SSSC

For the SSSC, it has only one control degree of freedom since the active power exchange with the DC link should be zero at any time. So the SSSC may be used to control only one of the following parameters, (a) the active power flow on the transmission line, (b) the reactive power flow on the transmission line, (c) the bus voltage, and (d) the impedance (precisely reactance) of the transmission line. A Newton power flow algorithm with simultaneous solution of power flow constraints and power flow control constraints of the SSSC may be represented by (2.54) as follows:

$$\begin{bmatrix} \frac{\partial F}{\partial \theta_{se}} & \frac{\partial F}{\partial V_{se}} & \frac{\partial F}{\partial \theta_i} & \frac{\partial F}{\partial V_i} & \frac{\partial F}{\partial \theta_j} & \frac{\partial F}{\partial V_j} \\ \frac{\partial PE}{\partial \theta_{se}} & \frac{\partial PE}{\partial V_{se}} & \frac{\partial PE}{\partial \theta_i} & \frac{\partial PE}{\partial V_i} & \frac{\partial PE}{\partial \theta_j} & \frac{\partial PE}{\partial V_j} \\ \frac{\partial P_i}{\partial \theta_{se}} & \frac{\partial P_i}{\partial V_{se}} & \frac{\partial P_i}{\partial \theta_i} & \frac{\partial P_i}{\partial V_i} & \frac{\partial P_i}{\partial \theta_j} & \frac{\partial P_i}{\partial V_j} \\ \frac{\partial Q_i}{\partial \theta_{se}} & \frac{\partial Q_i}{\partial V_{se}} & \frac{\partial Q_i}{\partial \theta_i} & \frac{\partial Q_i}{\partial V_i} & \frac{\partial Q_i}{\partial \theta_j} & \frac{\partial Q_i}{\partial V_j} \\ \frac{\partial P_j}{\partial \theta_{se}} & \frac{\partial P_j}{\partial V_{se}} & \frac{\partial P_j}{\partial \theta_i} & \frac{\partial P_j}{\partial V_i} & \frac{\partial P_j}{\partial \theta_j} & \frac{\partial P_j}{\partial V_j} \\ \frac{\partial Q_j}{\partial \theta_{se}} & \frac{\partial Q_j}{\partial V_{se}} & \frac{\partial Q_j}{\partial \theta_i} & \frac{\partial Q_j}{\partial V_i} & \frac{\partial Q_j}{\partial \theta_j} & \frac{\partial Q_j}{\partial V_j} \end{bmatrix} \begin{bmatrix} \Delta\theta_{se} \\ \Delta V_{se} \\ \Delta\theta_i \\ \Delta V_i \\ \Delta\theta_j \\ \Delta V_j \end{bmatrix} = \begin{bmatrix} -\Delta F \\ -PE \\ -\Delta P_i \\ -\Delta Q_i \\ -\Delta P_j \\ -\Delta Q_j \end{bmatrix} \quad (2.54)$$

In (2.54) the system Jacobian matrix is split into four blocks by the dotted line. The bottom diagonal block has the same structure as that of the system Jacobian matrix of conventional power flow. Though the terms of the former should consider the contributions from the SSSC. The other three blocks of the system Jacobian matrix in (2.54) are SSSC related.

### 2.3.2.2 Enforcement of Voltage and Current Constraints for SSSC

As discussed in Section 2.2.3, the basic constraint enforcement strategy is that, when there is an inequality constraint such as the current or voltage inequality constraint of the SSSC is violated, the constraint is enforced by being kept at its limit, while the control equality constraint of the SSSC given by one of the equality constraints (2.41)-(2.44) is released. The constraint enforcement equations of (2.46) and (2.48) can be generalized as:

$$\Delta G(x) = G(x) - G^{Spec} = 0 \qquad (2.55)$$

where $x = [\theta_i, V_i, \theta_j, V_j, \theta_{se}, V_{se}]^t$. $G(x) = V_{se}$ and $G^{Spec} = V_{se}^{max}$, or $G(x) = I_{se}$ and $G^{Spec} = I_{se}^{max}$.

Then when either the voltage constraint (2.46) or the current constraint (2.48) is violated, the Newton power flow equation becomes:

$$\begin{bmatrix} \frac{\partial G}{\partial \theta_{se}} & \frac{\partial G}{\partial V_{se}} & \frac{\partial G}{\partial \theta_i} & \frac{\partial G}{\partial V_i} & \frac{\partial G}{\partial \theta_j} & \frac{\partial G}{\partial V_j} \\ \frac{\partial PE}{\partial \theta_{se}} & \frac{\partial PE}{\partial V_{se}} & \frac{\partial PE}{\partial \theta_i} & \frac{\partial PE}{\partial V_i} & \frac{\partial PE}{\partial \theta_j} & \frac{\partial PE}{\partial V_j} \\ \frac{\partial P_i}{\partial \theta_{se}} & \frac{\partial P_i}{\partial V_{se}} & \frac{\partial P_i}{\partial \theta_i} & \frac{\partial P_i}{\partial V_i} & \frac{\partial P_i}{\partial \theta_j} & \frac{\partial P_i}{\partial V_j} \\ \frac{\partial Q_i}{\partial \theta_{se}} & \frac{\partial Q_i}{\partial V_{se}} & \frac{\partial Q_i}{\partial \theta_i} & \frac{\partial Q_i}{\partial V_i} & \frac{\partial Q_i}{\partial \theta_j} & \frac{\partial Q_i}{\partial V_j} \\ \frac{\partial P_j}{\partial \theta_{se}} & \frac{\partial P_j}{\partial V_{se}} & \frac{\partial P_j}{\partial \theta_i} & \frac{\partial P_j}{\partial V_i} & \frac{\partial P_j}{\partial \theta_j} & \frac{\partial P_j}{\partial V_j} \\ \frac{\partial Q_j}{\partial \theta_{se}} & \frac{\partial Q_j}{\partial V_{se}} & \frac{\partial Q_j}{\partial \theta_i} & \frac{\partial Q_j}{\partial V_i} & \frac{\partial Q_j}{\partial \theta_j} & \frac{\partial Q_j}{\partial V_j} \end{bmatrix} \begin{bmatrix} \Delta \theta_{se} \\ \Delta V_{se} \\ \Delta \theta_i \\ \Delta V_i \\ \Delta \theta_j \\ \Delta V_j \end{bmatrix} = \begin{bmatrix} -\Delta G \\ -PE \\ -\Delta P_i \\ -\Delta Q_i \\ -\Delta P_j \\ -\Delta Q_j \end{bmatrix} \quad (2.56)$$

It can be seen, that the formulation of Newton power flow with the constraint enforcement in (2.56) has exactly the same structure as that of Newton power flow without the constraint enforcement in (2.54). This property makes the implementation of the power flow algorithm easy and efficient.

### 2.3.2.3 Initialization of SSSC in Newton Power Flow

Basically, unlike that for the UPFC, there are no analytical solutions available, which can be used to initialize the values of the SSSC voltage variables in power flow analysis.

In the present implementation, for control modes 1, 2 and 3, the initial values of the voltage angle and magnitude of a SSSC may be set as follows:

$$\theta_{se}^0 = -\frac{\pi}{2} \quad (2.57)$$

$$V_{se}^0 = 0.1 \quad (2.58)$$

and for control mode 4, the initial values of the voltage angle and magnitude of a SSSC are set as follows:

$$\theta_{se}^0 = \begin{cases} \dfrac{\pi}{2}, & \text{if } X_{comp}^{Spec} > 0 \\ -\dfrac{\pi}{2}, & \text{if } X_{comp}^{Spec} < 0 \end{cases} \quad (2.59)$$

$$V_{se}^0 = |X_{comp}^{Spec}| \quad (2.60)$$

### 2.3.3 Numerical Results

Numerical results are carried out on the IEEE 30-bus system, IEEE118-bus system and IEEE 300-bus system. In the test, a convergence tolerance of 1.0e-12 p.u. (or 1.0e-10 *MW/MVAr*) for maximal absolute bus power mismatches and power flow control mismatches is utilized.

#### 2.3.3.1 Power Flow, Voltage and Reactance Control by the SSSC

In order to show the multi-control capabilities of the SSSC model and performance of the Newton power flow algorithm, the following cases based on the IEEE 30 bus system are carried out.

*Case 1*: This is a base case IEEE 30 bus system.

*Case 2*: This is similar to case 1 except that there is a SSSC installed for control of the active power flow of line 12-15. The active power flow control reference is set to $P_{15,12}^{Spec} = -30\ MW$, which is more than 60% of its corresponding base case active power flow.

*Case 3*: This is similar to case 2 except that the SSSC is used for control of the reactive power flow of line 12-15. The reactive power flow control reference is $Q_{15,12}^{Spec} = -1\ MVAr$.

*Case 4*: This is similar to case 2 except that the SSSC is used for control of the voltage magnitude at bus 15, and the voltage control reference is $V_{15}^{Spec} = 1.0$ p.u.

*Case 5*: This is similar to case 2 except that the SSSC is controlled to generate an equivalent reactance with a capacitive reactance control reference $X_{comp}^{Spec} = -0.2$ p.u.

*Case 6*: This is similar to case 1 except that there are three SSSCs installed on lines 12-15, 10-21 and 6-2, respectively. These SSSCs are used for control of the voltage at bus 15, reactive power flow of line 10-21, and active power flow of line 6-2, respectively. The control references are $V_{15}^{Spec} = 1.0$ p.u., $Q_{21,10}^{Spec} = -5\ MVAr$, and $P_{2,6}^{Spec} = 45\ MW$, respectively.

*Case 7*: This is similar to case 6 except that the SSSC on line 6-2 is controlled to generate an equivalent reactance with respect to the capacitive reactance control reference $X_{comp}^{Spec} = -0.1$ p.u., and the SSSC on line 12-15 is used to control the active power flow of that line.

The results of cases 1–7 are summarized in Table 2.5.

**Table 2.5.** Results of the IEEE 30 bus system

| Case No. | Solutions of the SSSCs parameters | Number of iterations |
|---|---|---|
| Case 1 | None | 4 |
| Case 2 | $\theta se_{12,15} = -109.64°$, $Vse_{12,15} = 0.06060$ p.u. | 5 |
| Case 3 | $\theta se_{12,15} = -97.03°$, $Vse_{12,15} = 0.09639$ p.u. | 5 |
| Case 4 | $\theta se_{12,15} = -101.78°$, $Vse_{12,15} = 0.08344$ p.u. | 6 |
| Case 5 | $\theta se_{12,15} = -109.93°$, $Vse_{12,15} = 0.05972$ p.u. | 5 |
| Case 6 | $\theta se_{12,15} = -101.05°$, $Vse_{12,15} = 0.08577$ p.u.<br>$\theta se_{10,21} = -105.13°$, $Vse_{10,21} = 0.05787$ p.u.<br>$\theta se_{6,2} = 75.77°$, $Vse_{6,2} = 0.03539$ p.u. | 6 |
| Case 7 | $\theta se_{12,15} = -108.02°$, $Vse_{12,15} = 0.06663$ p.u.<br>$\theta se_{10,21} = -105.27°$, $Vse_{10,21} = 0.05538$ p.u.<br>$\theta se_{6,2} = 77.79°$, $Vse_{6,2} = 0.04878$ p.u. | 5 |

In these cases above and the following discussions, the control references of active and reactive power flows are referred to $P_{ji}^{Spec}$, $Q_{ji}^{Spec}$, which are at the sending end of a transmission line. Active power flow and reactive power flows at the sending end of the line are referred to $-P_{ji}$, $-Q_{ji}$ since the sending end of the line is connected to the SSSC bus $j$.

The test results of the IEEE 118 bus system are described as follows,

*Case 8*: This is a base case of the IEEE 118 bus system.

*Case 9*: This is similar to case 8 except that there are three SSSCs installed for control of active power flow of line 21-20, reactive power flow of line 45-44 and voltage of bus 95, respectively.

*Case 10*: This is similar to case 9 except that the SSSC of line 94-95 is used to control the reactance of that line.

The test results of cases 8-10 are given by Table 2.6.

**Table 2.6.** Results of the IEEE 118 bus system

| Case No. | Number of SSSCs | Number of iterations |
|---|---|---|
| Case 8 | None | 4 |
| Case 9 | 3 | 5 |
| Case 10 | 3 | 5 |

Further cases are carried out on the IEEE 300 bus system, which are as follows,

*Case 11*: This is a base case of the IEEE 300 bus system

*Case 12*: Similar to case 11 except that there are four SSSCs installed. The first SSSC is installed for control of the reactive power flow of line 198-197. The second SSSC is used for control of the voltage of the SSSC at bus 49. The third SSSC is installed for control of the reactance of line 126-132. The fourth SSSC is used for control of the active power flow of line 140-137.

**Table 2.7.** Results of the IEEE 300 bus system

| Case No. | Number of SSSCs | Number of iterations |
|---|---|---|
| Case 11 | None | 6 |
| Case 12 | 4 | 7 |

From the results of Table 2.5 to Table 2.7, it can be seen that, in comparison with those cases of base power flow solutions, the Newton power flow solutions with the SSSCs need more iterations but can converge within 7 iterations with the convergence tolerance of 1e-12 *p.u.* (1e-10 *MW/MVAr*). The convergence characteristics of case 11 without the SSSCs and case 12 with four SSSCs on the IEEE 300-bus system are shown in Fig. 2.5. The quadratic convergence characteristics of the Newton's algorithm can be clearly observed.

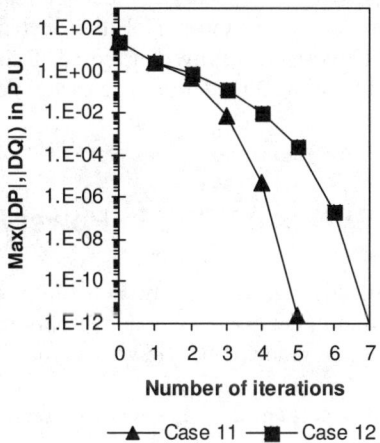

**Fig. 2.5.** Power mismatches as function of number of iterations

### 2.3.3.2 Enforcement of Voltage and Current Constraint of the SSSC

In the following, examples of the enforcement of the voltage and current constraints of the SSSC are given:

*Case* 13: This is similar to case 3 except that a voltage limit is applied to the SSSC.

*Case* 14: This is similar to case 2 except that a current limit is applied to the SSSC.

The test results of cases 13 and 14 are presented in Table 2.8. Test results of constraints enforcement of SSSC on the IEEE 118-bus system and the IEEE 300-bus system can be found in [34].

**Table 2.8.** Results of constraints enforcement of the SSSC for the IEEE 30 bus system

| Case No. | Actual voltage or current of the SSSC without constraint enforcement | Voltage or current limits of the SSSC for constraints enforcement | Number of iterations |
|---|---|---|---|
| Case 13 | $Vse_{12,15} = 0.09639\,p.u.$ | $Vse_{12,15}^{max} = 0.08\,p.u.$ | 6 |
| Case 14 | $Ise_{15,12} = 0.3\,p.u.$ | $Ise_{15,12}^{max} = 0.25\,p.u.$ | 7 |

This section has introduced a multi-control functional model for the Static Synchronous Series Compensator (SSSC) suitable for power flow analysis. The model has explored the multi-control options of the SSSC such as (a) the active power flow on the transmission line, (b) the reactive power flow on the transmission line, (c) the bus voltage, and (d) the impedance (precisely reactance) of the transmission line, etc. Furthermore, within the model, the operating voltage and current constraints of the SSSC have been fully considered. Detailed implementation of the novel multi-control functional model in the Newton power flow algorithm has been presented.

## 2.4 Modeling of SVC and TCSC in Power Flow Analysis

In the previous sections, the mathematical models of converter based FACTS devices have been discussed in details. The models proposed will be of great importance to develop production grade power flow programs. In this section, the modeling of SVC and TCSC will be addressed. Traditionally SVC and TCSC are considered as variable susceptance and reactance, respectively, in power flow analysis. In this section, SVC and TCSC will be equivalently represented as STATCOM and SSSC, respectively. Then SVC and TCSC can be incorporated in the Newton power flow program very easily.

### 2.4.1 Representation of SVC by STATCOM in Power Flow Analysis

SVC can provide voltage and reactive power control by varying its shunt reactance [33]. Two popular configurations of SVC are the combination of fixed capacitor and Thyristor Controlled Reactor (TCR) and the combination of Thyristor Switched Capacitor (TSC) and TCR. A SVC consisting of a fixed capacitor and a TCR is shown in Fig. 2.6a.

In power flow analysis, the total susceptance of the SVC may be taken as a variable and additional voltage or reactive power control equation should be included. In contrast to the SVC models used in power flow analysis, the SVC will be implemented in the way as for a STATCOM. So at first the SVC Fig. 2.6a should be converted into an equivalent STATCOM in the point of view of power flow analysis. The equivalent representation of the SVC is illustrated by Fig. 2.6b.

In Fig. 2.6b $Zsh$ is given by:

$$Zsh = j(X_{TCR}^{min} + X_{TCR}^{max})/2 \tag{2.61}$$

where $X_{TCR}^{min}$, $X_{TCR}^{max}$ are the lower and upper limits of the variable reactance of the TCR branch in Fig. 2.6. Now the variable reactance of the TCR branch in Fig. 2.6a can be equivalently represented by an impedance $Zsh$ in series with a variable voltage source $Vsh$ which can only inject reactive power into bus $i$. The equivalent of the branch is identical to that of STATCOM.

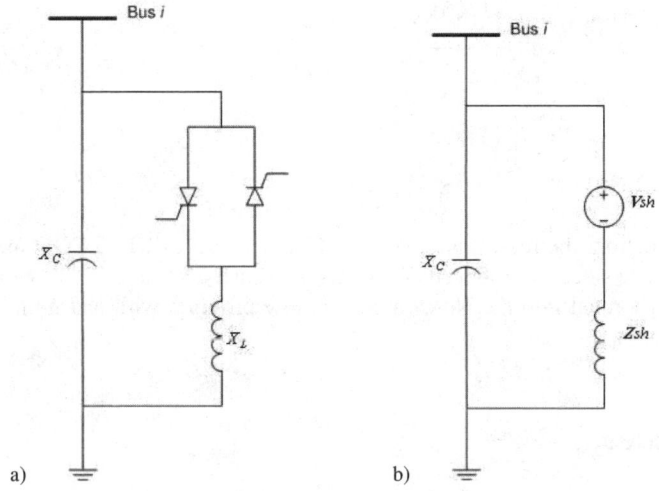

**Fig. 2.6.** a) A SVC with a fixed capacitor and TCR, b) Equivalent representation of SVC by STATCOM for power flow analysis

Hence, the STATCOM model can be applied directly to the SVC except that the following inequality should hold instead of (2.18):

$$X_{TCR}^{min} \leq X_{TCR} \leq X_{TCR}^{max} \tag{2.62}$$

where $X_{TCR}$ is given by:

$$X_{TCR} = |VshZsh/(V_i - Vsh)| \tag{2.63}$$

### 2.4.2 Representation of TCSC by SSSC in Power Flow Analysis

A typical TCSC, as shown in Fig. 2.7, can provide continuous control of power on the AC line with a variable series capacitive reactance. The TCSC consists of a fixed capacitor in parallel with a Thyristor Controlled Reactor (TCR). In principle, a TCSC is very similar to a SVC. The difference between them is that the former is usually series connected with a transmission line while the latter is usually shunt connected with a local bus.

Similarly, a TCSC can be represented by an equivalent SSSC for power flow analysis.

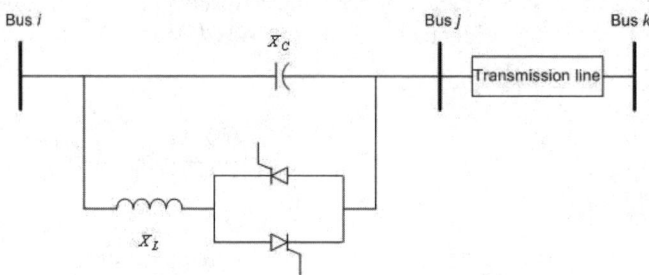

**Fig. 2.7.** A typical TCSC

In this section, the representation of SVC and TCSC as STATCOM and SSSC, respectively has been introduced. With this approach SVC and TCSC can be very easily incorporated into the Newton power flow program with just minor modifications of computer code.

## References

[1] Stott B (1974) Review of load-flow calculation methods. Proceedings of the IEEE, vol 62, no 7 pp 916-929
[2] Ward JB, Hale HW (1956) Digital computer solution of power flow problems. AIEE Trans. on Power App. Syst, vol 75, pp 398-404

[3] Brown HE, Carter GE, Happ HH, Person CE (1963) Power flow solution by impedance matrix iterative method. IEEE Transactions on Power App. Syst., vol 82, no 1, pp 1-10
[4] Van Ness JE, Griffin JH (1961) Elimination methods for load flow studies. AIEE Trans. on Power App. Syst., vol 80, pp 299-304
[5] Tinney WF, C.E. Hart (1967) Power flow solution by Newton's method. IEEE Trans. on Power App. Syst., vol 86, no 11, pp1449-1456
[6] Stott B (1972) Decoupled Newton power flow. IEEE Transactions on Power App. Syst, vol 91, pp. 1955-1959
[7] Stott B, Alsac O (1974) Fast decoupled load flow. IEEE Transactions on Power App. Syst, vol 93, no 3, pp 859-869
[8] Sasson AM, Trevino C, Aboytes F (1971) Improved Newton's load flow through a minimization technique. IEEE Transactions on Power Apparatus and Systems, vol 90, no 5, pp. 1974-1981
[9] Sachdev MS, Medicherla TKP (1977) A second order load flow technique. IEEE Transactions on Power Apparatus and Systems, vol 96, no 1.
[10] Iwamoto S, Tamura (1981) A load flow calculation method for ili-conditioned power systems. IEEE Transactions on Power Apparatus and Systems, vol 100, no 4, pp. 1736-1743
[11] Britton JP (1969) Improved area interchange control for Newton's method load flows. IEEE Transactions on Power Apparatus and Systems, vol 88, no 10, pp1577-1581
[12] Peterson N.M., Meyer WS (1971) Automatic adjustment of transformer and phase-shifter taps in the Newton power flow. IEEE Transactions on Power Apparatus and Systems, vol 90, no 1, pp103-108
[13] Britton JP (1971) Improved load flow performance through a more general equation form. IEEE Transactions on Power Apparatus and Systems, vol 90, no 1, pp.109-116
[14] Tinney W, Walker J (1967) Direct solution of sparse network equations by optimally ordered triangular factorization. Proceedings of the IEEE, vol 55, no 11, pp1801-1809
[15] Song YH, John AT (1999) Flexible AC Transmission Systems. IEE Press, London
[16] Hingorani NG, Gyugyi L (2000) Understanding FACTS – concepts and technology of flexible ac transmission systems. New York: IEEE Press
[17] Schauder C, Gernhardt M, Stacey E, Lemak T, Gyugyi L, Cease TW, Edris A (1995) Development of a ±100MVar static condenser for voltage control of transmission systems. IEEE Transactions on Power Delivery, vol 10, no 3, pp1486-1493
[18] Gyugyi L, Shauder CD, Sen KK (1997) Static synchronous series compensator: a solid-state approach to the series compensation of transmission lines. IEEE Transactions on Power Delivery; vol 12, no 1, pp 406-413
[19] Sen KK (1998) SSSC - Static synchronous series compensator: theory, modeling, and applications. IEEE Transactions on Power Delivery, vol. 13, no 1, pp 241-246
[20] Gyugyi L, Shauder CD, Williams SL, Rietman TR, Torgerson DR, Edris A (1995) The unified power flow controller: a new approach to power transmission control. IEEE Transactions on Power Delivery, vol 10, no 2, pp 1085-1093
[21] Sen KK, Stacey EJ (1998) UPFC – Unified power flow controller: theory, modeling and applications. IEEE Trans. on Power Delivery, vol 13, no 4, pp 1453-1460
[22] Fardanesh B, Henderson M, Shperling B, Zelingher S, Gyugyi L, Schauder C, Lam B, Mounford J, Adapa R, Edris A (1998) Convertible static compensator: application to the New York transmission system. CIGRE 14-103, CIGRE Session 1998, Paris, France, September.
[23] Fardanesh B, Shperling B, Uzunovic E, Zelingher S (2000) Multi-converter FACTS devices: the generalized unified power flow controller (GUPFC). Proceedings of IEEE 2000 PES Summer Meeting, Seattle, USA.

[24] Zhang XP, Handschin E, Yao MM (2001) Modeling of the generalized unified power flow controller in a nonlinear interior point OPF. IEEE Trans. on Power Systems, vol 16, no 3, pp 367-373.
[25] Zhang XP (2003) Modeling of the interline power flow controller and generalized unified power flow controller in Newton power flow. IEE Proc. - Generation, Transmission and Distribution, vol. 150, no. 3, pp 268-274.
[26] Asplund G, Eriksson K, Svensson K (1997) DC transmission based on voltage source converters. CIGRE SC14 Colloquium, South Africa
[27] Asplund G (2000) Application of HVDC light to power system enhancement. Proceedings of IEEE 2000 PES Winter Meeting, Sigapore
[28] Schetter F, Hung H, Christl N (2000) HVDC transmission system using voltage sourced converters – design and applications. Proceedings of IEEE 2000 PES Summer Meeting, Seattle, USA
[29] Lasson T, Edris A, Kidd D, Aboytes F (2001) Eagle pass back-to-back tie: a dual purpose application of voltage source converter technology. Proceedings of IEEE 2001 PES Summer Meeting, Vancouver, Canada
[30] Jiang H, Ekstrom A (1998) Multiterminal HVDC systems in urban areas of large cities. IEEE Transactions on Power Delivery, vol 13, no 4, pp 1278–1284
[31] Lu W, Ooi BT (2003) DC overvoltage control during loss of converter in multiterminal voltage-source converter-based HVDC (M-VSC-HVDC). IEEE Trans. on Power Delivery, vol 18, no 3, pp 915-920
[32] Zhang XP (2004) Multiterminal voltage-sourced converter based HVDC models for power flow analysis. IEEE Transactions on Power Systems, vol 18, no 4, 2004, pp1877-1884
[33] Miller TJE, Ed. (1982) Reactive Power Control in Electric Power Systems, John Wiley & Sons, New York
[34] Zhang XP, Handschin E, Yao M (2004) Multi-control functional static synchronous compensator (STATCOM) in power system steady state operations. Journal of Electric Power Systems Research, vol 72, no 3, pp 269-278

# 3 Modeling of Multi-Converter FACTS in Power Flow Analysis

This chapter discusses the recent developments in modeling of multi-functional multi-converter FACTS-devices in power flow analysis. The objectives of this chapter are:

1. to model not only the well-recognized two-converter shunt-series FACTS-device - UPFC, but also the latest multi-line FACTS-devices such as IPFC, GUPFC, VSC-HVDC and M-VSC-HVDC in power flow analysis,
2. to establish multi-control functional models of these multi-converter FACTS-devices to compare the control performance of these FACTS-devices.
3. to handle the small impedances of coupling transformers of FACTS-devices in power flow analysis.

## 3.1 Modeling of Multi-Control Functional UPFC

Among the converter based FACTS-devices, the Unified Power Flow Controller (UPFC) [10][11] is a versatile FACTS-device, which can simultaneously control a local bus voltage and power flows of a transmission line and make it possible to control circuit impedance, voltage angle and power flow for optimal operation performance of power systems. In recent years, there has been increasing interest in computer modeling of the UPFC in power flow and optimal power flow analysis [12],[15]-[24], However, in the most recent research work, the UPFC is primarily used to control a local bus voltage and active and reactive power flows of a transmission line. As reported in [24], in practice, the UPFC series converter may have other control modes such as direct voltage injection, phase angle shifting and impedance control modes, etc.

In contrast to the practical control possibilities of the UPFC, there has been a lack of modelling of the various control modes in power system analysis. In this section, besides the basic active and reactive power flow control mode, twelve new UPFC control modes are presented. The new modes include direct voltage injection, bus voltage regulation, line impedance compensation and phase angle regulation, etc. Mathematical modelling of these control modes is presented. Detailed implementation of the UPFC model with the twelve control modes in power flow analysis is given.

## 3.1.1 Advanced UPFC Models for Power Flow Analysis

### 3.1.1.1 Operating Principles of UPFC

The basic operating principle diagram of an UPFC is shown in Fig. 3.1 [10]. The UPFC consists of two switching converters based on VSC valves. The two converters are connected by a common DC link. The series inverter is coupled to a transmission line via a series transformer. The shunt inverter is coupled to a local bus $i$ via a shunt-connected transformer. The shunt inverter can generate or absorb controllable reactive power, and it can provide active power exchange to the series inverter to satisfy operating control requirements.

Based on the operating diagram of Fig. 3.1, an equivalent circuit shown in Fig. 3.2 can be established. In Fig. 3.2, the phasors $V_{sh}$ and $V_{se}$ represent the equivalent, injected shunt voltage and series voltage sources, respectively. $Z_{sh}$ and $Z_{se}$ are the UPFC series and shunt coupling transformer impedances, respectively.

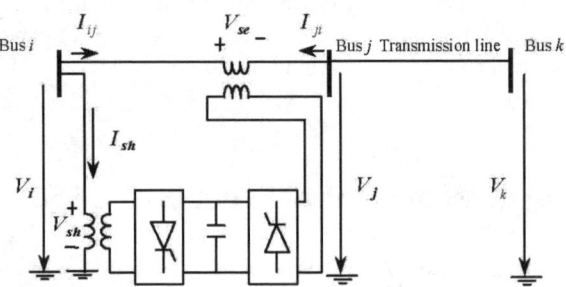

**Fig. 3.1.** Operating principle of UPFC

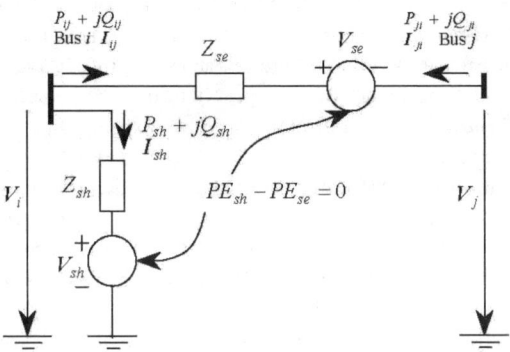

**Fig. 3.2.** Equivalent circuit of UPFC

$V_i$ and $V_j$ are voltages at buses $i$, $j$, respectively while $V_k$ is the voltage of bus $k$ of the receiving-end of the transmission line. $I_{sh}$ is the current through the UPFC shunt converter. $P_{sh}$ and $Q_{sh}$ are the shunt converter branch active and reactive power flows, respectively. The power flow direction of $P_{sh}$ and $Q_{sh}$ is leaving bus $i$. $I_{ij}$ and $I_{ji}$ are the currents through the UPFC series converter, and $I_{ij} = -I_{ji}$. $P_{ij}$ and $Q_{ij}$ are the UPFC series active and reactive power flows, respectively, leaving bus $i$. $P_{ji}$ and $Q_{ji}$ are the UPFC series branch active and reactive power flows, respectively, leaving bus $j$. $P_{sh}$ is the real power exchange of the shunt converter with the DC link. $P_{se}$ is the real power exchange of the series converter with the DC link.

### 3.1.1.2 Power Flow Constraints of UPFC

For the equivalent circuit of the UPFC shown in Fig. 3.2, suppose $V_{sh} = V_{sh} \angle \theta_{sh}$, $V_{se} = V_{se} \angle \theta_{se}$, $V_i = V_i \angle \theta_i$, $V_j = V_j \angle \theta_j$; then the power flow constraints of the UPFC shunt and series branches are:

$$P_{sh} = V_i^2 g_{sh} - V_i V_{sh} (g_{sh} \cos(\theta_i - \theta_{sh}) + b_{sh} \sin(\theta_i - \theta_{sh})) \tag{3.1}$$

$$Q_{sh} = -V_i^2 b_{sh} - V_i V_{sh} (g_{sh} \sin(\theta_i - \theta_{sh}) - b_{sh} \cos(\theta_i - \theta_{sh})) \tag{3.2}$$

$$\begin{aligned} P_{ij} = &V_i^2 g_{ij} - V_i V_j (g_{ij} \cos \theta_{ij} + b_{ij} \sin \theta_{ij}) \\ &- V_i V_{se} (g_{ij} \cos(\theta_i - \theta_{se}) + b_{ij} \sin(\theta_i - \theta_{se})) \end{aligned} \tag{3.3}$$

$$\begin{aligned} Q_{ij} = &-V_i^2 b_{ij} - V_i V_j (g_{ij} \sin \theta_{ij} - b_{ij} \cos \theta_{ij}) \\ &- V_i V_{se} (g_{ij} \sin(\theta_i - \theta_{se}) - b_{ij} \cos(\theta_i - \theta_{se})) \end{aligned} \tag{3.4}$$

$$\begin{aligned} P_{ji} = &V_j^2 g_{ij} - V_i V_j (g_{ij} \cos \theta_{ji} + b_{ij} \sin \theta_{ji}) \\ &+ V_j V_{se} (g_{ij} \cos(\theta_j - \theta_{se}) + b_{ij} \sin(\theta_j - \theta_{se})) \end{aligned} \tag{3.5}$$

$$\begin{aligned} Q_{ji} = &-V_j^2 b_{ij} - V_i V_j (g_{ij} \sin \theta_{ji} - b_{ij} \cos \theta_{ji}) \\ &+ V_j V_{se} (g_{ij} \sin(\theta_j - \theta_{se}) - b_{ij} \cos(\theta_j - \theta_{se})) \end{aligned} \tag{3.6}$$

where $g_{sh} + jb_{sh} = 1/Z_{sh}$, $g_{ij} + jb_{ij} = 1/Z_{se}$, $\theta_{ij} = \theta_i - \theta_j$, $\theta_{ji} = \theta_j - \theta_i$.

### 3.1.1.3 Active Power Balance Constraint of UPFC

The operating constraint of the UPFC (active power exchange between two inverters via the DC link) is:

$$\Delta P_\Sigma = PE_{sh} - PE_{se} = 0 \tag{3.7}$$

where $PE_{sh} = \text{Re}(V_{sh} I_{sh}^*)$ and $PE_{se} = \text{Re}(V_{se} I_{ji}^*)$ are active power exchange of the shunt converter and the series converter with the DC link, respectively. The symbol * represents conjugate.

### 3.1.1.4 Novel Control Modes of UPFC

For a UPFC, steady control for voltage and power flow is implemented as follows:

- The local voltage magnitude of bus $i$ is controlled;
- Active and reactive power flows, namely, $P_{ji}$ and $Q_{ji}$ (or $P_{jk}$ and $Q_{jk}$), of the transmission line are controlled.

The above voltage and power flow control has been used widely in UPFC models [15]-[22]. It has been recognised that besides the power flow control, UPFC has the ability to control angle, voltage and impedance or combination of those [23]. However, research work in the modeling of these controls is very limited [24]. In the following, the possibilities of alternative voltage, angle, impedance and power flow control modes or combination of these controls will be presented. Here we try to explore the control modes, discuss the similarities and differences between some of the control modes and those of traditional transformers and series compensation devices, and investigate the mathematical modeling of these control modes.

*Mode 1: Active and reactive power flow control*
The well-known independent active and reactive power flows control is:

$$P_{ji} - P_{ji}^{Spec} = 0 \tag{3.8}$$

$$Q_{ji} - Q_{ji}^{Spec} = 0 \tag{3.9}$$

where $P_{ji}^{Spec}$ is the specified active power flow control reference. $Q_{ji}^{Spec}$ is the specified reactive power flow control reference.

*Mode 2: Power flow control by voltage shifting*
In this control mode, the active power flow is controlled by voltage shifting between bus i and bus j while the voltage at bus j is equal to the voltage at bus $i$. The control constraints are:

$$P_{ji} - P_{ji}^{Spec} = 0 \tag{3.10}$$

$$V_i - V_j = 0 \tag{3.11}$$

where $P_{ji}^{Spec}$ is the specified active power flow control reference. For this control mode, the UPFC is very similar to a phase shifting transformer for active power flow control. However, the significant difference between them is that besides the power flow control, the UPFC also has powerful shunt reactive power or voltage control capability.

*Mode 3: General Direct Voltage Injection*

In this control mode, both the series voltage magnitude and angle are specified. The control mode is:

$$V_{se} - V_{se}^{Spec} = 0 \tag{3.12}$$

$$\theta_{se} - \theta_{se}^{Spec} = 0 \tag{3.13}$$

where $V_{se}^{Spec}$ and $\theta_{se}^{Spec}$ are the specified series voltage magnitude and angle control references, respectively.

*Mode 4: Direct Voltage Injection with $V_{se}$ in phase with $V_i$*

In this control mode, the series voltage magnitude is specified while $V_{se}$ is in phase with $V_i$. The control mode is:

$$V_{se} - V_{se}^{Spec} = 0 \tag{3.14}$$

$$\theta_{se} - \theta_i = 0 \text{ or } \theta_{se} - \theta_i - 180° = 0 \tag{3.15}$$

where $V_{se}^{Spec}$ is the specified series voltage magnitude control reference. This control mode is very similar to the function of a traditional ideal transformer.

The tap ratio of the above control mode is $V_i/(V_i + V_{se})$ or $V_i/(V_i - V_{se})$. The difference between the UPFC and a transformer is that the former also has the ability to control bus voltage to a control reference by the reactive power control of the shunt converter.

*Mode 5: Direct Voltage Injection with $V_{se}$ in Quadrature with $V_i$ (lead)*

In this control mode, the series voltage magnitude is specified while $V_{se}$ is in quadrature with $V_i$, and $V_{se}$ leads $V_i$. The control mode is:

$$V_{se} - V_{se}^{Spec} = 0 \tag{3.16}$$

$$\theta_{se} - \theta_i - \frac{\pi}{2} = 0 \tag{3.17}$$

where $V_{se}^{Spec}$ is the specified series voltage magnitude control reference. This control mode is to emulate the traditional Quadrature Boosting transformer.

*Mode 6: Direct Voltage Injection with $V_{se}$ in Quadrature with $V_i$ (lag)*

In this control mode, the series voltage magnitude is specified while $V_{se}$ is in quadrature with $V_i$, and $V_{se}$ lags $V_i$. The control mode is:

$$V_{se} - V_{se}^{Spec} = 0 \tag{3.18}$$

$$\theta_{se} - \theta_i + \frac{\pi}{2} = 0 \tag{3.19}$$

where $V_{se}^{Spec}$ is the specified series voltage magnitude control reference. This control mode is also to emulate the traditional Quadrature Boosting transformer.

*Mode 7: Direct Voltage Injection with $V_{se}$ in Quadrature with $I_{ij}$ (lead)*

In this control mode, the series voltage magnitude is specified while $V_{se}$ is in Quadrature with $I_{ij}$. $V_{se}$ leads $I_{ij}$. The control mode is:

$$V_{se} - V_{se}^{Spec} = 0 \tag{3.20}$$

$$\text{Im}[V_{se}(I_{ij}e^{j90°})] = 0 \tag{3.21}$$

where $V_{se}^{Spec}$ is the specified series voltage magnitude control reference.

*Mode 8: Direct Voltage Injection with $V_{se}$ in Quadrature with $I_{ij}$ (lag)*

In this control mode, the series voltage magnitude is specified while $V_{se}$ is in quadrature with $I_{ij}$. $V_{se}$ lags $I_{ij}$. The control mode is:

$$V_{se} - V_{se}^{Spec} = 0 \tag{3.22}$$

$$\text{Im}[V_{se}(I_{ij}e^{-j90°})] = 0 \tag{3.23}$$

where $V_{se}^{Spec}$ is the specified series voltage magnitude control reference.

*Mode 9: Voltage Regulation with $V_{se}$ in phase with $V_i$*

In this control mode, the $V_i$ magnitude is controlled while $V_{se}$ is in phase with $V_i$. The control mode is:

$$V_j - V_j^{Spec} = 0 \tag{3.24}$$

$$\theta_{se} - \theta_i = 0 \tag{3.25}$$

where $V_j^{Spec}$ is the voltage magnitude control reference at bus $j$.

## Mode 10: Phase Shifting Regulation

In this control mode, $V_{se}$ is regulated to control the voltage magnitudes at buses $i$ and $j$ to be equal while the phase shifting between $V_i$ and $V_j$ is controlled to a specified angle reference. The control mode is:

$$V_i - V_j = 0 \tag{3.26}$$

$$\theta_i - \theta_j - \theta_{ij}^{Spec} = 0 \tag{3.27}$$

where $\theta_{ij}^{Spec}$ is the specified phase angle control reference. This control mode is to emulate the function of a traditional phase shifting transformer.

## Mode 11: Phase Shifting and Quadrature Regulation (lead)

In this control mode, $V_{se}$ is regulated to control the voltage magnitudes at buses $i$ and $j$ to be equal while $V_{se}$ is in quadrature with $V_i$, and leads $V_i$. The control mode is:

$$V_i - V_j = 0 \tag{3.28}$$

$$\theta_{se} - \theta_i - \frac{\pi}{2} = 0 \tag{3.29}$$

## Mode 12: Phase Shifting and Quadrature Regulation (lag)

In this control mode, $V_{se}$ is regulated to control the voltage magnitudes at buses $i$ and $j$ to be equal while $V_{se}$ is in quadrature with $V_i$, and lags $V_i$. The control mode is:

$$V_i - V_j = 0 \tag{3.30}$$

$$\theta_{se} - \theta_i + \frac{\pi}{2} = 0 \tag{3.31}$$

## Mode 13: Line Impedance Compensation

In this control mode, $V_{se}$ is regulated to control the equivalent reactance of the UPFC series voltage source to a specified impedance reference. The control mode is:

$$R_{se} - Z_{se}^{Spec} \cos \gamma_{se}^{Spec} = 0 \tag{3.32}$$

$$X_{se} - Z_{se}^{Spec} \sin \gamma_{se}^{Spec} = 0 \tag{3.33}$$

where $R_{se} + jX_{se}$ is the equivalent impedance of the series voltage source. $Z_{se}^{Spec} \angle \gamma_{se}^{Spec}$ is the impedance control reference.

For the impedance control by the UPFC, the reactance may be either capacitive or inductive. Special cases of impedance compensation such as purely capacitive and inductive compensation can be emulated. These two cases are very similar to the traditional compensation techniques using a capacitor and a reactor. However, the impedance control by the UPFC is more powerful since not only the reactance but also the resistance can be compensated.

The control equations of any control mode above can be generally written as:

$$\Delta F(x, f^{Spec}) = 0 \tag{3.34}$$

$$\Delta G(x, g^{Spec}) = 0 \tag{3.35}$$

where $x = [\theta_i, V_i, \theta_j, V_j, \theta_{se}, V_{se}]^T$. $f^{Spec}$ and $g^{Spec}$ are control references.

In the multi-control functional model of UPFC, only the series control modes with two degrees of freedom have been described. It is imaginable that the shunt control modes of STATCOM discussed in chapter 2 are applicable to the shunt control of UPFC.

### 3.1.2 Implementation of Advanced UPFC Model in Newton Power Flow

#### 3.1.2.1 Modeling of UPFC in Newton Power Flow

Assuming that the shunt converter of the UPFC is used to control voltage magnitude at bus $i$, a Newton power flow algorithm with simultaneous solution of power flow constraints and power flow control constraints of the UPFC may be represented by:

$$\mathbf{J}\Delta\mathbf{X} = -\Delta\mathbf{R} \tag{3.36}$$

Here, $\mathbf{J}$ is the Jacobian matrix, $\Delta\mathbf{X}$ is the incremental vector of state variables and $\Delta\mathbf{R}$ is the power and control mismatch vector:

$$\Delta\mathbf{X} = [\Delta\theta_{se}, \Delta V_{se}, \Delta\theta_{sh}, \Delta V_{sh}, \Delta\theta_i, \Delta V_i, \Delta\theta_j, \Delta V_j]^T \tag{3.37}$$

$$\Delta\mathbf{R} = [\Delta F, \Delta G, \Delta P_\Sigma, V_i - V_i^{Spec}, \Delta P_i, \Delta Q_i, \Delta P_j, \Delta Q_j]^T \tag{3.38}$$

$$\mathbf{J} = \frac{\partial \Delta\mathbf{R}}{\partial \mathbf{X}} \tag{3.39}$$

where $\Delta P_i$ and $\Delta Q_i$ are power mismatches at bus $i$ while $\Delta P_j$ and $\Delta Q_j$ are power mismatches at bus $j$.

### 3.1.2.2 Modeling of Voltage and Current Constraints of the UPFC

The voltage and current constraints of the shunt branch of the UPFC are given by (2.18) and (2.20) while the voltage and current constraints of the series branch of the UPFC are given by (2.46) and (2.48).

As it was discussed in section 2.2 of chapter 2, the basic constraint enforcement strategy is that, when there is a voltage or current inequality constraint of the UPFC is violated, the constraint is enforced by being kept at its limit while the control equality constraint of the UPFC is released. In principle, a series inequality constraint is enforced by releasing a series control constraint; a shunt inequality constraint is enforced by releasing a shunt control constraint.

### 3.1.2.3 Initialization of UPFC Variables in Newton Power Flow

For the initialization of the series converter for power flow control mode, (3.8) and (3.9) can be applied. Assuming that shunt control is the control of the voltage magnitude of the local bus, $Vsh$ may be determined by:

$$Vsh = (Vsh^{max} + Vsh^{min})/2 \text{ or } Vsh = V^{Spec} \tag{3.40}$$

then $\theta sh$ can be found by solving (3.7):

$$\theta sh = -\sin^{-1}[B/(V_i Vsh\sqrt{(gsh^2 + bsh^2)})] + \tan^{-1}(-gsh/bsh) \tag{3.41}$$

where:

$$\begin{aligned} B = & Vsh^2 gsh + Vse^2 g_{ij} \\ & + V_i Vse(g_{ij} \cos(\theta_j - \theta se) - b_{ij} \sin(\theta_j - \theta se)) \\ & - V_i Vse(g_{ij} \cos(\theta_i - \theta se) - b_{ij} \sin(\theta_i - \theta se)) \end{aligned} \tag{3.42}$$

For other control modes, similar initialization may be derived.

### 3.1.3 Numerical Results

Numerical results are given for tests carried out on the IEEE 30-bus system and the IEEE 118-bus system. In the tests, a convergence tolerance of $10^{-12}$ p.u. (or $10^{-10}$ MW/MVAr) for maximal absolute bus power mismatches and power flow control mismatches is utilized.

In order to show the capabilities of the UPFC model and the performance of the Newton power flow algorithm, 14 cases including the base case have been investigated. In case 2–14, a UPFC is installed between bus 12 and the sending end of the transmission line 12-15.

The computational results are summarized in Table 3.1. In the simulations, the bus voltage control reference is $V_{12}^{Spec} = 1.05$ p.u.

**Table 3.1.** Results of the IEEE 30 bus system

| Case No. | Control mode | UPFC series control reference | Solution of the UPFC series voltage | Number of iterations |
|---|---|---|---|---|
| 1 | Base Case | None | None | 4 |
| 2 | 1 | $P^{Spec} = -30e^{-2}\ p.u.$ <br> $Q^{Spec} = -5e^{-2}\ p.u.$ | $\theta_{se} = -90.48°$ <br> $V_{se} = 0.0681\ p.u.$ | 7 |
| 3 | 2 | $P^{Spec} = -30e^{-2}\ p.u.$ | $\theta_{se} = -103.13°$ <br> $V_{se} = 0.12916\ p.u.$ | 8 |
| 4 | 3 | $\theta_{se}^{Spec} = 45°$ <br> $V_{se}^{Spec} = 0.2$ | $\theta_{se} = 45°$ <br> $V_{se} = 0.2\ p.u.$ | 5 |
| 5 | 4 | $V_{se}^{Spec} = 0.2$ | $\theta_{se} = -9.63°$ <br> $V_{se} = 0.2\ p.u.$ | 5 |
| 6 | 5 | $V_{se}^{Spec} = 0.2$ | $\theta_{se} = 81.42°$ <br> $V_{se} = 0.2\ p.u.$ | 5 |
| 7 | 6 | $V_{se}^{Spec} = 0.2$ | $\theta_{se} = 101.86°$ <br> $V_{se} = 0.2\ p.u.$ | 5 |
| 8 | 7 | $V_{se}^{Spec} = 0.1$ | $\theta_{se} = 18.10°$ <br> $V_{se} = 0.1\ p.u.$ | 10 |
| 9 | 8 | $V_{se}^{Spec} = 0.1$ | $\theta_{se} = -107.92°$ <br> $V_{se} = 0.1\ p.u.$ | 7 |
| 10 | 9 | $\gamma_{se}^{Spec} = -80°$ <br> $Z_{se}^{Spec} = 0.1$ | $\theta_{se} = -117.67°$ <br> $V_{se} = 0.02538\ p.u.$ | 7 |
| 11 | 10 | $\theta_{ij}^{Spec} = 10°$ | $\theta_{se} = -78.95°$ <br> $V_{se} = 0.18790\ p.u.$ | 5 |
| 12 | 11 | No explicit control reference | $\theta_{se} = 81.10°$ <br> $V_{se} = 0.11724\ p.u.$ | 8 |

**Table 3.1.** (cont.)

| | | | | |
|---|---|---|---|---|
| 13 | 12 | No explicit control reference | $\theta_{se} = -100.25°$<br>$V_{se} = 0.06334\ p.u.$ | 6 |
| 14 | 13 | $V_j^{Spec} = 1.02\ p.u.$ | $\theta_{se} = -9.60°$<br>$V_{se} = 0.02829\ p.u.$ | 4 |

In the cases above and the following discussions, the control references of active and reactive power flows are referred to $P_{ji}^{Spec}, Q_{ji}^{Spec}$, which are at the sending end of a transmission line. It can be seen that the Newton power flow algorithm can converge for all the control modes with very tight tolerance.

Further test cases were also carried out on the IEEE 118-bus system, which are presented as follows:

*Case 15*: This is the base case system without UPFCs.

*Case 16*: In this case, three UPFCs are installed on the IEEE 118-bus system. The three UPFCs are installed, one each on line 21-20, line 45-44 and line 94-95. The three UPFCs are using series control mode 1, 7, 13, respectively when the shunt control of the three UPFC is to control the voltages at buses 21, 45 and 94, respectively.

The quadratic convergence characteristics of case 15 (Base case without UPFC) and case 16 (with three UPFCs) are shown in Fig. 3.3. This shows that the Newton power flow algorithm can converge in 6 iterations.

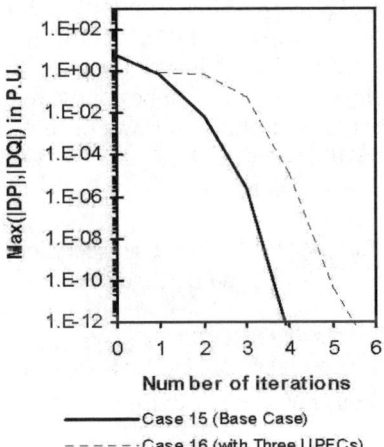

**Fig. 3.3.** Power and control mismatches as functions of number of iterations for the IEEE 118-bus system

Besides the well-known active and reactive power flow control mode, some twelve-control modes for the UPFC have been proposed. Mathematical modeling of these control modes has been described. The new control modes proposed are complementary to the well-known independent active and reactive power control mode by the UPFC, and will be helpful to fully understand the control capabilities of the UPFC. The new control modes of the UPFC have been successfully implemented in a Newton power flow program. Furthermore, the similarities and differences between the traditional angle, voltage and impedance control devices and the UPFC have also been discussed.

## 3.2 Modeling of Multi-Control Functional IPFC and GUPFC

As discussed in chapter 1, a series-series FACTS device called Interline Power Flow Controller (IPFC) was recently installed at NYPA's Marcy Substation, which can increase power transfer capability and maximize the use of the existing transmission network. The salient features of the IPFC are its convertibility and expandability, which are becoming increasingly important as electric utilities are being transformed into highly competitive marketplaces. The functional convertibility enables the IPFC to adapt to changing system operating requirements and changing power flow patterns. The expandability of the IPFC is that a number of voltage-source converters coupled with a common DC bus can be operated. Additional compatible converter or converters can be connected to the common DC bus to expand the functional capabilities of the IPFC. The convertibility and expandability of the CSC enables it to be operated in various configurations. The IPFC installed at NYPA consists of two converters, and it can operate as a Static Synchronous Shunt Compensator (STATCOM) [7], Static Synchronous Series Compensator (SSSC) [8], Unified Power Flow Controller (UPFC) [10] or the innovative Interline Power Flow Controller (IPFC) [2][3][6]. In principle, with an extra shunt converter, a Generalized Unified Power Flow Controller (GUPFC) [4], [5], which requires at least three converters, can be configured The IPFC and GUPFC are significantly extended to control power flows of multi-lines or a sub-network beyond that achievable by the UPFC or SSSC or STATCOM. In principle, with at least two converters, an IPFC can be configured. With at least three converters, a GUPFC can be configured.

The GUPFC model for EMTP simulation has been proposed [4]. A model of the GUPFC with voltage and active and reactive power flow control has been proposed and successfully implemented in an optimal power flow algorithm [5]. The power flow model of the IPFC and GUPFC is presented in [6]. The detailed modeling of novel and versatile FACTS-devices – the IPFC and GUPFC under practical operating inequality constraints in power flow analysis will be presented here.

## 3.2.1 Mathematical Modeling of IPFC in Newton Power Flow under Practical Constraints

### 3.2.1.1 Mathematical Model of the IPFC

The IPFC obtained by combining two or more series-connected converters working together extends the concept of power flow control beyond what is achievable with the known one-converter series FACTS-device - SSSC [8][9][14]. A simplest IPFC, with three FACTS buses – $i$, $j$ and $k$ shown functionally in Fig. 3.4, is used to illustrate the basic operation principle [2][3][6]. The IPFC consists of two converters being series-connected with two transmission lines via transformers. It can control three power system quantities - independent three power flows of the two lines. It can be seen that the sending-ends of the two transmission lines are series-connected with the FACTS buses $j$ and $k$, respectively.

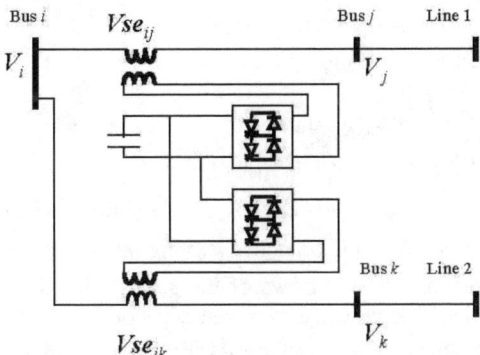

**Fig. 3.4.** Operational principle of the simplest IPFC with two converters

An equivalent circuit of the IPFC with two controllable series injected voltage sources is shown in Fig. 3.5. The real power can be exchanged between or among the series converters via the common DC link while the sum of the real power exchange should be zero.

Suppose in Fig. 3.5 the series transformer impedance is $Zse_{in}$, and the controllable injected voltage source is $Vse_{in} = Vse_{in} \angle \theta se_{in}$ ($n = j, k$). Active and reactive power flows of the FACTS branches leaving buses $i$, $j$, $k$ are given by:

$$P_{in} = V_i^2 g_{in} - V_i V_n (g_{in} \cos\theta_{in} + b_{in} \sin\theta_{in}) \\ - V_i Vse_{in}(g_{in}\cos(\theta_i - \theta se_{in}) + b_{in}\sin(\theta_i - \theta se_{in})) \quad (3.43)$$

$$Q_{in} = -V_i^2 b_{in} - V_i V_n (g_{in} \sin\theta_{in} - b_{in} \cos\theta_{in}) \\ - V_i Vse_{in}(g_{in}\sin(\theta_i - \theta se_{in}) - b_{in}\cos(\theta_i - \theta se_{in})) \quad (3.44)$$

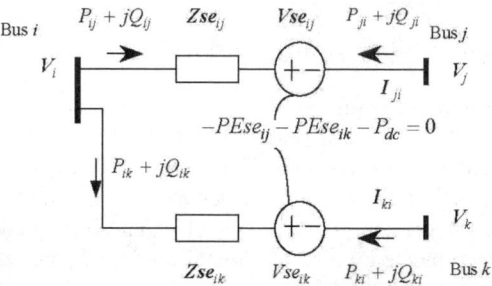

**Fig. 3.5.** Equivalent circuit of the simplest IPFC with two converters

$$P_{ni} = V_n^2 g_{in} - V_i V_n (g_{in} \cos(\theta_n - \theta_i) + b_{in} \sin(\theta_n - \theta_i)) \\ + V_n V se_{in} (g_{in} \cos(\theta_n - \theta se_{in}) + b_{in} \sin(\theta_n - \theta se_{in})) \quad (3.45)$$

$$Q_{ni} = -V_n^2 b_{nn} - V_i V_n (g_{in} \sin(\theta_n - \theta_i) - b_{in} \cos(\theta_n - \theta_i)) \\ + V_n V se_{in} (g_{in} \sin(\theta_n - \theta se_{in}) - b_{in} \cos(\theta_n - \theta se_{in}) \quad (3.46)$$

where $g_{in} = \text{Re}(1/Zse_{in})$, $b_{in} = \text{Im}(1/Zse_{in})$. $P_{in}$, $Q_{in}$ ($n=j, k$) are the active and reactive power flows of two IPFC branches leaving bus $i$ while $P_{ni}$, $Q_{ni}$ ($n = j, k$) are the active and reactive power flows of the series FACTS branch $n$-$i$ leaving bus $n$ ($n = j, k$), respectively. Since two transmission lines are series connected with the FACTS branches $i$-$j$, $i$-$k$ via the FACTS buses $j$ and $k$, respectively, $P_{ni}$, $Q_{ni}$ ($n = j, k$) are equal to the active and reactive power flows at the sending-end of the transmission lines, respectively.

For the IPFC, the power mismatches at buses $i, j, k$ should hold:

$$\Delta P_m = Pg_m - Pd_m - P_m = 0 \quad (3.47)$$

$$\Delta Q_m = Qg_m - Qd_m - Q_m = 0 \quad (3.48)$$

where, without loss of generality, $Pg_m$, $Qg_m$ ($m=i, j, k$) are the real and reactive power generation entering the bus $m$, and $Pd_m$, $Qd_m$ ($m=i, j, k$) are the real and reactive power load leaving bus $m$. $P_m$, $Q_m$ ($m=i, j, k$) are the sum of real and reactive power flows of the circuits connected to bus $m$, which include the power flow contributions of the FACTS branches given by equations (3.47), (3.48).

According to the operating principle of the IPFC, the operating constraint representing the active power exchange between or among the series converters via the common DC link is:

$$PEx = -\sum PEse_{in} - P_{dc} = 0 \quad (3.49)$$

where $PEse_{in} = \text{Re}(Vse_{in}I_{ni}^{*})$ ($n = j, k$). * means complex conjugate. $I_{ni}$ ($n = j, k$) is the current through the series converter.

The IPFC shown in Fig. 3.4. and Fig. 3.5. can control both active and reactive power flows of primary line 1 but only active power flow (or reactive power flow) of secondary line 2. The active and reactive power flow control constraints of the IPFC are:

$$\Delta P_{ni} = P_{ni} - P_{ni}^{Spec} = 0 \tag{3.50}$$

$$\Delta Q_{ni} = Q_{ni} - Q_{ni}^{Spec} = 0 \tag{3.51}$$

where $n = j, k$. $P_{ni}^{Spec}$, $Q_{ni}^{Spec}$ are specified active and reactive power flow control references and $P_{ni} = \text{Re}(V_n I_{ni}^{*})$, $Q_{ni} = \text{Im}(V_n I_{ni}^{*})$

The constraints of each series converter are:

$$0 \leq \theta se_{in} \leq 2\pi \tag{3.52}$$

$$Vse_{in}^{min} \leq Vse_{in} \leq Vse_{in}^{max} \tag{3.53}$$

$$-PEse_{in}^{max} \leq PEse_{in} \leq PEse_{in}^{max} \tag{3.54}$$

$$I_{ni} \leq I_{ni}^{max} \quad (n = j, k) \tag{3.55}$$

where $n = j, k$. $PEse_{in}^{max}$ is the maximum limit of the power exchange of the series converter with the DC link. $PEse_{in} = \text{Re}(Vse_{ni}I_{in}^{*})$ ($n = j, k$). $I_{ni}^{max}$ is the current rating of the series converter.

### 3.2.1.2 Modeling of IPFC in Newton Power Flow

For the IPFC shown in Fig. 3.4. and Fig. 3.5., the primary series converter $i$-$j$ has two control degrees of freedom while the secondary series converter $i$-$k$ has one control degree of freedom since another control degree of freedom of the converter is used to balance the active power exchange between the two series converters. Combining power flow mismatch equations (3.47), (3.48), and operating and control equations (3.49)-(3.51), the Newton power flow solution may be given by

$$\mathbf{J}\Delta\mathbf{X} = -\Delta\mathbf{R} \tag{3.56}$$

where

$\Delta\mathbf{X}$ - the incremental vector of state variables, and $\Delta\mathbf{X} = [\Delta\mathbf{X}_1, \Delta\mathbf{X}_2]^T$

$\Delta\mathbf{X}_1 = [\Delta\theta_i, \Delta V_i, \Delta\theta_j, \Delta V_j, \Delta\theta_k, \Delta V_k]^T$ - the incremental vector of bus voltage magnitudes and angles.

$\Delta\mathbf{X}_2 = [\Delta\theta se_{ij}, \Delta Vse_{ij}, \Delta\theta se_{ik}, \Delta Vse_{ik}]^T$ - the incremental vector of the state variables of the IPFC.

$\Delta \mathbf{R}$ - the bus power mismatch and IPFC control mismatch vector, and $\Delta \mathbf{R} = [\Delta \mathbf{R}_1, \Delta \mathbf{R}_2]^T$.

$\Delta \mathbf{R}_1 = [\Delta P_i, \Delta Q_i, \Delta P_j, \Delta Q_j, \Delta P_k, \Delta Q_k]^T$ - bus power mismatch vector.

$\Delta \mathbf{R}_2 = [P_{ji} - P_{ji}^{Spec}, Q_{ji} - Q_{ji}^{Spec}, P_{ki} - P_{ki}^{Spec}, PEx]^T$ operating control mismatch vector of the IPFC

$\mathbf{J} = \dfrac{\partial \Delta \mathbf{R}}{\partial \mathbf{X}}$ - System Jacobian matrix

It is worth pointing out that for the reason mentioned above, for the secondary series converter $i$-$k$, there is only one associated active power flow control equation considered in (3.56). The Jacobian matrix in (3.56) can be partitioned into four blocks. The bottom diagonal block has very similar structure to that of conventional power flow. The other three blocks are FACTS related. The (3.56) can be solved by first eliminating $\Delta\theta se$, $\Delta Vse$ of the IPFC. However, this will result in new fill-in elements in the bottom diagonal block. Then the resulting reduced bottom diagonal block Newton equation can be solved by block sparse matrix techniques.

It should be pointed out that the multi-control modes of UPFC is applicable to IPFC. In addition, the techniques for the handling of the violated functional inequalities of STATCOM and SSSC [13] is applicable to IPFC.

### 3.2.1.3 Initialization of IPFC Variables in Newton Power Flow

With setting bus voltage $V_i$, $V_j$, $V_k$, $\theta_i$, $\theta_j$, $\theta_k$ to the flat start values, say $V_i = V_j = V_k = 1.0$ if buses $i$, $j$, $k$ are not voltage controlled buses, and $\theta_i = \theta_j = \theta_k = 0$, the initial values of $Vse_{ij}$, $\theta se_{ij}$ for the primary series converter $i$-$j$ can be found by solving two simultaneous equations (3.50) and (3.51):

$$Vse_{ij} = \sqrt{A/(g_{ij}^2 + b_{ij}^2)} / V_j \qquad (3.57)$$

$$\theta se_{ij} = \tan^{-1}[(P_{ji}^{Spec} - V_j^2 g_{jj} + V_i V_j g_{ij})/(Q_{ji}^{Spec} + V_j^2 b_{jj} - V_i V_j b_{ij})] \\ - \tan^{-1}(-g_{ij}/b_{ij}) \qquad (3.58)$$

where A is given by:

$$A = (P_{ji}^{Spec} - V_j^2 g_{jj} + V_i V_j g_{ij})^2 + (Q_{ji}^{Spec} + V_j^2 b_{jj} - V_i V_j b_{ij})^2 \qquad (3.59)$$

For the secondary series converter $i$-$k$, assume that $Vse_{ik}$ is chosen to a value between $Vse_{ik}^{min}$ and $Vse_{ik}^{max}$, then $\theta se_{ik}$ can be determined by solving (3.49):

$$\theta se_{ik} = \sin^{-1}[(P_{ki}^{Spec} - V_k^2 g_{ik} + V_i V_k g_{ik})/(V_k V se_{ki}\sqrt{g_{ik}^2 + b_{ik}^2})]$$
$$- \tan^{-1}(-g_{ik}/b_{ik}) \qquad (3.60)$$

## 3.2.2 Mathematical Modeling of GUPFC in Newton Power Flow under Practical Constraints

### 3.2.2.1 Mathematical Model of GUPFC

The GUPFC by combining three or more converters working together extends the concepts of voltage and power flow control of the known two-converter UPFC controller to multi-line voltage and power flow control [4][5]. A simplest GUPFC shown in Fig. 3.6 consists of three converters.

One converter is shunt-connected with a bus and the other two series-connected with two transmission lines via transformers in a substation. The GUPFC can explicitly control total five power system quantities such as the voltage magnitude of bus $i$ and independent active and reactive power flows of the two lines.

The equivalent circuit of the GUPFC including one controllable shunt injected voltage source and two controllable series injected voltage sources is shown in Fig. 3.7. Real power can be exchanged among the shunt and series converters via the common DC link, and the sum of the real power exchange should be zero.

$Zsh_i$ in Fig. 3.7 is the shunt transformer impedance, and $Vsh_i$ is the controllable shunt injected voltage of the shunt converter; $Psh_i$ is the power exchange of the shunt converter via the common DC link. Other variables and parameters are the same as those of Fig. 3.6 and Fig. 3.7. The controllable shunt injected voltage source is defined as $Vsh_i = Vsh_i \angle \theta sh_i$.

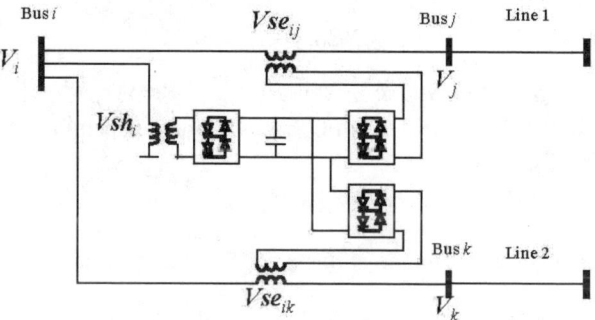

**Fig. 3.6.** Operational principle of the GUPFC with three converters

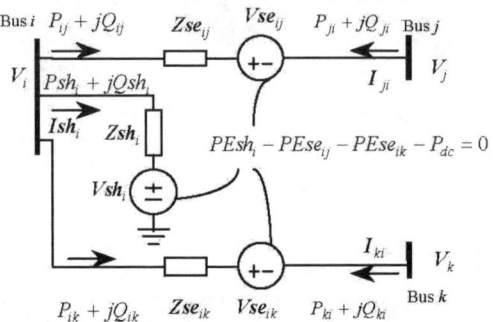

**Fig. 3.7.** The equivalent circuit of the GUPFC

Based on the equivalent circuit of the GUPFC shown in Fig. 3.7, the power flows of the shunt converter $Psh_i$, $Qsh_i$ leaving bus $i$ can be derived as:

$$Psh_i = V_i^2 gsh_i - V_i Vsh_i (gsh_i \cos(\theta_i - \theta sh_i) + bsh_i \sin(\theta_i - \theta sh_i)) \quad (3.61)$$

$$Qsh_i = -V_i^2 bsh_i - V_i Vsh_i (gsh_i \sin(\theta_i - \theta sh_i) - bsh_i \cos(\theta_i - \theta sh_i)) \quad (3.62)$$

while the active and reactive power flows $P_{in}$, $Q_{in}$ ($n=j, k$) are the same as those given by (3.50) and (3.51), respectively, and the active and reactive power flows $P_{ni}$, $Q_{ni}$ ($n=j, k$) are the same as those given by (3.45) and (3.46). In (3.50) and (3.51), $gsh_i = \text{Re}(1/\mathbf{Zsh}_i)$, $bsh_i = \text{Im}(1/\mathbf{Zsh}_i)$.

The bus power mismatch equations for the GUPFC are similar to (3.47), (3.48). The active and reactive power flow control constraints of the GUPFC are the same as those given by (3.50) and (3.51).

The operating constraint representing the active power exchange among converters via the common DC link is:

$$PEx = PEsh_i - \sum PEse_{in} - P_{dc} = 0 \quad (3.63)$$

where $n =j, k$. $PEsh_i = \text{Re}(\mathbf{Vsh}_i \mathbf{Ish}_i^*)$, and $PEse_{in} = \text{Re}(\mathbf{Vse}_{in} \mathbf{I}_{ni}^*)$

In contrast to the IPFC, the GUPFC has additional capability to control the voltage magnitude of bus $i$:

$$V_i - V_i^{Spec} = 0 \quad (3.64)$$

where $V_i$ is the voltage magnitude at bus $i$. $V_i^{Spec}$ is the specified bus voltage control reference at bus $i$.

For the operation of the GUPFC, the power flow equality constraints (3.47), (3.48), and operation and control constraints (3.50), (3.51), (3.63) and (3.64) should hold. Besides, the GUPFC is also constrained by its operating inequality constraints such as voltage, power and thermal constraints.

Similarly to the IPFC, the equivalent controllable injected voltage source of each series converter of the GUPFC is constrained by the voltage limits given by (3.53) (3.54) and (3.55).

The constraints of the shunt converter of the GUPFC are:

$$0 \le \theta sh_i \le 2\pi \tag{3.65}$$

$$Vsh_i^{\min} \le Vsh_i \le Vsh_i^{\max} \tag{3.66}$$

$$-PEsh_i^{\max} \le PEsh_i \le PEsh_i^{\max} \tag{3.67}$$

$$Ish_i \le Ish_i^{\max} \tag{3.68}$$

where $PEsh_i^{\max}$ is the maximum limit of the power exchange of the shunt converter with the DC link, and $PEsh_i = \text{Re}(Vsh_i Ish_i^*)$. $Ish_i^{\max}$ is the current rating.

### 3.2.2.2 Modeling of the GUPFC in Newton Power Flow

For the GUPFC in Fig. 3.6 and Fig. 3.7, the control degrees of freedom of any of the two series converters $i$-$j$ and $i$-$k$ are two except the shunt converter has one control degree of freedom since the power exchange among the three series-shunt converters should be balanced. Combining power flow mismatch equations (3.47), (3.48), and operating and control equations (3.50), (3.51), (3.63) and (3.64), the Newton power flow solution may be given by:

$$\mathbf{J}\Delta\mathbf{X} = -\Delta\mathbf{R} \tag{3.69}$$

where

$\Delta\mathbf{X}$ - the incremental vector of state variables, and $\Delta\mathbf{X} = [\Delta\mathbf{X}_1, \Delta\mathbf{X}_2]^T$

$\Delta\mathbf{X}_1 = [\Delta\theta_i, \Delta V_i, \Delta\theta_j, \Delta V_j, \Delta\theta_k, \Delta V_k]^T$ - the incremental vector of bus voltage magnitudes and angles.

$\Delta\mathbf{X}_2 = [\Delta\theta se_{ij}, \Delta Vse_{ij}, \Delta\theta se_{ik}, \Delta Vse_{ik}, \Delta\theta sh_i, \Delta Vsh_i]^T$ - the incremental vector of the state variables of the GUPFC.

$\Delta\mathbf{R}$ - the bus power mismatch and GUPFC control mismatch vector, and $\Delta\mathbf{R} = [\Delta\mathbf{R}_1, \Delta\mathbf{R}_2]^T$.

$\Delta\mathbf{R}_1 = [\Delta P_i, \Delta Q_i, \Delta P_j, \Delta Q_j, \Delta P_k, \Delta Q_k]^T$ - bus power mismatch vector.

$\Delta\mathbf{R}_2 = [P_{ji} - P_{ji}^{Spec}, Q_{ji} - Q_{ji}^{Spec}, P_{ki} - P_{ki}^{Spec}, Q_{ki} - Q_{ki}^{Spec}, V_i - V_i^{Spec}, PEx]^T$ - operating and control mismatch vector of the GUPFC

$\mathbf{J} = \dfrac{\partial \Delta\mathbf{R}}{\partial \mathbf{X}}$ - System Jacobian matrix

The multi-control modes of STATCOM and UPFC are applicable to GUPFC. The techniques for the handling of the violated functional inequalities of STATCOM and SSSC [13] are applicable to GUPFC.

### 3.2.2.3 Initialization of GUPFC Variables in Newton Power Flow

For the initialization of the series converters, (3.57) and (3.58) can be applied. Assume $Vsh_i$ is given by:

$$Vsh_i = (Vsh_i^{max} + Vsh_i^{min})/2 \quad \text{or} \quad Vsh_i = V_i^{Spec} \tag{3.70}$$

then $\theta sh_i$ can be found by solving (3.63):

$$\theta sh_i = -\sin^{-1}[B/(V_i Vsh_i \sqrt{(gsh_i^2 + bsh_i^2)})] + \tan^{-1}(-gsh_i/bsh_i) \tag{3.71}$$

where:

$$\begin{aligned}B = &Vsh_i^2 gsh_i + \sum Vse_{in}^2 g_{in} \\ &+ \sum V_i Vse_{in}(g_{in}\cos(\theta_n - \theta se_{in}) - b_{in}\sin(\theta_n - \theta se_{in})) \\ &- \sum V_i Vse_{in}(g_{in}\cos(\theta_i - \theta se_{in}) - b_{in}\sin(\theta_i - \theta se_{in}))\end{aligned} \tag{3.72}$$

### 3.2.3 Numerical Examples

Numerical tests are carried out on the IEEE 118-bus system, IEEE 300-bus system and a 1000-bus system. In the tests, a convergence tolerance of 1.0e-11 p.u. (or 1.0e-9 *MW/MVAr*) for maximal absolute bus power mismatches and power flow control mismatches is utilized. The test cases are described as follows,

*Case 1*: This is a base case of the IEEE 118-bus system.

*Case 2*: This is similar to case 1 except that there are an IPFC and two GUPFCs installed. The IPFC is used to control the active and reactive power flows of line 12-11 and the active power flow of line 12-3. The first GUPFC is used to control the voltage at bus 45 and active and reactive power flows of line 45-44 and line 45-46. The second GUPFC is used to control the voltage of bus 94 and power flows of line 94-95, line 94-93, line 94-100, respectively.

*Case 3*: This is a base case of the IEEE 300-bus system.

*Case 4*: This is similar to case 3 except that there are an IPFC and three GUPFCs installed. The IPFC is used to control the active and reactive power flows of line 198-197 and active power flow of line 198-211. The first GUPFC is used to control the voltage at bus 37 and active and reactive power flows of line 37-49, line 37-74, and line 37-34. The second GUPFC is used to control the voltage of bus 126 and power flows of line 126-132, line 126-169, line 126-127. The third GUPFC is used to control the voltage at bus 140 and power flows of line 140-137, line 140-141, and line 140-145.

*Case 5*: This is a base case of the 1000-bus system.

*Case 6*: This is similar to case 5 except that there are an IPFC and three GUPFCs installed. The IPFC is used to control the active and reactive power flows of line 142-388 and active power flow of line 142-376. The first GUPFC is used to control the voltage at bus 82 and active and reactive power flows of line 82-200 and line 82-203. The second GUPFC is used to control the voltage at bus 126 and power flows of line 126-132, line 126-169, line 126-127. The third GUPFC is for control the voltage at bus 142 and power flows of lines 142-146, 142-141, and 140-170, respectively.

### 3.2.3.1 Initialization of the Power Flow with FACTS-Devices

Table 3.2 gives the number of iterations for cases 1-6. From the results, it can be seen that in comparison with those cases of base power flow solutions the Newton power flow with the IPFC and GUPFC needs more iterations but can converge within 9 iterations with the special initialization procedure. It can also be seen that without the special initialization procedure for the FACTS controllers, the Newton power flow may need more iterations or even diverge.

The convergence characteristics of case 5 without the FACTS controllers and case 6 with the FACTS controllers on the 1000-bus system are shown in Fig. 3.8, which demonstrate the quadratic convergence characteristics of the Newton power flow algorithm when the algorithm is approaching the final solution.

**Table 3.2.** Iteration count of the test systems (Tolerance $1.0e^{-11}$ p.u.)

| System | | IEEE 118 Bus | IEEE 300 Bus | 1000 Bus |
|---|---|---|---|---|
| Base cases | Case No. | Case 1 | Case 3 | Case 5 |
| | Number of iterations | 4 | 5 | 6 |
| FACTS cases without the special initialization | Case No. | Case 2 | Case 4 | Case 6 |
| | Number of iterations | 9 | Diverge | 11 |
| FACTS cases with the special initialization | Case No. | Case 2 | Case 4 | Case 6 |
| | Number of iterations | 8 | 8 | 9 |

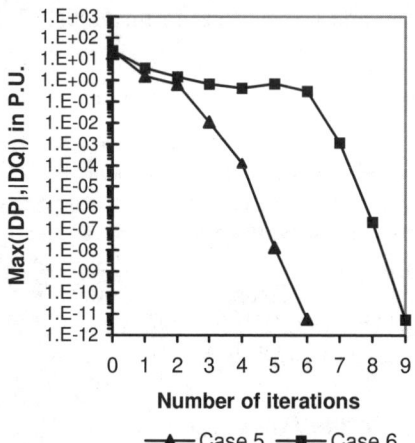

**Fig. 3.8.** Absolute power mismatches as function of number of iterations for the 1000 bus system

### 3.2.3.2 Enforcement of Practical Constraints of FACTS

The practical current and active power exchange of the FACTS-devices of case 6 of the 1000-bus system without the constraint enforcement are shown in the first column of Table 3.3. The second column of Table 3.3 shows the current and power limits of the FACTS-devices, which will be used in the following examples.

**Table 3.3.** Current and power through the converter and their operating limits

| Current and power without constraint enforcement | Current and power limits with constraint enforcement | Constraint violation percentage |
|---|---|---|
| $Pse_{82,200} = -6.37\text{e-}2$ p.u. | $Pse_{82,200}^{max} = 5.00\text{e-}2$ p.u. | 27% |
| $Ise_{200,82} = 4.76$ p.u. | $Ise_{200,82}^{max} = 4.00$ p.u. | 19% |
| $Psh_{82} = 8.44\text{e-}2$ p.u. | $Psh_{82}^{max} = 6.00\text{e-}2$ p.u. | 40% |
| $Ish_{82} = 4.39$ p.u. | $Ish_{82}^{max} = 4.00$ p.u. | 10% |

### 3.2.3.3 Enforcement of Practical Constraints of Series Converters

Two examples based on case 6 are given here to show the enforcement of the two practical functional inequality constraints of the series converter 82-200: the active power exchange constraint and current constraint, respectively.

## 3.2 Modeling of Multi-Control Functional IPFC and GUPFC 81

In the first example, when only the active power exchange limit $Pse_{82,200}^{max} = 5.00e-2$ p.u. is applied, the corresponding active power constraint is violated. This constraint is enforced while the reactive power flow control constraint $Q_{200,82} - Q_{200,82}^{Spec} = 0$ is released. The Newton power flow can converge in 9 iterations.

In the second example, when only the current limit $I_{200,82}^{max} = 4.0$ p.u. is applied, the corresponding current constraint $I_{200,82} < I_{200,82}^{max}$ is violated. Then the constraint is enforced while the reactive power flow control constraint $Q_{200,82} - Q_{200,82}^{Spec} = 0$ is released. The Newton power flow can converge in 9 iterations.

### 3.2.3.5 Enforcement of Practical Constraints of the Shunt Converter

There are two examples based on case 6 to show the enforcement of the current and active power exchange constraints of the shunt converter of the GUPFC at bus 82.

In the first example here, when the active power exchange limit $Psh_{82}^{max} = 6.0e-2$ p.u. is applied, the corresponding active power exchange constraint is violated. Then the constraint is enforced while the active power flow control constraint $P_{200,82} - P_{200,82}^{Spec} = 0$ is released. The Newton power flow can converge in 9 iterations.

In the second example, when the current limit $Ish_{82}^{max} = 4.0$ p.u. is applied, the corresponding current constraint is violated. Then the constraint is enforced while the voltage control constraint of bus 82 is released. The Newton power flow can converge in 10 iterations.

### 3.2.3.6 Enforcement of Series and Shunt Converter Constraints

For case 6, when the active power exchange limit $Pse_{82,200}^{max} = 5.00e-2$ p.u. and current limit $Ish_{82}^{max} = 4.0$ p.u. are applied, the corresponding power and current constraints are violated. Then they are enforced. The Newton power flow can converge in 10 iterations.

For case 6 when the voltage magnitude limit $Vse_{82,200}^{max}$ and current limit $Ish_{82}^{max}$ are applied, the corresponding voltage and current constraints are violated. Then they are enforced. The Newton power flow can converge in 9 iterations.

In conclusion, this section has proposed mathematical models for the Interline Power Flow Controller (IPFC) and Generalized Unified Power Flow Controller (GUPFC). The implementation of these models in Newton power flow with the particular consideration of the practical functional inequality constraints, including the current and active power limits of the series and shunt converters of the FACTS-devices, has been reported. Furthermore, a special initialization procedure for the FACTS models has been presented.

## 3.3 Multi-Terminal Voltage Source Converter Based HVDC

With the advance of voltage source converter (VSC) technologies, a number of VSC based FACTS-devices such as the STATCOM, SSSC, UPFC, IPFC and GUPFC [1]-[11] have been proposed. Among these VSC-FACTS-devices, the IPFC and GUPFC [2]-[6], which have the ability to control active and reactive power on two or more transmission lines, may be used in a major network substation in a power system to effectively control power flows on specified transmission paths.

Along with the success of application of VSC technologies in FACTS, application of such technologies in HVDC transmission has gained great success. It was reported recently that VSC based HVDC systems have been successfully installed in several electric utilities in European Countries and U.S. [25]-[28]. However, it should be pointed out that these HVDC systems are basically used for two-terminal HVDC power transmission. Simultaneous power flow and voltage control serves for various applications such as back-to-back or cable installations.

In contrast to the traditional Thyristor based HVDC system, the VSC-HVDC system has the following features: (a) it is very easy to make multiterminal connections; (b) it has the ability to independently control active and reactive flows at its terminals; (c) in addition, it has the option to control its terminal bus voltages instead of reactive powers; (d) The costs for filtering of harmonics may be significantly reduced if suitable PWM techniques are used; (e) construction and commissioning of a VSC-HVDC system takes less time than that for a traditional Thyristor based HVDC-system.

Beyond two terminal VSC-Back-to-Back-HVDC systems [25]-[28], there is the option for multi-terminal VSC-HVDC (M-VSC-HVDC) [29] in electric transmission and distribution systems. Before the practical application of the M-VSC-HVDC in electric power systems, extensive research and development need to be carried out to understand the steady and dynamic characteristics of the M-VSC-HVDC. In the light of this, the dynamic over-voltage problem of the M-VSC-HVDC was investigated [30]. Due to the fact that power flow analysis is one of the fundamental power system calculations for power system operation, control and planning, in this section, the steady-state modeling of the M-VSC-HVDC for power flow analysis [31] will be presented.

## 3.3.1 Mathematical Model of M-VSC-HVDC with Converters Co-located in the same Substation

### 3.3.1.1 Operating Principles of M-VSC-HVDC

A M-VSC-HVDC shown in Fig. 3.9. consists of three converters, which are connected with three buses $i, j, k$ via three coupling transformers. The three converters are directly connected with a common DC link and co-located in a substation. Basically, the converters at bus $j$ and $k$, which are considered as primary converters, can provide independent active and reactive power flow control. Alternatively a primary converter can also provide active power flow and voltage control. The converter at bus $i$ is considered as secondary converter, which can provide voltage control at bus $i$ and power exchange balance among the converters.

It should be pointed out that for simplifying the presentation, the M-VSC-HVDC given in Fig. 3.9 consists of only three terminals. However, in principle, the following mathematical derivation will be applicable to a M-VSC-HVDC with any number of terminals.

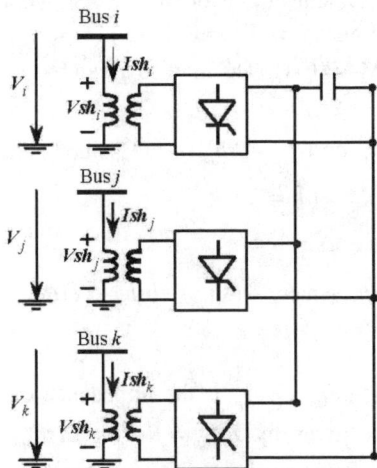

**Fig. 3.9.** Schematic setup of M-VSC-HVDC

### 3.3.1.2 Power Flow Constraints of M-VSC-HVDC

Based on the operating principles shown in Fig. 3.9, the equivalent circuit of the M-VSC-HVDC can be derived, which is given in Fig. 3.10.

In the derivation, we assume that (a) harmonics generated by the converters are neglected and (b) the system as well as the VSC are three phase balanced. Then each converter can be equivalently represented at the fundamental (power system) frequency by voltage phasor $\mathbf{Vsh}_m = Vsh_m \angle \theta sh_m$ ($m = i, j, k$).

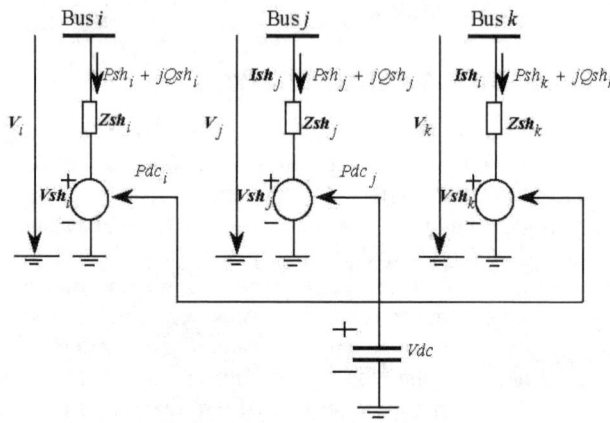

**Fig. 3.10.** M-VSC-HVDC equivalent circuit

According to the equivalent circuit of the M-VSC-HVDC shown in Fig. 3.9, suppose the bus voltage phasor is $V_m = V_m \angle \theta_m$, ($m = i, j, k$), then the AC terminal power flows of the M-VSC-HVDC may be given by:

$$Psh_m = V_m^2 gsh_m \\ - V_m Vsh_m (gsh_m \cos(\theta_m - \theta sh_m) + bsh_m \sin(\theta_m - \theta sh_m)) \quad (3.73)$$

$$(m = i, j, k)$$

$$Qsh_m = -V_m^2 bsh_m \\ -V_m Vsh_m (gsh_m \sin(\theta_m - \theta sh_m) - bsh_m \cos(\theta_m - \theta sh_m)) \quad (3.74)$$

$$(m = i, j, k)$$

where $gsh_m + jbsh_m = 1/Zsh_m$. $Zsh_m$ is the impedance of the converter coupling transformer, and may be given by $Zsh_m = Rsh_m + jXsh_m$. $Rsh_m$ and $Xsh_m$ are the resistance and reactance, respectively, of the coupling transformer.

### 3.3.1.3 Active Power Balance of M-VSC-HVDC

The active power exchange among the converters via the DC link should be balanced at any instant, which is described by:

$$Pdc_\Sigma = Pdc_i + Pdc_j + Pdc_k + Ploss = 0 \quad (3.75)$$

where *Ploss* represents losses in converter circuits. Each converter losses consist of two terms. The first term is proportional to its AC terminal current squared, and the second term is a constant. The former may be represented by an equivalent resistance, and can be included into its coupling transformer impedance. The second term may be represented by an equivalent resistance in parallel with the DC bus.

However, Considering that there is no explicit DC network being represented in the M-VSC-HVDC formulation here, the second terms of all the converters can be combined and represented by *Ploss* which is included in the power balance equation (3). $Pdc_m$ ($m = i, j, k$) as shown in Fig. 3.10 is the power exchange of the converter with the DC link and given by:

$$Pdc_m = \text{Re}(-Vsh_m Ish_m^*)$$
$$= Vsh_m^2 gsh_m - V_i Vsh_m (gsh_m \cos(\theta_i - \theta sh_m) - bsh_m \sin(\theta_i - \theta sh_m)) \quad (3.76)$$
$$(m = i, j, k)$$

### 3.3.1.4 Voltage and Power Flow Control of M-VSC-HVDC

**Primary converters.** Each primary converter has two control modes such as PQ and PV, which are presented as follows.

*Control mode 1: PQ control*

In principle, the primary converters at buses *j* and *k* can be used to control the independent active and reactive power of terminals *j* and *k*, respectively. In the PQ control mode, the independent active and reactive power control constraints are:
at bus *j*:

$$Psh_j - Psh_j^{Spec} = 0 \quad (3.77)$$

$$Qsh_j - Qsh_j^{Spec} = 0 \quad (3.78)$$

at bus *k*:

$$Psh_k - Psh_k^{Spec} = 0 \quad (3.79)$$

$$Qsh_k - Qsh_k^{Spec} = 0 \quad (3.80)$$

where $Psh_j^{Spec}$, $Qsh_j^{Spec}$ are the specified active and reactive power control references at bus *j* while $Psh_k^{Spec}$, $Qsh_k^{Spec}$ are the specified active and reactive power control references at bus *k*.

*Control mode 2: PV control*

In the PV control mode, alternatively, the primary converters at buses *j* and *k* may control voltage rather than reactive power. In other words, the reactive control constraints of (6) and (8) may be replaced by the following voltage control constraints, respectively:
at bus *j*:

$$V_j - V_j^{Spec} = 0 \quad (3.81)$$

at bus *k*:

$$V_k - V_k^{Spec} = 0 \qquad (3.82)$$

where $V_j^{Spec}$ and $V_k^{Spec}$ are the bus voltage control references at buses $j$ and $k$, respectively. It should be pointed out that the voltage at a remote bus instead of a local bus may be controlled.

**Secondary converter.** In operation of the M-VSC-HVDC, the secondary converter at bus $i$ can be used to control the voltage magnitude at its terminal bus $i$. Such a control is given by:

$$V_i - V_i^{Spec} = 0 \qquad (3.83)$$

where $V_i^{Spec}$ is the bus voltage control reference.

In addition to the voltage control constraint (3.83), the secondary converter is also used to balance the active power exchange among the converters. Such an active power balance constraint is given by (3.75).

### 3.3.1.5 Voltage and Current Constraints of M-VSC-HVDC

The voltage constraint of each converter is:

$$Vsh_m^{\min} \le Vsh_m \le Vsh_m^{\max} \quad (m=i,\,j,\,k) \qquad (3.84)$$

where $Vsh_m^{\max}$ is the voltage rating of the converter while $Vsh_m^{\min}$ is the minimal limit for the injected VSC voltage. $Vsh_m$ is the actual voltage of the converter.

The current through each VSC should be within its thermal capability:

$$Ish_m \le Ish_m^{\max} \quad (m=i,\,j,\,k) \qquad (3.85)$$

where $Ish_m^{\max}$ is the current rating of the VSC converter while $Ish_m$ is the actual current through the converter, which is given by:

$$Ish_m = \sqrt{V_m^2 + Vsh_m^2 - 2V_m Vsh_m \cos(\theta_m - \theta sh_m)} \,/\, \sqrt{Rsh_m^2 + Xsh_m^2} \qquad (3.86)$$

### 3.3.1.6 Modeling of M-VSC-HVDC in Newton Power Flow

For the three-terminal VSC-HVDC shown in Fig. 3.10, the Newton equation including power mismatches at buses $i$, $j$ and $k$ and control mismatches may be written as:

$$\mathbf{J}\Delta\mathbf{X} = -\Delta\mathbf{R} \qquad (3.87)$$

where

$\Delta\mathbf{X}$ - the incremental vector of state variables, and $\Delta\mathbf{X} = [\Delta\mathbf{X}_1, \Delta\mathbf{X}_2]^T$

$\Delta \mathbf{X}_1 = [\Delta\theta_i, \Delta V_i, \Delta\theta_j, \Delta V_j, \Delta\theta_k, \Delta V_k]^T$ - the incremental vector of bus voltage angles and magnitudes.

$\Delta \mathbf{X}_2 = [\Delta\theta sh_i, \Delta Vsh_i, \Delta\theta sh_j, \Delta Vsh_j, \Delta\theta sh_k, \Delta Vsh_k]^T$ - the incremental vector of state variables of the M-VSC-HVDC.

$\Delta \mathbf{R}$ - the M-VSC-HVDC bus power mismatch and control mismatch vector, and $\Delta \mathbf{R} = [\Delta \mathbf{R}_1, \Delta \mathbf{R}_2]^T$.

$\Delta \mathbf{R}_1 = [\Delta P_i, \Delta Q_i, \Delta P_j, \Delta Q_j, \Delta P_k, \Delta Q_k]^T$ - power mismatches.

$\Delta \mathbf{R}_2 = [Pdc_\Sigma, V_i - V_i^{Spec}, Psh_j - Psh_j^{Spec}, Qsh_j - Qsh_j^{Spec}, Psh_k - Psh_k^{Spec}, Qsh_k - Qsh_k^{Spec}]^T$
control mismatches of the M-VSC-HVDC

$\mathbf{J} = \dfrac{\partial \Delta \mathbf{R}}{\partial \mathbf{X}}$ - System Jacobian matrix

In the above formulation, PQ control mode is applied to the primary converters $j$ and $k$. If however, PV control mode is applied to the primary converters, the reactive power flow control mismatch equations such as (3.78) and (3.80) in $\Delta \mathbf{R}_2$ should be replaced by the voltage control mismatch equations (3.81) and (3.82), respectively. It can be found that in the Newton formulation of (3.87), the implementation of PQ control mode for a primary converter has the same dimension and the similar Jacobian matrix structure as that of PV control mode for that converter. In addition, handling of international voltage and current limits, as will be discussed in the next, will not affect the dimension and basic structure of the Jacobian matrix in (3.87). The above features are in particular desirable for incorporation of the M-VSC-HVDC in production grade program since the complexity of such a multi-converter HVDC.

### 3.3.1.7 Handling of Internal Voltage and Current Limits of M-VSC-HVDC

**Primary converters.** If the voltage or current limit of a primary converter is violated, the voltage or current is simply kept at the limit while the reactive power control (for PQ control mode) or voltage control (for PV control mode) is released.

**Secondary converter.** If the voltage or current limit of a secondary converter is violated, the voltage or current is simply kept at the limit while the voltage control is released.

### 3.3.1.8 Comparison of M-VSC-HVDC and GUPFC

In principle, the M-VSC-HVDC with all converters being co-located in the same substation can be used to replace a GUPFC for voltage and power flow control purposes. Here we try to discuss the different characteristics of the M-VSC-

HVDC and the GUPFC when they can be used interchangeably. Modeling of the GUPFC for steady state voltage and power flow control is referred to [5], [6].

First, the power rating of a primary converter of the M-VSC-HVDC may be higher than that of a corresponding series converter of the GUPFC since the voltage rating of the former is higher than that of the latter. Hence, the power rating of the secondary converter of the M-VSC-HVDC may be higher than that of the shunt converter of the GUPFC. It can be anticipated that the investment for the M-VSC-HVDC is higher than that for the GUPFC.

Second, the GUPFC can be used to control bus voltage by its shunt converter, and it can provide independent active and reactive power flows by its series converters. The M-VSC-HVDC can control bus voltage by its secondary converter, and it can provide independent active and reactive power flows by its primary converters. In addition, the primary converters of the M-VSC-HVDC can alternatively control bus voltages instead of reactive powers. In contrast, the series converters of the GUPFC have relatively limited voltage control capability. Hence, generally speaking, the M-VSC-HVDC may have stronger voltage control capability than that of the GUPFC.

### 3.3.2 Generalized M-VSC-HVDC Model with Incorporation of DC Network Equation

#### *3.3.2.1 Generalized M-VSC-HVDC*

In section 3.3.1, the M-VSC-HVDC model is only applicable to situations when the converters are co-located in a substation. In the mathematical model and the Newton power flow algorithm, explicit representation of the DC link is not required. Instead, the active power balance equation (3.75) is applied to represent the effect of the DC link. However, if the M-VSC-HVDC converters are not co-located in a substation, then the DC network of the M-VSC-HVDC needs to be explicitly represented.

For the sake of simplicity, we assume that the M-VSC-HVDC shown in Fig. 3.10 is extended to the generalized M-VSC-HVDC in Fig. 3.11. In Fig. 3.11, a DC network is explicitly represented, which consists of three DC buses and three DC lines. The VSC converters $i, j, k$ are coupled with the DC buses $i, j, k$, respectively, and the DC buses are interconnected via the DC lines. The DC bus voltages $Vdc_i$, $Vdc_j$ and $Vdc_k$ are state variables of the DC network.

A VSC converter may not be lossless. As has been discussed in section 3.3.1, each VSC converter losses consist of two terms. The first term, which is proportional to the AC terminal current squared, can be included into the transformer impedance as an equivalent resistance. There are two approaches to represent the losses of the second term. The first approach is that the second term of each converter may be represented by an equivalent resistance in parallel with its DC coupling capacitor.

3.3 Multi-Terminal Voltage Source Converter Based HVDC    89

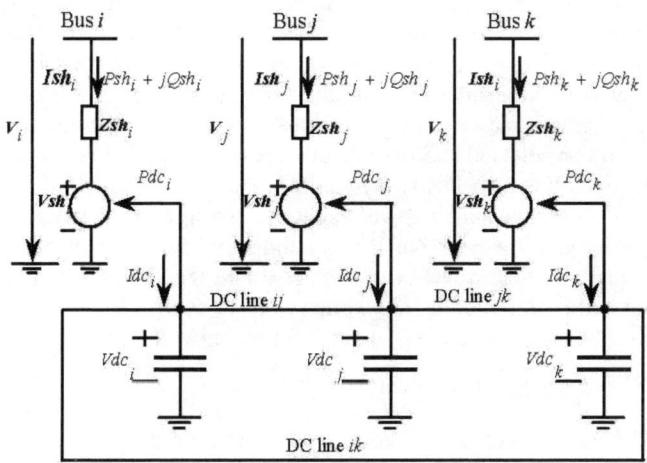

**Fig. 3.11.** Generalized M-VSC-HVDC with a DC network

The second approach is that the second term of each converter, which is almost constant, may be represented by $Ploss_m$ ($m = i, j, k$). $Ploss_m$ can be represented as a power injection to the DC bus $m$ and the direction of $Ploss_m$ is leaving the positive terminal of the DC bus $m$. It should be pointed out here that in the Generalized M-VSC-HVDC model here, the equivalent resistance approach for representing the losses of the second terms is preferred since there is an explicit DC network representation for the Generalized M-VSC-HVDC here and the equivalent resistances can be directly included in the DC network equation that will be introduced next.

### 3.3.3.1  DC Network Equation

Assuming that the DC lines can be represented by equivalent DC resistances, and power losses are represented by equivalent resistances in parallel with the DC buses, the voltage and current relationships of the DC network may be represented by:

$$\mathbf{Y}_{dc}\mathbf{V}_{dc} = \mathbf{I}_{dc} \tag{3.88}$$

where $\mathbf{Y}_{dc}$ is the DC network Y-bus matrix. $\mathbf{V}_{dc}$ is the DC bus voltage vector, given by $\mathbf{V}_{dc} = [Vdc_i, Vdc_j, Vdc_k]^T$. $\mathbf{I}_{dc}$ is the DC network bus current injection vector, given by $\mathbf{I}_{dc} = [-Pdc_i/Vdc_i, -Pdc_j/Vdc_j, -Pdc_k/Vdc_k]^T$. $Pdc_m$ ($m = i, j, k$), as defined in (3.76), is the power exchange of the VSC converter with the coupling DC link.

In AC power flow analysis, a slack bus should be selected and usually the voltage magnitude and angle at that bus should be kept constant. AC slack bus usually serves two roles such as (a) keeping slack bus voltage constant; (b) providing the balance between generation and load. In the Newton power mismatch equation of power flow analysis, the relevant row and column of slack bus are usually removed. Due to similar reasons, for the DC network voltage equation (3.88), a DC slack bus should be selected and the DC bus voltage should be kept constant. The selected DC slack bus also has two functions such as (a) providing DC voltage control; (b) balancing the active power exchange among the converters via the DC network. However, different from the handling technique for slack bus in AC power flow analysis, here a DC bus is selected and the voltage of the DC bus is kept constant and represented by an explicit voltage control equation. If DC bus $i$ is selected as the DC slack bus, we have the following voltage control equation:

$$Vdc_i - Vdc_i^{Spec} = 0 \qquad (3.89)$$

where $Vdc_i^{Spec}$ is the specified DC voltage control reference.

Equations (3.88) and (3.89) are the basic operating constraints of the DC network. The DC network is mathematically coupled with the AC terminals of the M-VSC-HVDC via the DC powers $Pdc_i$, $Pdc_j$, $Pdc_k$, respectively. Due to the fact that the DC network is represented by the DC nodal voltage equation in (3.88), any topologies of the DC network may be modeled without difficulty. In addition, if an energy storage system is connected with the DC network, it can be included into the DC network equation.

### 3.3.2.2 Incorporation of DC Network Equation into Newton Power Flow

With incorporation of the DC network into the generalized M-VSC-HVDC model, the Newton power flow equation (3.87) may be augmented as:

$$\mathbf{J}\Delta\mathbf{X} = -\Delta\mathbf{R} \qquad (3.90)$$

where

$\Delta\mathbf{X}$ - the incremental vector of state variables, and $\Delta\mathbf{X} = [\Delta\mathbf{X}_1, \Delta\mathbf{X}_2, \Delta\mathbf{X}_3]^T$

$\Delta\mathbf{X}_1 = [\Delta\theta_i, \Delta V_i, \Delta\theta_j, \Delta V_j, \Delta\theta_k, \Delta V_k]^T$ - the incremental vector of bus voltage magnitudes and angles.

$\Delta\mathbf{X}_2 = [\Delta\theta sh_i, \Delta Vsh_i, \Delta\theta sh_j, \Delta Vsh_j, \Delta\theta sh_k, \Delta Vsh_k]^T$ - the incremental vector of the state variables of the M-VSC-HVDC.

$\Delta\mathbf{X}_3 = [\Delta Vdc_i, \Delta Vdc_j, \Delta Vdc_k]^T$ - the incremental vector of DC state variables.

$\Delta\mathbf{R}$ - the bus power mismatch and M-VSC-HVDC control mismatch vector, and
$\Delta\mathbf{R} = [\Delta\mathbf{R}_1, \Delta\mathbf{R}_2, \Delta\mathbf{R}_3]^T$.

$\Delta\mathbf{R}_1 = [\Delta P_i, \Delta Q_i, \Delta P_j, \Delta Q_j, \Delta P_k, \Delta Q_k]^T$ - bus power mismatch vector.

$$\Delta \mathbf{R}_2 = [V_i - V_i^{Spec}, Vdc_i - Vdc_i^{Spec}, Psh_j - Psh_j^{Spec}, Qsh_j - Qsh_j^{Spec},$$
$$Psh_k - Psh_k^{Spec}, Qsh_k - Qsh_k^{Spec}]^T$$

control mismatch vector of the M-VSC-HVDC

$\Delta \mathbf{R}_3 = \mathbf{Y}_{dc}\mathbf{V}_{dc} - \mathbf{I}_{dc}$ - DC network bus mismatch vector

$\mathbf{J} = \dfrac{\partial \Delta \mathbf{R}}{\partial \mathbf{X}}$ - System Jacobian matrix

### 3.3.3 Numerical Examples

Numerical results are presented on the IEEE 30-bus system, IEEE118-bus system and IEEE 300-bus system. In the tests, a convergence tolerance of 1.0e-12 p.u. (or 1.0e-10 MW/MVAr) for maximum absolute bus power mismatches and power flow control mismatches is used. In order to simplify the following presentation, the M-VSC-HVDC model proposed in section 3.3.1 is referred to Model I while the Generalized M-VSC-HVDC model with incorporation of the DC network in section 3.3.2 is referred to Model II.

#### 3.3.3.1 Comparison of the M-VSC-HVDC to the GUPFC

Three cases are given on the IEEE 30 bus systems to compare the M-VSC-HVDC with the GUPFC:

*Case 1*: A GUPFC is installed for control of the voltage at bus 12 and control of active and reactive power flows in line 12-15 and line 12-16. Suppose two FACTS buses 15' and 16' are created, and assume that the sending ends of the two transmission lines 12-15 and 12-16 are now connected with the FACTS buses 15' and 16', respectively while the series converters are installed between buses 12 and 15', and buses 12 and 16', respectively. The active power flows transferred on the two transmission lines are over 70% of their corresponding base case active power flows.

*Case 2*: A M-VSC-HVDC (Model I) is used to replace the GUPFC in case 1 while the control settings for voltage and power flows are as the same as that of case 1. This also means the two primary converters are using the PQ control mode.

*Case 3*: This is similar to case 2. But the two primary VSC converters are using the PV control mode.

In the above cases, the impedances of all the converter coupling transformers are set to $0 + j0.025$ p.u. The power flow solutions of cases 1, 2 and 3 are summarized in Table 3.4. In Table 3.4, the actual power through a converter is defined as the equivalent voltage of the converter times the current through the converter (i.e. $S = V I$).

**Table 3.4.** Power flow solutions for the IEEE 30-bus system

| Case No. | FACTS-Type | Control mode & converter state variables | Actual power of converter in p.u. | Iterations |
|---|---|---|---|---|
| 1 | GUPFC | Shunt converter: (V control) $Vsh_{12} = 0.9950$ p.u. $\theta sh_{12} = -10.57°$ | Shunt converter: $Ssh_{12} = 0.201$ p.u. | 7 |
| | | Series converters: (PQ control) $Vse_{12,15'} = 0.0737$ p.u. $\theta se_{12,15'} = -107.19°$ $Vse_{12,16'} = 0.0558$ p.u. $\theta se_{12,16'} = -94.01°$ | Series converters: $Sse_{12,15'} = 0.022$ p.u. $Sse_{12,16'} = 0.006$ p.u. | |
| 2 | M-VSC-HVDC (Model I) | Converter at bus 12: (V control) $Vsh_{12} = 0.9939$ p.u. $\theta sh_{12} = -11.15°$ | Converter at bus 12: $Ssh_{12} = 0.467$ p.u. | 5 |
| | | Converter at bus 15': (PQ control): $Vsh_{15'} = 1.0111$ p.u. $\theta sh_{15'} = -6.42°$ | Converter at bus 15': $Ssh_{15'} = 0.305$ p.u. | |
| | | Converter at bus 16' (PQ control): $Vsh_{16'} = 0.9952$ p.u. $\theta sh_{16'} = -7.38°$ | Converter at bus 16': $Ssh_{16'} = 0.102$ p.u. | |
| 3 | M-VSC-HVDC (Model I) | Converter at bus 12: (V control) Converter at bus 15': (PV control) Converter at bus 16': (PV control) | - | 5 |

## 3.3 Multi-Terminal Voltage Source Converter Based HVDC

From the results, it can be seen:

1. The Newton power flow with incorporation of the M-VSC-HVDC converges in 5 iterations for cases 2 and 3 with a flat start. Special initialization of VSC state variables for the M-VSC-HVDC is not needed. In contrast, the Newton power flow with the incorporation of GUPFC converges in 7 iterations for case 1 with a special initialization procedure for VSC state variables.
2. The actual power of a VSC converter for the M-VSC-HVDC is much higher than that of a corresponding VSC converter for the GUPFC. This supports the observations made in section 3.3.1.
3. It can be anticipated that the investment for the M-VSC-HVDC is higher than that for the GUPFC. However, any VSC converter of the M-VSC-HVDC has strong voltage control capability. In contrast, only the shunt converter of the GUPFC has strong voltage control capability. This indicates that in terms of control capability, the M-VSC-HVDC may be more powerful than the GUPFC. In principle, the M-VSC-HVDC and the GUPFC may be used interchangeably.

### *3.3.3.2 Power Flow and Voltage Control by M-VSC-HVDC*

The following test cases on the IEEE 118-bus system are presented:

*Case 4*: A three-terminal VSC-HVDC (Model I) is installed at bus 45 and the sending-ends of line 45-44 and line 45-46. Suppose two FACTS buses 44' and 46' are created, and assume that the sending ends of the two transmission lines 45-44 and 45-46 are now connected with the FACTS buses 44' and 46', respectively while the three AC terminals of the M-VSC-HVDC are buses 45, 44' and 46'. A four-terminal VSC-HVDC is placed at bus 94 and the sending-ends of line 94-95, line 94-93 and line 94-100. For this four-terminal VSC-HVDC, the terminal buses are 94, 95', 93' and 100' are created. It is assumed that PQ control mode is applied to all the primary converters of the two M-VSC-HVDCs.

*Case 5*: Similar to case 4. But PV control mode is applied to all the primary converters of the two M-VSC-HVDCs.

*Case 7*: A generalized three-terminal VSC-HVDC (Model II) is placed at bus 45, bus 44 and bus 46 to replace the ac transmission line 45-44 and line 45-46. A generalized four-terminal VSC-HVDC is placed at bus 94, bus 95, bus 93 and bus 100 to replace the transmission line 94-95, line 94-93 and line 94-100. PQ control mode is applied to all the primary converters.

*Case 8*: Similar to case 7, except that PV control mode is applied all the primary converters.

*Case 9*: Similar to case 8, except that PQ control mode is applied to the primary converter at bus 93.

The test results of cases 4-9 are shown in Table 3.5 while the detailed power flow solution for case 6 is given by Table 3.6. The tests have also been carried out on the IEEE 300-bus system [31].

**Table 3.5.** Results on the IEEE 118-bus system

| Case No. | M-VSC-HVDC model | Control mode | Number of iterations |
|---|---|---|---|
| 4 | I | PQ control for all primary converters | 5 |
| 5 | I | PV control for all primary converters | 4 |
| 6 | I | PV control for the primary converter at bus 44'; PQ for the primary converter at bus 46' | 5 |
| 7 | II | PQ control for all primary converters | 5 |
| 8 | II | PV control for all primary converters | 4 |
| 9 | II | PV control for the primary converters at buses 95' and 100'; PQ for the primary converter at bus 93' | 5 |

**Table 3.6.** Power flow solution for case 6

| Location of M-VSC-HVDC | Control mode & converter state variables |
|---|---|
| The three-terminal M-VSC-HVDC at bus 45 | Secondary converter at bus 45 (V control):<br>$Vsh_{45} = 1.0095$ p.u.  $\theta sh_{45} = -13.65°$<br>Primary converter at bus 44' (PV control):<br>$Vsh_{44'} = 1.0000$ p.u.  $\theta sh_{44'} = -9.74°$<br>Primary converter at bus 46' (PQ Control):<br>$Vsh_{46'} = 0.96775$ p.u.  $\theta sh_{46'} = -16.99°$ |
| The four-terminal M-VSC-HVDC at bus 94 | Secondary converter at bus 94 (V control):<br>$Vsh_{94} = 1.0077$ p.u.  $\theta sh_{94} = 0.69°$<br>Primary converter at bus 95' (PV control):<br>$Vsh_{95'} = 1.0064$ p.u.  $\theta sh_{95'} = 0.53°$<br>Primary converter at bus 93' (PQ Control):<br>$Vsh_{93'} = 0.9253$ p.u.  $\theta sh_{93'} = -2.32°$<br>Primary converter at bus 100' (PV Control):<br>$Vsh_{100'} = 1.0021$ p.u.  $\theta sh_{100'} = -5.44°$ |

In conclusion, two M-VSC-HVDC models suitable for power flow analysis have been proposed. The first M-VSC-HVDC model (Model I) assumes that all converters of the M-VSC-HVDC are co-located in the same substation while the second M-VSC-HVDC model (Model II) is a general one, in which a DC network can be explicitly represented. For both the M-VSC-HVDC models proposed, the primary converter can use either PQ or PV control mode while the secondary converter can provide voltage control (V control). The Newton power flow algorithm with incorporation of the M-VSC-HVDC models proposed performs well with a flat start. Hence, unlike that for the GUPFC [5], [6], a special initialization procedure for VSC state variables is not needed.

In principle, M-VSC-HVDC (Model I) and GUPFC can be used interchangeably while the power ratings of converters of the former should be higher than that of converters of the latter. This theoretic conclusion has been further confirmed by numerical results. However, in comparison to the GUPFC, it has been found that the M-VSC-HVDC has not only strong active and reactive power flow control capability (PQ control mode) but also strong voltage control capability (PV control mode).

## 3.4 Handling of Small Impedances of FACTS in Power Flow Analysis

### 3.4.1 Numerical Instability of Voltage Source Converter FACTS Models

It has been found that: 1) the voltage source model for the IPFC and GUPFC may not be numerically stable when the coupling transformer impedances are too small; 2) the voltage source models have difficulties to be directly included in the Newton power flow algorithm when the IPFC and GUPFC are transformer-less devices [32]. For the former case, even with the advanced initialization procedure for the IPFC and the GUPFC derived in previous sections, the Newton power flow algorithm may not be able to converge. For the latter, the IPFC and GUPFC converters will become pure voltage sources. As a matter of fact, dealing with pure voltage sourced branches is extremely difficult in the Newton power flow calculations if not impossible. In a general power flow analysis tool, the above two situations need to be considered.

### 3.4.2 Impedance Compensation Model

In order to deal with the difficulties mentioned above, an impedance compensation method will be introduced here. Suppose that the branch $ij$ of the GUPFC shown in Fig. 3.7 is depicted in Fig. 3.12 while the branch $ij$ of the GUPFC with an impedance compensation is shown in Fig. 3.13.

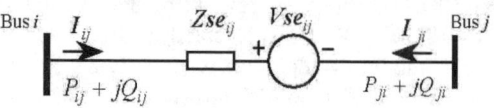

**Fig. 3.12.** Original branch *ij* of the GUPFC in Fig. 3.7

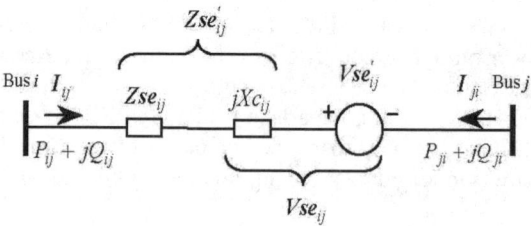

**Fig. 3.13.** Branch *ij* of the GUPFC with an impedance compensation

In the equivalent circuit shown in Fig. 3.12, the GUPFC branch *ij* can be represented by a new equivalent voltage source $Vse'_{ij}$ in series with a new equivalent impedance $Zse'_{ij}$.

The equivalent circuits in Fig. 3.12 and in Fig. 3.13 are mathematically identical if the following equations hold:

$$Zse'_{ij} = Zse_{ij} + jXc_{ij} \qquad (3.91)$$

$$Vse_{ij} = Vse'_{ij} + jXc_{ij}I_{ij} \qquad (3.92)$$

where $jXc_{ij}$ is the compensation impedance (precisely pure reactance). The current $I_{ij}$ in Fig.3.12 is:

$$I_{ij} = -I_{ji} = (V_i - V_j - Vse'_{ij})/Zse'_{ij} \qquad (3.93)$$

In the equivalent circuit shown in Fig.3.12, the active power exchange of the series converter *ij* with the DC link is $PEse_{ij} = \text{Re}(Vse_{ij}I^*_{ji})$. We can substitute equation (3.93) into this equation, we get:

$$PEse_{ij} = \text{Re}(Vse_{ij}I^*_{ji}) = \text{Re}((Vse'_{ij} + jXc_{ij}I_{ij})I^*_{ji}) = \text{Re}(Vse'_{ij}I^*_{ji}) \qquad (3.94)$$

Equation (3.94) indicates that the active power exchange can be represented directly in terms of the new voltage source state variable $Vse'_{ij}$ and the new series impedance $Zse'_{ij}$. The conveter model in Fig. 3.12 can be equivalently represented by the converter model shown in Fig.3.13 if we replace:

1. in the power equations (3.43)-(3.46), $Vse_{ij}$ and $Zse_{ij}$ by the $Vse'_{ij}$ and $Zse'_{ij}$, respectively.
2. in the power balance equation (3.63) $Vse_{ij}$ and $Zse_{ij}$ by the $Vse'_{ij}$ and $Zse'_{ij}$, respectively.
3. in the current and power inequalities (3.54) and (3.55), $Vse_{ij}$ and $Zse_{ij}$ by the $Vse'_{ij}$ and $Zse'_{ij}$, respectively.

The simple voltage inequality constraint (3.53) now becomes the following functional inequality constraint

$$Vse_{ij}^{\min} \leq |Vse_{ij}| \leq Vse_{in}^{\max} \qquad (3.95)$$

where $Vse_{ij}$ can be determined by (3.92).

It should be pointed out that the impedance compensation method is also applicable to a shunt converter.

In this chapter the recent developments in modeling of multi-functional multi-converter FACTS-devices in power flow analysis have been discussed. Not only the two-converter shunt-series FACTS-device - UPFC, but also the latest multi-line FACTS-devices such as IPFC, GUPFC, and HVDC-devices such as VSC HVDC and M-VSC-HVDC in power flow analysis have been proposed. The control performance of different FACTS-devices have also been presented. In addition, handling of the small impedances of coupling transformers of FACTS-devices in power flow analysis has also been discussed. Further work would investigate novel control modes and possible new configurations of FACTS-devices.

## References

[1] Song YH, John AT (1999) Flexible AC Transmission Systems. IEE Press, London
[2] Hingorani NG, Gyugyi L (2000) Understanding FACTS – concepts and technology of flexible ac transmission systems. New York: IEEE Press
[3] Fardanesh B, Henderson M, Shperling B, Zelingher S, Gyugyi L, Schauder C, Lam B, Mounford J, Adapa R, Edris A (1998) Convertible static compensator: application to the New York transmission system. CIGRE 14-103, Paris, France
[4] Fardanesh B, Shperling B, Uzunovic E, Zelingher S (2000) Multi-converter FACTS devices: the generalized unified power flow controller (GUPFC). Proceedings of IEEE 2000 PES Summer Meeting, Seattle, USA
[5] Zhang XP, Handschin E, Yao MM (2001) Modeling of the generalized unified power flow controller in a nonlinear interior point OPF. IEEE Trans. on Power Systems, vol 16, no 3, pp. 367-373

[6] Zhang XP (2003) Modelling of the interline power flow controller and generalized unified power flow controller in Newton power flow. IEE Proc. - Generation, Transmission and Distribution, vol 150, no 3, pp 268-274
[7] Schauder C, Gernhardt M, Stacey E, Lemak T, Gyugyi L, Cease TW, Edris A (1995) Development of a ±100MVar static condenser for voltage control of transmission systems. IEEE Transactions on Power Delivery, vol 10, no 3, pp1486-1493
[8] Gyugyi L, Shauder CD, Sen KK (1997) Static synchronous series compensator: a solid-state approach to the series compensation of transmission lines. IEEE Transactions on Power Delivery; vol 12, no 1, pp 406-413
[9] Sen KK (1998) SSSC - Static synchronous series compensator: theory, modeling, and applications. IEEE Transactions on Power Delivery, vol. 13, no 1, pp 241-246
[10] Gyugyi L, Shauder CD, Williams SL, Rietman TR, Torgerson DR, Edris A (1995) The unified power flow controller: a new approach to power transmission control. IEEE Transactions on Power Delivery, vol 10, no 2, pp 1085-1093
[11] Sen KK, Stacey EJ (1998) UPFC – Unified power flow controller: theory, modeling and applications. IEEE Trans. on Power Delivery, vol 13, no 4, pp 1453-1460
[12] Zhang XP, Handschin E, (2001) Optimal power flow control by converter based FACTS controllers. 7th International Conference on AC-DC Power Transmission, 28-30 November 2001
[13] Zhang XP, Handschin E, Yao M (2004) Multi-control functional static synchronous compensator (STATCOM) in power system steady state operations. Journal of Electric Power Systems Research, vol 72, no 3, pp 269-278
[14] Zhang XP (2003) Advanced Modeling of the multi-control functional static synchronous series compensator (SSSC) in Newton power flow. *IEEE Transactions on Power Systems*, vol 18, no 4, pp 1410-1416
[15] Nabavi-Niaki A, and Iravani MR (1996) Steady state and dynamic models of unified power flow controller (UPFC) for power system studies. IEEE Trans. on Power Systems, vol 11, no 4, pp 1937-1943
[16] Raman M, Ahmed M, Gutman R, O'Keefe, RJ, Nelson RJ, Bian J (1997) UPFC application on the AEP system: planning considerations. IEEE Transactions on Power Systems, vol 12, no 4, pp 1695-1701
[17] Noroozian M, Angquist L, Ghandhari M, Andersson G (1997) Use of UPFC for optimal power flow control. IEEE Transactions on Power Delivery, vol 12, no 4, pp 1629-1634
[18] Fuerte CR, Acha E, H. Ambriz-Perez H (200) A comprhensive Newton-Raphson UPFC model for the quadratic power flow solution of practical power networks. IEEE Transactions on Power Systems, vol 15, no 1, pp 102-109
[19] Handschin E, Lehmkoester C (1999) Optimal power flow for deregulated systems with FACTS-Devices. 13th PSCC, Trondheim, Norway, pp 1270-1276
[20] Acha E, H. Ambriz-Perez H (1999) FACTS devices modelling in optimal power flow using Newton's method. 13th PSCC, Trondheim, Norway, pp 1277-1284
[21] Zhang XP, Handschin E (2001) Advanced implementation of UPFC in a nonlinear interior point OPF. IEE Proceedings– Generation, Transmission & Distribution, vol 148, no 3, pp 489-496
[22] Lehmkoster C (2002) Security constrained optimal power flow for an economical operation of FACTS-devices in liberalized energy markets. IEEE Transactions on Power Delivery, vol 17 no 2, pp 603–608
[23] Schauder CD, Gyugyi L, Lund MR, Hamai DM, Rietman TR, Torgerson DR, Edris A (1998) Operation of the unified power flow controller (UPFC) under practical constraints. IEEE Trans. on Power Delivery, vol 13:630-637

[24] Zhang XP (2005) Comprehensive modelling of the unified power flow controller for power system control. Electrical Engineering - Archiv für Elektrotechnik, DOI: 10.1007/S00202-004-0280-0, published online
[25] Asplund G, Eriksson K, Svensson K (1997) DC transmission based on voltage source converters. CIGRE SC14 Colloquium, South Africa
[26] Asplund G (2000) Application of HVDC light to power system enhancement. Proceedings of IEEE 2000 PES Winter Meeting, Singapore
[27] Schetter F, Hung H, Christl N (2000) HVDC transmission system using voltage sourced converters – design and applications. Proceedings of IEEE 2000 PES Summer Meeting, Seattle, USA
[28] Lasson T, Edris A, Kidd D, Aboytes F (2001) Eagle pass back-to-back tie: a dual purpose application of voltage source converter technology. Proceedings of IEEE 2001 PES Summer Meeting, Vancouver, Canada
[29] Jiang H, Ekstrom A (1998) Multiterminal HVDC systems in urban areas of large cities. IEEE Transactions on Power Delivery, vol 13, no 4, pp 1278-1284
[30] Lu W, Ooi BT (2003) DC overvoltage control during loss of converter in multiterminal voltage-source converter-based HVDC (M-VSC-HVDC). IEEE Trans. on Power Delivery, vol 18, no 3, pp 915-920
[31] Zhang XP (2004) Multiterminal voltage-sourced converter based HVDC models for power flow analysis. IEEE Transactions on Power Systems, vol 18, no 4, 2004, pp1877-1884
[32] Hochgraf C, Lasseter RH (1997) A transformer-less static synchronous compensator employing a multi-level inverter. IEEE Trans. on Power Delivery, vol 12, no 2, pp 881-887

# 4 Modeling of FACTS-Devices in Optimal Power Flow Analysis

In recent years, energy, environment, deregulation of power utilities have delayed the construction of both generation facilities and new transmission lines. Better utilisation of existing power system capacities by installing new FACTS-devices has become imperative. FACTS-devices are able to change, in a fast and effective way, the network parameters in order to achieve a better system performance. FACTS-devices, such as phase shifter, shunt or series compensation and the most recent developed converter-based power electronic devices, make it possible to control circuit impedance, voltage angle and power flow for optimal operation of power systems, facilitate the development of competitive electric energy markets, and stipulate the unbundling the power generation from transmission and mandate open access to transmission services, etc.

However, in contrast to the practical applications of the STATCOM, SSSC and UPFC in power systems, very few publications have been focused on the mathematical modeling of these converter based FACTS-devices in optimal power flow analysis. This chapter covers

- Review of optimal power flow (OPF) solution techniques.
- Introduction of OPF solution by the nonlinear interior point methods.
- Mathematical modeling of FACTS-devices including STATCOM, SSSC, UPFC, IPFC, GUPFC, and VSC HVDC.
- The detailed models of the multi-converter FACTS-devices GUPFC, and VSC HVDC and their implementation into the nonlinear interior point OPF.
- Comparison of UPFC and VSC HVDC, and GUPFC and multiterminal VSC HVDC.
- Numerical examples for demonstration of FACTS controls

## 4.1 Optimal Power Flow Analysis

### 4.1.1 Brief History of Optimal Power Flow

The Optimal Power Flow (OPF) problem was initiated by the desire to minimize the operating cost of the supply of electric power when load is given [1][2]. In 1962 a generalized nonlinear programming formulation of the economic dispatch problem including voltage and other operating constraints was proposed by Carpentier [3]. The Optimal Power Flow (OPF) problem was defined in early 1960's as an expansion of conventional economic dispatch to determine the optimal settings for control variables in a power network considering various operating and control constraints [4]. The OPF method proposed in [4] has been known as the reduced gradient method, which can be formulated by eliminating the dependent variables based on a solved load flow. Since the concept of the reduced gradient method for the solution of the OPF problem was proposed, continuous efforts in the developments of new OPF methods have been found. Several review papers were published [5]-[13]. Among the various OPF methods proposed, it has been recognized that the main techniques for solving the OPF problems are the gradient method [4], linear programming (LP) method [15][16], successive sparse quadratic programming (QP) method [18], successive non-sparse quadratic programming (QP) method [20], Newton's method [21] and Interior Point Methods [27]-[32]. Each method has its own advantages and disadvantages. These algorithms have been employed with varied success.

### 4.1.2 Comparison of Optimal Power Flow Techniques

It has been well recognised that the OPF problems are very complex mathematical programming problems. In the past, numerous papers on the numerical solution of the OPF problems have been published [7][10][11][13]. In this section, a review of several OPF methods is given.

#### *4.1.2.1 Gradient Methods*

The widely used gradient methods for the OPF problems include the reduced gradient method [4] and the generalised gradient method [14]. Gradient methods basically exhibit slow convergence characteristics near the optimal solution. In addition, the methods are difficult to solve in the presence of inequality constraints.

#### *4.1.2.2 Linear Programming Methods*

LP methods have been widely used in the OPF problems. The main strengths of LP based OPF methods are summarised as follows:

1. Efficient handling of inequalities and detection of infeasible solutions;
2. Dealing with local controls;
3. Incorporation of contingencies.

Noting the fact that it is quite common in the OPF problems, the nonlinear equalities and inequalities and objective function need to be handled. In this situation, all the nonlinear constraints and objective function should be linearized around the current operating point such that LP methods can be applied to solve the linear optimal problems. For a typical LP based OPF, the solution can be found through the iterations between load flow and linearized LP subproblem. The LP based OPF methods have been shown to be effective for problems where the objectives are separable and convex. However, the LP based OPF methods may not be effective where the objective functions are non-separable, for instance in the minimization of transmission losses.

### 4.1.2.3 Quadratic Programming Methods

QP based OPF methods [17]-[20] are efficient for some OPF problems, especially for the minimization of power network losses. In [20], the non-sparse implementation of the QP based OPF was proposed while in [17][18][19], the sparse implementation of the QP based OPF algorithm for large-scale power systems was presented. In [17][18], the successive QP based OPF problems are solved through a sequence of linearly constrained subproblems using a quasi-Newton search direction. The QP formulation can always find a feasible solution by adding extra shunt compensation. In [19], the QP method, which is a direct solution method, solves a set of linear equations involving the Hessian matrix and the Jacobian matrix by converting the inequality constrained quadratic program (IQP) into the equality constrained quadratic program (EQP) with an initial guess at the correct active set. The computational speed of the QP method in [19] has been much improved in comparison to those in [17][18]. The QP methods in [17]-[19] are solved using MINOS developed by Stanford University.

### 4.1.2.4 Newton's Methods

The development of the OPF algorithm by Newton's method [21]-[24], is based on the success of the Newton's method for the power flow calculations. Sparse matrix techniques applied to the Newton power flow calculations are directly applicable to the Newton OPF calculations. The major idea is that the OPF problems are solved by the sequence of the linearized Newton equations where inequalities are being treated as equalities when they are binding. However, most critical aspect of the Newton's algorithm is that the active inequalities are not known prior to the solution and the efficient implementations of the Newton's method usually adopt the so-called trial iteration scheme where heuristic constraints enforcement/release is iteratively performed until acceptable convergence is achieved. In [22][25], alternative approaches using linear programming techniques have been proposed to identify the active set efficiently in the Newton's OPF.

In principle, the successive QP methods and Newton's method both using the second derivatives, which are considered as second order optimization method, are theoretically equivalent.

### 4.1.2.5 Interior Point Methods

Since Karmarkar published his paper on an interior point method for linear programming in 1984 [26], a great interest on the subject has arisen. Interior point methods have proven to be a promising alternative for the solution of power system optimization problems. In [27] and [28], a Security-Constrained Economic

Dispatch (SCED) is solved by sequential linear programming and the IP Dual-Affine Scaling (DAS). In [29], a modified IP DAS algorithm was proposed. In [30], an interior point method was proposed for linear and convex quadratic programming. It is used to solve power system optimization problems such as economic dispatch and reactive power planning. In [31]-[36], nonlinear primal-dual interior point methods for power system optimization problems were developed. The nonlinear primal-dual methods proposed can be used to solve the nonlinear power system OPF problems efficiently. The theory of nonlinear primal-dual interior point methods has been established based on three achievements: Fiacco & McCormick's barrier method for optimization with inequalities, Lagrange's method for optimization with equalities and Newton's method for solving nonlinear equations [37]. Experience with application of interior point methods to power system optimization problems has been quite positive.

### 4.1.3 Overview of OPF-Formulation

The OPF problem may be formulated as follows:

$$\text{Minimize: } f(\mathbf{x},\mathbf{u}) \tag{4.1}$$

subject to:

$$\mathbf{g}(\mathbf{x},\mathbf{u}) = 0 \tag{4.2}$$

$$\mathbf{h}_{min} \leq \mathbf{h}(\mathbf{x},\mathbf{u}) \leq \mathbf{h}_{max} \tag{4.3}$$

where

$\mathbf{u}$ - the set of control variables
$\mathbf{x}$ - the set of dependent variables
$f(\mathbf{x},\mathbf{u})$ - a scalar objective function
$\mathbf{g}(\mathbf{x},\mathbf{u})$ - the power flow equations
$\mathbf{h}(\mathbf{x},\mathbf{u})$ - the limits of the control variables and operating limits of power system components.

The objectives, controls and constraints of the OPF problems are summarized in Table 4.1. The limits of the inequalities in Table 4.1 can be classified into two categories: (a) physical limits of control variables; (b) operating limits of power system. In principle, physical limits on control variables can not be violated while operating limits representing security requirements can be violated or relaxed temporarily.

In addition to the steady state power flow constraints, for the OPF formulation, stability constraints, which are described by differential equations, may be considered and incorporated into the OPF. In recent years, stability constrained OPF problems have been proposed [38]-[42].

**Table 4.1.** Objectives, Constraints and Control Variables of the OPF Problems

| | |
|---|---|
| Objectives | • Minimum cost of generation and transactions <br> • Minimum transmission losses <br> • Minimum shift of controls <br> • Minimum number of controls shifted <br> • Mininum number of controls rescheduled <br> • Minimum cost of VAr investment |
| Equalitiy constraints | • Power flow constraints <br> • Other balance constraints |
| Inequalitiy constraints | • Limits on all control variables <br> • Branch flow limits (amps, MVA, MW, MVAr) <br> • Bus voltage variables <br> • Transmission interface limits <br> • Active/reactive power reserve limits |
| Controls | • Real and reative power generation <br> • Transformer taps <br> • Generator voltage or reactive control settings <br> • MW interchange transactions <br> • HVDC link MW controls <br> • FACTS voltage and power flow controls <br> • Load shedding |

## 4.2 Nonlinear Interior Point Optimal Power Flow Methods

### 4.2.1 Power Mismatch Equations

The power mismatch equations in rectangular coordinates at a bus are given by:

$$\Delta P_i = Pg_i - Pd_i - P_i \qquad (4.4)$$

$$\Delta Q_i = Qg_i - Qd_i - Q_i \qquad (4.5)$$

where $Pg_i$ and $Qg_i$ are real and reactive powers of generator at bus $i$, respectively; $Pd_i$ and $Qd_i$ the real and reactive load powers, respectively; $P_i$ and $Q_i$ the power injections at the node and are given by:

$$P_i = V_i \sum_{j=1}^{N} V_j (G_{ij} \cos\theta_{ij} + B_{ij} \sin\theta_{ij}) \qquad (4.6)$$

$$Q_i = V_i \sum_{j=1}^{N} V_j (G_{ij} \sin\theta_{ij} - B_{ij} \cos\theta_{ij}) \qquad (4.7)$$

where $V_i$ and $\theta_i$ are the magnitude and angle of the voltage at bus $i$, respectively; $Y_{ij} = G_{ij} + jB_{ij}$ is the system admittance element while $\theta_{ij} = \theta_i - \theta_j$. $N$ is the total number of system buses.

### 4.2.2 Transmission Line Limits

The transmission MVA limit may be represented by:

$$(P_{ij})^2 + (Q_{ij})^2 \leq (S_{ij}^{max})^2 \tag{4.8}$$

where $S_{ij}^{max}$ is the MVA limit of the transmission line $ij$. $P_{ij}$ and $Q_{ij}$ are given by:

$$P_{ij} = -V_i^2 G_{ij} + V_i V_j (G_{ij} \cos\theta_{ij} + B_{ij} \sin\theta_{ij}) \tag{4.9}$$

$$Q_{ij} = V_i^2 b_{ii} + V_i V_j (G_{ij} \sin\theta_{ij} - B_{ij} \cos\theta_{ij}) \tag{4.10}$$

where $b_{ii} = -B_{ij} + bc_{ij}/2$. $bc_{ij}$ is the shunt admittance of transmission line $ij$.

### 4.2.3 Formulation of the Nonlinear Interior Point OPF

Mathematically, as an example the objective function of an OPF may minimize the total operating cost as follows:

$$\text{Minimize } f(x) = \sum_i^{Ng} (\alpha_i * Pg_i^2 + \beta_i * Pg_i + \gamma_i) \tag{4.11}$$

while being subject to the following constraints:

*Nonlinear equality constraints:*

$$\Delta P_i(x) = Pg_i - Pd_i - P_i(t,e,f) = 0 \tag{4.12}$$

$$\Delta Q_i(x) = Qg_i - Qd_i - Q_i(t,e,f) = 0 \tag{4.13}$$

*Nonlinear inequality constraints*

$$h_j^{min} \leq h_j(x) \leq h_j^{max} \tag{4.14}$$

where
$x = [Pg, Qg, t, \theta, V]^T$ is the vector of variables
$\alpha_i, \beta_i, \gamma_i$      coefficients of production cost functions of generator
$\Delta P(x)$      bus active power mismatch equations
$\Delta Q(x)$      bus reactive power mismatch equations

| | |
|---|---|
| $h(x)$ | functional inequality constraints including line flow and voltage magnitude constraints, simple inequality constraints of variables such as generator active power, generator reactive power, transformer tap ratio |
| $Pg$ | the vector of active power generation |
| $Qg$ | the vector of reactive power generation |
| $t$ | the vector of transformer tap ratios |
| $\theta$ | the vector of bus voltage magnitude |
| $V$ | the vector of bus voltage angle |
| $Ng$ | the number of generators |

By applying Fiacco and McCormick's barrier method, the OPF problem equations (4.11)-(4.14) can be transformed into the following equivalent OPF problem:

$$\text{Objective:} \quad \text{Min}\{f(x) - \mu \sum_{j=1}^{M} \ln(sl_j) - \mu \sum_{j=1}^{M} \ln(su_j)\} \quad (4.15)$$

subject to the following constraints:

$$\Delta P_i = 0 \quad (4.16)$$

$$\Delta Q_i = 0 \quad (4.17)$$

$$h_j - sl_j - h_j^{\min} = 0 \quad (4.18)$$

$$h_j + su_j - h_j^{\max} = 0 \quad (4.19)$$

where $sl > 0$ and $su > 0$.

Thus the Lagrangian function for equalities optimisation of equations (4.15)-(4.19) is given by:

$$\begin{aligned}L = f(x) &- \mu \sum_{j=1}^{M} \ln(sl_j) - \mu \sum_{j=1}^{M} \ln(su_j) - \sum_{i=1}^{N} \lambda p_i \Delta P_i - \sum_{i=1}^{N} \lambda q_i \Delta Q_i \\ &- \sum_{j=1}^{M} \pi l_j (h_j - sl_j - h_j^{\min}) - \sum_{j=1}^{M} \pi u_j (h_j + su_j - h_j^{\max})\end{aligned} \quad (4.20)$$

where $\lambda p_i$, $\lambda q_i$, $\pi l_j$, $\pi u_j$ are Langrage multipliers for the constraints of equations (4.16)-(4.19), respectively. $N$ represents the number of buses and $M$ the number of inequality constraints. Note that $\mu > 0$. The Karush-Kuhn-Tucker (KKT) first order conditions for the Lagrangian function shown in equation (4.20) are as follows:

$$\nabla_x L_\mu = \nabla f(x) - \nabla \Delta P^T \lambda p - \nabla \Delta Q^T \lambda q - \nabla h^T \pi l - \nabla h^T \pi u = 0 \quad (4.21)$$

$$\nabla_{\lambda p} L_\mu = -\Delta P = 0 \quad (4.22)$$

$$\nabla_{\lambda q} L_\mu = -\Delta Q = 0 \qquad (4.23)$$

$$\nabla_{\pi l} L_\mu = -(h - sl - h^{\min}) = 0 \qquad (4.24)$$

$$\nabla_{\pi u} L_\mu = -(h + su - h^{\max}) = 0 \qquad (4.25)$$

$$\nabla_{sl} L_\mu = \mu - Sl\Pi l = 0 \qquad (4.26)$$

$$\nabla_{su} L_\mu = \mu + Su\Pi u = 0 \qquad (4.27)$$

where $Sl = diag(sl_j)$, $Su = diag(su_j)$, $\Pi l = diag(\pi l_j)$, $\Pi u = diag(\pi u_j)$.

As suggested in [31], the above equations can be decomposed into the following three sets of equations:

$$\begin{bmatrix} -\Pi l^{-1} Sl & 0 & -\nabla h & 0 \\ 0 & \Pi u^{-1} Su & -\nabla h & 0 \\ -\nabla h^T & -\nabla h^T & H & -J^T \\ 0 & 0 & -J & 0 \end{bmatrix} \begin{bmatrix} \Delta \pi l \\ \Delta \pi u \\ \Delta x \\ \Delta \lambda \end{bmatrix} = \begin{bmatrix} -\nabla_{\pi l} L_\mu - \Pi l^{-1} \nabla_{Sl} L_\mu \\ -\nabla_{\pi u} L_\mu - \Pi u^{-1} \nabla_{Su} L_\mu \\ -\nabla_x L_\mu \\ -\nabla_\lambda L_\mu \end{bmatrix} \qquad (4.28)$$

$$\Delta sl = \Pi l^{-1} (\nabla_{sl} L_\mu - Sl\Delta \pi l) \qquad (4.29)$$

$$\Delta su = \Pi u^{-1} (-\nabla_{su} L_\mu - Su\Delta \pi u) \qquad (4.30)$$

where $H(x, \lambda, \pi l, \pi u) = \nabla^2 f(x) - \sum \lambda \nabla^2 g(x) - \sum (\pi l + \pi u) \nabla^2 h(x)$,

$J(x) = \left[ \dfrac{\partial \Delta P(x)}{\partial x}, \dfrac{\partial \Delta Q(x)}{\partial x} \right]$, $g(x) = \begin{bmatrix} \Delta P(x) \\ \Delta Q(x) \end{bmatrix}$, and $\lambda = \begin{bmatrix} \lambda_p \\ \lambda_q \end{bmatrix}$.

The elements corresponding to the slack variables $sl$ and $su$ have been eliminated from equation (4.28) using analytical Gaussian elimination. By solving equation (4.28), $\Delta \pi l$, $\Delta \pi u$, $\Delta x$, $\Delta \lambda$ can be obtained, then by solving equations (4.29) and (4.30), respectively, $\Delta sl$, $\Delta su$ can be obtained. With $\Delta \pi l$, $\Delta \pi u$, $\Delta x$, $\Delta \lambda$, $\Delta sl$, $\Delta su$ known, the OPF solution can be updated using the following equations:

$$sl^{(k+1)} = sl^{(k)} + \sigma \alpha_p \Delta sl \qquad (4.31)$$

$$su^{(k+1)} = su^{(k)} + \sigma \alpha_p \Delta su \qquad (4.32)$$

$$x^{(k+1)} = x^{(k)} + \sigma \alpha_p \Delta x \qquad (4.33)$$

$$\pi l^{(k+1)} = \pi l^{(k)} + \sigma \alpha_d \Delta \pi l \qquad (4.34)$$

$$\pi u^{(k+1)} = \pi u^{(k)} + \sigma \alpha_d \Delta \pi u \qquad (4.35)$$

$$\pi u^{(k+1)} = \pi u^{(k)} + \sigma \alpha_d \Delta \pi u \qquad (4.36)$$

$$\lambda^{(k+1)} = \lambda^{(k)} + \sigma \alpha_d \Delta \lambda \qquad (4.37)$$

where $k$ is the iteration count, parameter $\sigma \in [0.995 - 0.99995]$ and $\alpha_p$ and $\alpha_d$ are the primal and dual step-length parameters, respectively. The step-lengths are determined as follows:

$$\alpha_p = \min\left[\min\left(\frac{sl}{-\Delta sl}\right), \min\left(\frac{su}{-\Delta su}\right), 1.00\right] \qquad (4.38)$$

$$\alpha_d = \min\left[\min\left(\frac{\pi l}{-\Delta \pi l}\right), \min\left(\frac{\pi u}{-\Delta \pi u}\right), 1.00\right] \qquad (4.39)$$

for those sl<0, $\Delta$su<0, $\Delta\pi$l<0 and $\Delta\pi$u>0.

The barrier parameter $\mu$ can be evaluated by:

$$\mu = \frac{\beta \times Cgap}{2 \times M} \qquad (4.40)$$

where $\beta \in [0.01-0.2]$ and $Cgap$ is the complementary gap for the nonlinear interior point OPF and can be determined using:

$$Cgap = \sum_{j=1}^{M}(sl_j \pi l_j - su_j \pi u_j) \qquad (4.41)$$

### 4.2.4 Implementation of the Nonlinear Interior Point OPF

Equations (4.28)-(4.30) are the basic formulation of the nonlinear interior point OPF that has been well reported in [31][51]. Equation (4.28) is the reduced equation with respect to the original OPF problem. However, equation (4.28) can be further reduced by eliminating all the dual variables of the inequalities, generator output variables and transformer tap ratios. The elimination will result in new fill-in elements. For example, eliminating a transformer tap ratio will result in sixteen new elements in the reduced Newton equation. Such a significant reduction means that the reduced Newton equation only involves the state variables of $\theta_i$, $V_i$, $\lambda p_i$, $\lambda q_i$. The details will be discussed in the next section.

#### 4.2.4.1 Eliminating Dual Variables $\pi l$, $\pi u$ of the Inequalities

In order to obtain the final reduced Newton equation consisting of only the variables $\theta$, $V$, $\lambda p$, $\lambda q$, the following Gaussian elimination steps can be applied.

The dimension of the Newton equation (4.28) can be reduced using analytical Gaussian elimination techniques. Basically, the dual variables $\pi l$, $\pi u$ in equation (4.28) can be eliminated to obtain:

$$H'\Delta x - J_P^T \Delta \lambda p - J_Q^T \Delta \lambda q = -\nabla_x L'_\mu \qquad (4.42)$$

$$-J_P \Delta \lambda p = -\nabla_{\lambda p} L_\mu \qquad (4.43)$$

$$-J_Q \Delta \lambda q = -\nabla_{\lambda q} L_\mu \qquad (4.44)$$

$$H'\Delta x - J_P^T \Delta \lambda p - J_Q^T \Delta \lambda q = -\nabla_x L'_\mu \qquad (4.45)$$

where:

$$H' = H + \nabla h^T \nabla h \left( \Pi l Sl^{-1} - \Pi u Su^{-1} \right) \qquad (4.46)$$

$$\nabla_x L'_\mu = \nabla_x L_\mu - \nabla h^T Sl^{-1} \left( \nabla_{sl} L_\mu + \Pi l \nabla_{\pi l} L_\mu \right) + \nabla h^T Su^{-1} \left( \nabla_{su} L_\mu + \Pi u \nabla_{\pi u} L_\mu \right) \qquad (4.47)$$

$$J_P = \frac{\partial \Delta P(x)}{\partial x} \qquad (4.48)$$

$$J_Q = \frac{\partial \Delta Q(x)}{\partial x} \qquad (4.49)$$

The equations (4.42)-(4.45) can be written as the following compact form:

$$\begin{bmatrix} H' & -Jp^T & -Jq^T \\ -Jp & 0 & 0 \\ -Jq & 0 & 0 \end{bmatrix} \cdot \begin{bmatrix} \Delta x \\ \Delta \lambda p \\ \Delta \lambda q \end{bmatrix} = \begin{bmatrix} -\nabla_x L'_\mu \\ -\nabla_{\lambda p} L_\mu \\ -\nabla_{\lambda q} L_\mu \end{bmatrix} \qquad (4.50)$$

By solving equation (4.50), $\Delta x$ can be obtained, then the dual variables $\Delta \pi l$ and $\Delta \pi u$ can be found by solving the following equations:

$$\Delta \pi l = -\Pi l Sl^{-1} (\nabla h \Delta x - \nabla_{\pi l} L_\mu) + Sl^{-1} \nabla_{Sl} L_\mu \qquad (4.51)$$

$$\Delta \pi u = \Pi u Su^{-1} (\nabla h \Delta x - \nabla_{\pi u} L_\mu) - Su^{-1} \nabla_{Su} L_\mu \qquad (4.52)$$

Up to now, equation (4.28) has been reduced to three lower dimension equations (4.50), (4.51) and (4.52). In (4.50), all inequalities have been eliminated, while equations (4.51) and (4.52) are relatively simple to solve.

### 4.2.4.2 Eliminating Generator Variables $P_g$ and $Q_g$

In equation (4.50), generator variables $P_g$, $Q_g$ can be further eliminated. The equation (4.50) may be written in the following form, in which only the relevant major diagonal block of bus $i$ is displayed:

$$\begin{bmatrix} H'P_{g_i}P_{g_i} & 0 & 0 & 0 & -1 & 0 \\ 0 & H'Q_{g_i}Q_{g_i} & 0 & 0 & 0 & -1 \\ 0 & 0 & H'\theta_i\theta_i & H'\theta_iV_i & -Jp_i,\theta_i & -Jq_i,\theta_i \\ 0 & 0 & H'\theta_iV_i & H'V_iV_i & -Jp_i,V_i & -Jq_i,V_i \\ -1 & 0 & -Jp_i,\theta_i & -Jp_i,V_i & 0 & 0 \\ 0 & -1 & -Jq_i,\theta_i & -Jq_i,V_i & 0 & 0 \end{bmatrix} \times \begin{bmatrix} \Delta P_{g_i} \\ \Delta Q_{g_i} \\ \Delta \theta_i \\ \Delta V_i \\ \Delta \lambda p_i \\ \Delta \lambda q_i \end{bmatrix} = \begin{bmatrix} -\nabla_{P_{g_i}}L'_\mu \\ -\nabla_{Q_{g_i}}L'_\mu \\ -\nabla_{\theta_i}L'_\mu \\ -\nabla_{V_i}L'_\mu \\ -\nabla_{\lambda p_i}L'_\mu \\ -\nabla_{\lambda q_i}L'_\mu \end{bmatrix} \quad (4.53)$$

Eliminating $\Delta P_{g_i}$ and $\Delta Q_{g_i}$ from the above equation, we have:

$$\begin{bmatrix} H'\theta_i\theta_i & H'\theta_iV_i & -Jp_i,\theta_i & -Jq_i,\theta_i \\ H'V_i\theta_i & H'V_iV_i & -Jp_i,V_i & -Jq_i,V_i \\ -Jp_i,\theta_i & -Jp_i,V_i & -J\lambda p_i & 0 \\ -Jq_i,\theta_i & -Jq_i,V_i & 0 & -J\lambda q_i \end{bmatrix} \begin{bmatrix} \Delta \theta_i \\ \Delta V_i \\ \Delta \lambda p_i \\ \Delta \lambda q_i \end{bmatrix} = \begin{bmatrix} -\nabla_{\theta_i}L'_\mu \\ -\nabla_{V_i}L'_\mu \\ -\nabla_{\lambda p_i}L'_\mu \\ -\nabla_{\lambda q_i}L'_\mu \end{bmatrix} \quad (4.54)$$

where:

$$J\lambda p_i = (H'P_{g_i}P_{g_i})^{-1}$$

$$J\lambda q_i = (H'Q_{g_i}Q_{g_i})^{-1}$$

$$\nabla_{\lambda p_i}L'_\mu = \nabla_{\lambda p_i}L_\mu + (H'P_{g_i}P_{g_i})^{-1}\nabla_{P_{g_i}}L'_\mu$$

$$\nabla_{\lambda q_i}L'_\mu = \nabla_{\lambda q_i}L_\mu + (H'Q_{g_i}Q_{g_i})^{-1}\nabla_{Q_{g_i}}L'_\mu$$

By solving equation (4.54), $\Delta \lambda p_i$ and $\Delta \lambda q_i$ can be obtained, then $\Delta P_{g_i}$ and $\Delta Q_{g_i}$ can be found by the following equations:

$$\Delta P_{g_i} = (H'P_{g_i}P_{g_i})^{-1}(\Delta \lambda p_i - \nabla_{P_{g_i}}L'_\mu) \quad (4.55)$$

$$\Delta Q_{g_i} = (H'Q_{g_i}Q_{g_i})^{-1}(\Delta \lambda q_i - \nabla_{Q_{g_i}}L'_\mu) \quad (4.56)$$

Similarly, the elements corresponding to the transformer tap ratio can be eliminated using the same principle resulting in a reduced Newton equation consisting of only the variables $\theta$, $V$, $\lambda p$ and $\lambda q$. Since the reduced Newton equation has the similar structure as that of (4.54), in the next discussions the final reduced Newton equation (without considering FACTS) is still referred to (4.54). However, it should be pointed out that the elimination of the transformer tap ratio will affect the sparsity of the matrix because new fill in elements are resulting.

### 4.2.5 Solution Procedure for the Nonlinear Interior Point OPF

The solution of the nonlinear interior point OPF may be summarised as follows:

*Step 0:*  *Formulation of equation* (4.28)

*Step 1:*  *Forward substitution*
- (a) Eliminating the dual variables $\pi l$, $\pi u$ of the inequalities from equation (4.28), obtain equation (4.50);
- (b) Eliminating generator variables $Pg$, $Qg$ from equation (4.50), obtain equation (4.54);
- (c) Eliminating the transformer tap ratio $t$ from equation (4.54), obtain highly reduced Newton matrix equation.

*Step 2:*  *Solution of the final highly reduced Newton equation by sparse matrix techniques*
- (a) The highly reduced system matrix has a dimension of $4N$ where $N$ is the total number of buses.
- (b) Having been grouped into 4 by 4 blocks, solution to the final matrix is produced by sparse matrix techniques.

*Step 3:*  *Back substitution*
- (a) First substitution for transformer tap ratio. After solving the final matrix equation, $\Delta\theta$, $\Delta V$ $\Delta\lambda p$ and $\Delta\lambda q$ are known, then $\Delta t$ can be found by back substitution.
- (b) Second substitution for generator output variables: $\Delta Pg$ and $\Delta Qg$ can be found from equations (4.55) and (4.56);
- (c) Third substitution for all the dual variables of the inequalities: The dual variables $\pi l$, $\pi u$ of the inequalities can be found by equation (4.51) and equation (4.52);
- (d) Fourth substitution for all slack variables: All slack variables can be found by equations (4.29) and (4.30).

## 4.3 Modeling of FACTS in OPF Analysis

Very few publications have been focused on the mathematical modeling of FACTS-devices in optimal power flow analysis. In [44][45], a UPFC model has been proposed, and the model has been implemented in a Successive QP. In [46], mathematical models for TCSC, IPC and UPFC have been established, and the OPF problem with these FACTS-devices is solved by Newton's method. In [47], a versatile model for UPFC in OPF analysis has been proposed and the model has been implemented into the nonlinear interior point methods. In this model, explicit controls such local voltage and power flow controls can be explicitly represented. Furthermore, in this model, global controls of UPFC can be achieved without explicit controls. In [48], the modeling techniques in [47] have been extended to general mathematical models for the converter based FACTS-devices such as STATCOM, SSSC, and UPFC suitable for optimal power flow analysis. Applying the techniques in chapter 3, the Thyristor controlled FACTS-devices such as SVC and TCSC can be modeled in OPF analysis. The detailed models of STATCOM, SSSC, UPFC are referred to [47][48]. In the next sections, novel models for IPFC, GUPFC, multi-terminal VSC HVDC will be discussed, the modeling techniques of which are applicable to STATCOM, SSSC and UPFC.

### 4.3.1 IPFC and GUPFC in Optimal Voltage and Power Flow Control

An innovative approach to utilization of FACTS-devices providing a multifunctional power flow management device was proposed in [43]. There are several possibilities of operating configurations by combing two or more converter blocks with flexibility. Among them, there are two novel operating configurations, namely the Interline Power Flow Controller (IPFC) and the Generalized Unified Power Flow Controller (GUPFC) [43][49][50], which are significantly extended to control power flows of multi-lines or a sub-network rather than controlling the power flow of a single line by a UPFC or SSSC.

In contrast to the practical applications of the GUPFC in power systems, very few publications have been focused on the mathematical modeling of this new FACTS-device in power system analysis. A fundamental frequency model of the GUPFC consisting of one shunt converter and two series converters for EMTP study was proposed quite recently in [50]. The modeling of IPFC and GUPFC in power flow and optimal power flow (OPF) analysis has been reported [51][52]. In the next, novel model for GUPFC will be proposed, which are very convenient to consider various control constraints and control modes. The model for IPFC can be very easily derived once the model for GUPFC has been established.

### 4.3.2 Operating and Control Constraints of GUPFC

As discussed in chapter 3, the GUPFC combining three or more converters working together extends the concepts of voltage and power flow control beyond what is achievable with the known two-converter UPFC. The simplest GUPFC consists of three converters, one connected in shunt and the other two in series with two transmission lines in a substation. It can control total five power system quantities such as a bus voltage and independent active and reactive power flows of two lines. The equivalent circuit of such a GUPFC, which is shown in Fig. 4.1, is used to show the basic operation principle for the sake of simplicity. However, the mathematical derivation is applicable to a GUPFC with an arbitrary number of series converters.

In the steady state operation, the main objective of the GUPFC is to control voltage and power flow. Real power can be exchanged among these shunt and series converters via the common DC link. The sum of the real power exchange should be zero if we neglect the losses of the converter circuits.

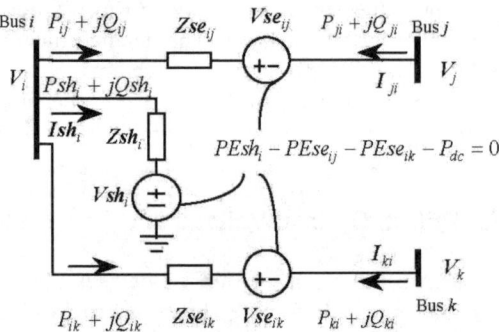

**Fig. 4.1.** The equivalent circuit of the GUPFC

For the GUPFC shown in Fig. 4.1, it has total 5 degrees of control freedom, that means it can control five power system quantities such as one bus voltage, and 4 active and reactive power flows of two lines. It can be seen that with more series converters included within the GUPFC, more degrees of control freedom can be introduced and hence more control objectives can be achieved.

In Fig. 4.1 $Zsh_i$ and $Zse_{in}$ are the shunt and series transformer impedances, respectively. $Vsh_i = Vsh_i \angle \theta sh_i$ and $Vse_{in} = Vse_{in} \angle \theta se_{in}$ ($n = j, k$) are the controllable injected shunt and series voltage sources. $PEsh_i$ and $PEse_{in}$ are the power exchange of the shunt converter and series converter, respectively, via the common DC link.

### 4.3.2.1 Power Flow Constraints of GUPFC

The power flow constraints of the GUPFC are summarized as follows:

*Shunt power flows:*

$$Psh_i = V_i^2 gsh_i - V_i Vsh_i (gsh_i \cos(\theta_i - \theta sh_i) + bsh_i \sin(\theta_i - \theta sh_i)) \qquad (4.57)$$

$$Qsh_i = -V_i^2 bsh_i - V_i Vsh_i (gsh_i \sin(\theta_i - \theta sh_i) - bsh_i \cos(\theta_i - \theta sh_i)) \qquad (4.58)$$

*Series power flows:*

$$\begin{aligned} P_{in} = &V_i^2 g_{in} - V_i V_n (g_{in} \cos\theta_{in} + b_{in} \sin\theta_{in}) \\ &- V_i Vse_{in} (g_{in} \cos(\theta_i - \theta se_{in}) + b_{in} \sin(\theta_i - \theta se_{in})) \end{aligned} \qquad (4.59)$$

$$\begin{aligned} Q_{in} = &-V_i^2 b_{in} - V_i V_n (g_{in} \sin\theta_{in} - b_{in} \cos\theta_{in}) \\ &- V_i Vse_{in} (g_{in} \sin(\theta_i - \theta se_{in}) - b_{in} \cos(\theta_i - \theta se_{in})) \end{aligned} \qquad (4.60)$$

$$P_{ni} = V_n^2 g_{in} - V_i V_n (g_{in} \cos(\theta_n - \theta_i) + b_{in} \sin(\theta_n - \theta_i))$$
$$+ V_n Vse_{in}(g_{in} \cos(\theta_n - \theta se_{in}) + b_{in} \sin(\theta_n - \theta se_{in})) \quad (4.61)$$

$$Q_{ni} = -V_n^2 b_{nn} - V_i V_n (g_{in} \sin(\theta_n - \theta_i) - b_{in} \cos(\theta_n - \theta_i))$$
$$+ V_n Vse_{in}(g_{in} \sin(\theta_n - \theta se_{in}) - b_{in} \cos(\theta_n - \theta se_{in})) \quad (4.62)$$

where $g_{in} = \text{Re}(1/Zse_{in})$, $b_{in} = \text{Im}(1/Zse_{in})$. $P_{in}$, $Q_{in}$ ($n=j, k$) are the active and reactive power flows of two GUPFC series branches leaving bus $i$ while $P_{ni}$, $Q_{ni}$ ($n = j, k$) are the active and reactive power flows of the GUPFC series branch $n$-$i$ leaving bus $n$ ($n = j, k$), respectively. Since two transmission lines are series connected with the FACTS branches $i$-$j$, $i$-$k$ via the GUPFC buses $j$ and $k$, respectively, $P_{ni}$, $Q_{ni}$ ($n = j, k$) are equal to the active and reactive power flows at the sending-end of the transmission lines, respectively.

The operating constraint representing the active power exchange among converters via the common DC link is:

$$PEx = PEsh_i - \sum PEse_{in} - P_{dc} = 0 \quad (4.63)$$

where $n =j, k$. $P_{dc}$ is the power loss of the DC circuit of the GUPFC. $PEsh_i$ and $PEse_{in}$ satisfy the following equalities:

$$PEsh_i - \text{Re}(Vsh_i Ish_i^*) = 0 \quad (4.64)$$

$$PEse_{in} - \text{Re}(Vse_{in} I_{ni}^*) = 0 \quad (4.65)$$

### 4.3.2.2 Operating Control Equalities of GUPFC

The GUPFC shown in Fig. 4.1 can control both active and reactive power flows of the two transmission lines. The active and reactive power flow control constraints of the GUPFC are given by (3.50) and (3.51)

The GUPFC has additional capability to control the voltage magnitude of bus $i$:

$$V_i - V_i^{Spec} = 0 \Leftrightarrow V_i^{Spec} - \varepsilon \leq V_i \leq V_i^{Spec} + \varepsilon \quad (4.66)$$

where $V_i$ is the voltage magnitude at bus $i$. $V_i^{Spec}$ is the specified bus voltage control reference at bus $i$. In the point of view of the implementation, the inequality is preferred since incorporation of the simple variable inequality is very easy.

### 4.3.2.3 Operating Inequalities of GUPFC

For the operation of the GUPFC, the injected voltage sources should be within their operating ratings while the currents through the converters should be within the current ratings:

*Shunt converter*

$$\theta sh_i^{min} \leq \theta sh_i \leq \theta sh_i^{max} \tag{4.67}$$

$$Vsh_i^{min} \leq Vsh_i \leq Vsh_i^{max} \tag{4.68}$$

$$-PEsh_i^{max} \leq PEsh_i \leq PEsh_i^{max} \tag{4.69}$$

$$Ish_i \leq Ish_i^{max} \tag{4.70}$$

*Series converter*

$$\theta se_{in}^{min} \leq \theta se_{in} \leq \theta se_{in}^{max} \tag{4.71}$$

$$Vse_{in}^{min} \leq Vse_{in} \leq Vse_{in}^{max} \tag{4.72}$$

$$-PEse_{in}^{max} \leq PEse_{in} \leq PEse_{in}^{max} \tag{4.73}$$

$$I_{ni} \leq I_{ni}^{max} \tag{4.74}$$

where $n = j, k$. $PEsh_i^{max}$ is the maximum limit of the power exchange of the shunt converter with the DC link. $Ish_i^{max}$ is the current rating. $PEse_{in}^{max}$ is the maximum limit of the power exchange of the series converter with the DC link ($n = j, k$). $I_{ni}^{max}$ is the current rating of the series converter.

### 4.3.3 Incorporation of GUPFC into Nonlinear Interior Point OPF

#### 4.3.3.1 Constraints of GUPFC

The GUPFC consists of the power flow constraints (4.57)-(4.62), the internal power exchange balance constraint (4.63), the operating inequality constraints (4.67)-(4.74), and the power flow control constraints and voltage control constraint (4.66). In the formulation of the nonlinear interior point OPF algorithm, the power flow constraints (4.57)-(4.62) can be directly incorporated into the power mismatch equations at bus $i$, $j$ and $k$. By introducing slack variables and barrier parameter, the inequalities (4.67)-(4.74) can be converted into equalities. Then all the transformed equalities of the GUPFC can be incorporated into the Lagrangian function of the OPF problem.

#### 4.3.3.2 Variables of GUPFC

The state variables of the GUPFC are $\theta sh_i$, $Vsh_i$, $PEsh_i$, $\theta se_{in}$, $Vse_{in}$, $PEse_{in}$. With incorporation of $PEsh_i$, $PEse_{in}$ into the state variables of the GUPFC, the formulation of the Newton OPF equation and the implementation of multi-control

## 4.3 Modeling of FACTS in OPF Analysis

functional model become simple and straightforward. In addition to the state variables, dual variables should be introduced for all the equalities and inequalities while slack variables should be introduced for all the inequalities. In the implementation, the angle constraints (4.67) and (4.71) are optional since they are usually allowed to move around 360°. For the simplicity of presentation, the angle constraints (4.67) and (4.71) are not discussed here.

The dual variables of the GUPFC inequality constraints are defined as follows:

$$\pi d V sh_i : \quad V sh_i - V sh_i^{min} - Sl V sh_i = 0 \tag{4.75}$$

$$\pi u V sh_i : \quad V sh_i - V sh_i^{max} + Su V sh_i = 0 \tag{4.76}$$

$$\pi d P Esh_i : \quad P Esh_i - P Esh_i^{min} - Sl P Esh_i = 0 \tag{4.77}$$

$$\pi u P Esh_i : \quad P Esh_i - P Esh_i^{max} + Su P Esh_i = 0 \tag{4.78}$$

$$\pi u Ish_i : \quad Ish_i - Ish_i^{max} + Su Ish_i = 0 \tag{4.79}$$

$$\pi d V se_{in} : \quad V se_{in} - V se_{in}^{min} - Sl V se_{in} = 0 \tag{4.80}$$

$$\pi u V se_{in} : \quad V se_{in} - V se_{in}^{max} + Su V se_{in} = 0 \tag{4.81}$$

$$\pi d P Ese_{in} : \quad P Ese_{in} - P Ese_{in}^{min} - Sl P Ese_{in} = 0 \tag{4.82}$$

$$\pi u P Ese_{in} : \quad P Ese_i - P Ese_i^{max} + Su P Ese_i = 0 \tag{4.83}$$

$$\pi u Ise_{ni} : \quad Ise_{ni} - Ise_{ni}^{max} + Su Ise_{ni} = 0 \tag{4.84}$$

In the above equations, all the variables that start with 'S' are slack variables and they are positive values while all the variables that start with '$\pi$' are dual variables.

The dual variables of the equalities of the GUPFC are defined as follows:

$$\lambda PEx : \quad PEx = PEsh_i + \sum PEse_{in} = 0 \tag{4.85}$$

$$\lambda PEsh_i : \quad PEsh_i - \text{Re}(\mathbf{V sh}_i \mathbf{Ish}_i^*) = 0 \tag{4.86}$$

$$\lambda PEse_{in} : \quad PEse_{in} - \text{Re}(\mathbf{V se}_{ni} \mathbf{I}_{in}^*) = 0 \tag{4.87}$$

$$\lambda P_{ni} : \quad \Delta P_{ni} = P_{ni} - P_{ni}^{Spec} = 0 \tag{4.88}$$

$$\lambda Q_{ni} : \quad \Delta Q_{ni} = Q_{ni} - Q_{ni}^{Spec} = 0 \tag{4.89}$$

$$\lambda p_m : \quad \Delta P_m = 0 \quad (m = i, j, k) \tag{4.90}$$

$$\lambda q_m: \qquad \Delta Q_m = 0 \qquad (m = i, j, k) \qquad (4.91)$$

where $\Delta P_m$ and $\Delta Q_m$ are power mismatch equations at bus $m$.

### 4.3.3.3 Augmented Lagrangian Function of GUPFC in Nonlinear Interior OPF

The augmented Lagrangian function of the equalities (4.75)-(4.84) is as follows:

$$\pi * \text{equality} - \mu \ln(S) \qquad (4.92)$$

The augmented Lagrangian function of the equalities (4.85)-(4.91) is defined as follows:

$$\begin{aligned}
&- \lambda PEx(PEsh_i - \sum_{n=j,k} PEse_{in}) - \lambda PEsh_i(PEsh_i - \text{Re}(Vsh_i Ish_i^*)) \\
&- \lambda PEse_{in}(PEse_{in} - \text{Re}(Vse_{in} Ise_{in}^*)) \\
&- \lambda P_{ni}(P_{ni} - P_{ni}^{Spec}) - \lambda Q_{ni}(Q_{ni} - Q_{ni}^{Spec}) \\
&- \lambda p_m \Delta P_m - \lambda q_m \Delta Q_m
\end{aligned} \qquad (4.93)$$

$$(n = j, k; \; m = i, j, k)$$

### 4.3.3.4 Newton Equation of Nonlinear Interior OPF with GUPFC

With the incorporation of the augmented Lagrangian functions above into the OPF problem in section 4.2, a reduced Newton equation can be derived:

$$\begin{bmatrix} \mathbf{A} & \mathbf{C} \\ \mathbf{C}^T & \mathbf{B} \end{bmatrix} \begin{bmatrix} \Delta \mathbf{X}^{gupfc} \\ \Delta \mathbf{X}^{sys} \end{bmatrix} = \begin{bmatrix} \mathbf{a} \\ \mathbf{b} \end{bmatrix} \qquad (4.94)$$

where

$\Delta \mathbf{X}^{gupfc} = [\Delta \mathbf{X}_{ik}^{gufc}, \Delta \mathbf{X}_{ij}^{gupfc}, \Delta \mathbf{X}_{i}^{gupfc}]^T$ - the incremental vector of the GUPFC variables, and

$\Delta \mathbf{X}_{in}^{gupfc} = [\Delta \pi u Ise_{ni}, \Delta \theta se_{in}, \Delta Vse_{in}, \Delta PEse_{in}, \lambda PEse_{in}, \Delta \lambda Pse_{ni}, \Delta \lambda Qse_{ni}]^T$ - the incremental vector of the variables of the GUPFC series branch $in$.

$\Delta \mathbf{X}_{i}^{gupfc} = [\Delta \pi u Ish_i, \Delta \theta sh_i, \Delta Vsh_i, \Delta PEsh_i, \Delta \lambda PEsh_i, \Delta \lambda PEx]^T$ - the incremental vector of the variables of the GUPFC shunt branch $i$.

$\Delta \mathbf{X}^{sys} = [\Delta \mathbf{X}_i^{sys}, \Delta \mathbf{X}_j^{sys}, \Delta \mathbf{X}_k^{sys}]^T$ - the incremental vector of the variables of the system buses.

$\mathbf{a} = [\mathbf{a}_{ij}, \mathbf{a}_{ik}, \mathbf{a}_i]^T$ - the right hand vector of the GUPFC.

$\Delta \mathbf{X}_m^{sys} = [\Delta \theta_m, \Delta V_m, \Delta \lambda p_m, \Delta \lambda q_m]^T$ $(m = i, j, k)$ - the incremental vector of the variables of system bus $m$.

## 4.3 Modeling of FACTS in OPF Analysis

In (4.94), all the slack and dual variables of the simple variable inequalities have been eliminated from the formulation. **B** and **b** are the system matrix and right hand vector, which have similar structure to the system matrix and right hand of (4.54), respectively except that in calculating the former, the contributions from the GUPFC should be considered. $\mathbf{a}_{in}$ and $\mathbf{a}_i$ are given by

$$\mathbf{a}_{in} = \begin{bmatrix} -\nabla_{\pi u I se_{in}} L_\mu - (\pi u I se_{in})^{-1} \nabla_{S u I se_{in}} L_\mu \\ -\nabla_{\theta se_{in}} L_\mu \\ -\nabla_{V se_{in}} L_\mu + \mu(1/SlVse_{in} - 1/SuVse_{in}) \\ -\nabla_{PEse_{in}} L_\mu + \mu(1/SlPEse_{in} - 1/SuPEse_{in}) \\ -\nabla_{\lambda PEse_{in}} L_\mu \\ -\nabla_{\lambda Pse_{in}} L_\mu \\ -\nabla_{\lambda Qse_{in}} L_\mu \end{bmatrix} \quad (n = j, k) \quad (4.95)$$

$$\mathbf{a}_i = \begin{bmatrix} -\nabla_{\pi u I sh_i} L_\mu - (\pi u I sh_i)^{-1} \nabla_{S u I sh_i} L_\mu \\ -\nabla_{\theta sh_i} L_\mu \\ -\nabla_{V sh_i} L_\mu + \mu(1/SlVsh_i - 1/SuVsh_i) \\ -\nabla_{PEsh_i} L_\mu ++\mu(1/SlPEsh_i - 1/SuPEsh_i) \\ -\nabla_{\lambda PEsh_i} L_\mu \\ -\nabla_{\lambda PEx} L_\mu \end{bmatrix} \quad (4.96)$$

In (4.54), **A** and **C** are given by:

$$\mathbf{A} = \left[ \frac{\partial^2 L}{\partial \mathbf{X}^{gupfc} \partial \mathbf{X}^{gupfc}} \right] + \mathbf{D} \quad (4.97)$$

$$\mathbf{C} = \left[ \frac{\partial^2 L}{\partial \mathbf{X}^{gupfc} \partial \mathbf{X}^{sys}} \right] \quad (4.98)$$

where **D** is given by:

$$\mathbf{D} = Diag\lfloor \mathbf{d}_{ij}, \mathbf{d}_{ik}, \mathbf{d}_i \rfloor \quad (4.99)$$

$$\mathbf{d}_{in} = Diag \begin{bmatrix} 0,0,(\pi dVse_{in}/SlVse_{in} - \pi uVse_{in}/SuVse_{in}) \\ (\pi dPEse_{in}/SlPEse_{in} - \pi uPEse_{in}/SuPEse_{in}),0,0,0 \end{bmatrix} \quad (4.100)$$

$$(n = j,k)$$

$$\mathbf{d}_i = Diag\begin{bmatrix} 0,0,(\pi dVse_i / SlVse_i - \pi uVse_i / SuVse_i) \\ (\pi dPEse_i / SlPEse_i - \pi uPEse_i / SuPEse_i),0,0 \end{bmatrix} \quad (4.101)$$

Some of the first and second terms of the Newton equation of the nonlinear interior point OPF are given in the Appendix of this chapter.

### 4.3.3.5 Implementation of Multi-Configurations and Multi-Control Functions of GUPFC

**Multi-configurations of GUPFC.** The GUPFC may have the configurations or topologies such as GUPFC or SSSCs plus STATCOM.

For the GUPFC configuration, the operation of GUPFC is mainly constrained by its voltage and current ratings of the converters and the active power exchange of the converters with the DC link. For the SSSCs plus STATCOM configuration the series converters are operated as SSSCs while the shunt converter is operated as a STATCOM, there is active power exchange between the SSSCs and STATCOM. The control configuration is simulated by setting the power exchange limits to zero.

**Multi-control functions of GUPFC.** As discussed in chapter 3, there are a number of control objectives that can be achieved by the series and shunt control. For instance, in the implementation of other series control functions other than the active and reactive power flow control, the latter can be simply replaced by the new control equations. For the case of power flow control, the following control modes may be adopted:

1. Active and reactive power flow control.
2. Active power flow control only.
3. Reactive power flow control only.
4. Without explicit active and reactive power control objectives.

For the implementation of 1, the dual variable of the reactive power flow control constraint should be dummied in the Newton equation of (4.94). Similarly by dummying the relevant dual variable in the Newton equation, the corresponding control equation can be removed from the equation.

Noting the fact that an OPF is to optimize the system globally, the control mode 4. is more practical and useful. However, for power flow analysis, the active and reactive power flow control equations must be retained.

### 4.3.3.6 Initialization of GUPFC Variables in Nonlinear Interior OPF

If the GUPFC has explicit series control objectives, then the initialization of the series converters can be done in the same way as discussed in chapter 3 for the GUPFC in power flow calculations. However, if there are no explicit objectives applied, then the injected series voltage magnitude may be set to a value, say $Vse_{in} = 0.1 p.u.$ $(n=j, k)$ while $Vsh_i$ is given by:

$$Vsh_i = (Vsh_i^{max} + Vsh_i^{min})/2 \text{ or } Vsh_i = V_i^{Spec} \qquad (4.102)$$

### 4.3.4 Modeling of IPFC in Nonlinear Interior Point OPF

In comparison to the GUPFC, the IPFC has no shunt converter and the associated control. For the IPFC shown in Fig. 3.4 and Fig. 3.5, the primary series converter $i$-$j$ has two control degrees of freedom while the secondary series converter $i$-$k$ has one control degree of freedom since another control degree of freedom of the converter is used to balance the active power exchange between the two series converters. Very similar to that for the GUPFC, A reduced Newton equation for the IPFC can be derived as follows:

$$\begin{bmatrix} \mathbf{A} & \mathbf{C} \\ \mathbf{C}^T & \mathbf{B} \end{bmatrix} \begin{bmatrix} \Delta \mathbf{x}^{gupfc} \\ \Delta \mathbf{x}^{sys} \end{bmatrix} = \begin{bmatrix} \mathbf{a} \\ \mathbf{b} \end{bmatrix} \qquad (4.103)$$

where

$\Delta \mathbf{X}^{ipfc} = [\Delta \mathbf{X}_{ij}^{ipfc}, \Delta \mathbf{X}_{ik}^{ipfc}]^T$ - the incremental vector of the IPFC variables, and

$\Delta \mathbf{X}_{ij}^{ipfc} = [\Delta \pi u Ise_{ji}, \Delta \theta se_{ij}, \Delta V se_{ij}, \Delta PEse_{ij}, \Delta \lambda PEse_{ij}, \Delta \lambda Pse_{ji}, \Delta \lambda Qse_{ji}]^T$ - the incremental vector of the variables of the IPFC primary series branch $ij$.

$\Delta \mathbf{X}_{ik}^{ipfc} = [\Delta \pi u Ise_{ik}, \Delta \theta se_{ik}, \Delta V se_{ik}, \Delta PEse_{ik}, \Delta \lambda PEse_{ik}, \Delta \lambda Pse_{ki}, \Delta \lambda PEx]^T$ - the incremental vector of the variables of the IPFC secondary series branch $ik$.

$\Delta \mathbf{X}^{sys} = [\Delta \mathbf{X}_i^{sys}, \Delta \mathbf{X}_j^{sys}, \Delta \mathbf{X}_k^{sys}]^T$ - the incremental vector of the variables of the system buses.

$\mathbf{a} = [\mathbf{a}_{ij}, \mathbf{a}_{ik}]^T$ - the right hand vector of the IPFC

$\Delta \mathbf{X}_m^{sys} = [\Delta \theta_m, \Delta V_m, \Delta \lambda p_m, \Delta \lambda q_m]^T$ ($m = i, j, k$) - the incremental vector of the variables of system bus $m$.

In (4.103), all the slack and dual variables of the simple variable inequalities have been eliminated from the formulation. $\mathbf{B}$ and $\mathbf{b}$ are the system matrix and right hand vector, which have similar structure to the system matrix and right hand of (4.54), respectively except that in calculating the former, the contributions from the IPFC should be considered. $\mathbf{a}_{ij}$ and $\mathbf{a}_{ik}$ are given by:

$$\mathbf{a}_{ij} = \begin{bmatrix} -\nabla_{\pi\iota Ise_{ij}} L_\mu - (\pi\iota Ise_{ij})^{-1}\nabla_{SuIse_{ij}} L_\mu \\ -\nabla_{\theta se_{ij}} L_\mu \\ -\nabla_{Vse_{ij}} L_\mu + \mu(1/SlVse_{ij} - 1/SuVse_{ij}) \\ -\nabla_{PEse_{ij}} L_\mu + \mu(1/SlPEse_{ij} - 1/SuPEse_{ij}) \\ -\nabla_{\lambda PEse_{ij}} L_\mu \\ -\nabla_{\lambda Pse_{ji}} L_\mu \\ -\nabla_{\lambda Qse_{ji}} L_\mu \end{bmatrix} \qquad (4.104)$$

$$\mathbf{a}_{ik} = \begin{bmatrix} -\nabla_{\pi\iota Ise_{ik}} L_\mu - (\pi\iota Ise_{ik})^{-1}\nabla_{SuIse_{ik}} L_\mu \\ -\nabla_{\theta se_{ik}} L_\mu \\ -\nabla_{Vse_{ik}} L_\mu + \mu(1/SlVse_{ik} - 1/SuVse_{ik}) \\ -\nabla_{PEse_{ik}} L_\mu + \mu(1/SlPEse_{ik} - 1/SuPEse_{ik}) \\ -\nabla_{\lambda PEse_{ik}} L_\mu \\ -\nabla_{\lambda Pse_{ki}} L_\mu \\ -\nabla_{\lambda PEx} L_\mu \end{bmatrix} \qquad (4.105)$$

Similar to that of the GUPFC, **A** and **C** are given by:

$$\mathbf{A} = \left[\frac{\partial^2 L}{\partial \mathbf{X}^{ipfc} \partial \mathbf{X}^{ipfc}}\right] + \mathbf{D} \qquad (4.106)$$

$$\mathbf{C} = \left[\frac{\partial^2 L}{\partial \mathbf{X}^{ipfc} \partial \mathbf{X}^{sys}}\right] \qquad (4.107)$$

where **D** is given by

$$\mathbf{D} = Diag[\mathbf{d}_{ij}, \mathbf{d}_{ik}] \qquad (4.108)$$

$$\mathbf{d}_{ij} = Diag\begin{bmatrix} 0,0,(\pi dVse_{ij}/SlVse_{ij} - \pi\iota Vse_{ij}/SuVse_{ij}) \\ (\pi dPEse_{ij}/SlPEse_{ij} - \pi\iota PEse_{ij}/SuPEse_{ij}),0,0,0 \end{bmatrix} \qquad (4.109)$$

$$\mathbf{d}_{ik} = Diag\begin{bmatrix} 0,0,(\pi dVse_{ik}/SlVse_{ik} - \pi\iota Vse_{ik}/SuVse_{ik}) \\ (\pi dPEse_{ik}/SlPEse_{ik} - \pi\iota PEse_{ik}/SuPEse_{ik}),0,0,0 \end{bmatrix} \qquad (4.110)$$

## 4.4 Modeling of Multi-Terminal VSC-HVDC in OPF

### 4.4.1 Multi-Terminal VSC-HVDC in Optimal Voltage and Power Flow

The multi-terminal VSC-HVDC models for power flow analysis have been presented in chapter 3. A multi-terminal VSC-HVDC model suitable for optimal power flow analysis will be discussed here. The multi-terminal VSC-HVDC has not only power flow but also voltage control capability. It is useful to compare the GUPFC with the M-VSC-HVDC and investigate different control capabilities of these two FACTS-devices.

The equivalent circuit of the multi-terminal VSC-HVDC (M-VSC-HVDC) is shown in Fig. 4.2. In this figure, for the sake of simplicity, the VSC HVDC consists of three terminals. However, the derivation is applicable to a M-VSC-HVDC with any number of terminals.

As discussed in chapter 3, the M-VSC-HVDC combining three or more converters working together extends the concepts of voltage and power flow control beyond what is achievable with the known two-converter VSC-HVDC-device. The simplest M-VSC-HVDC consists of three converters connected in shunt with two buses in a substation. It can control total five power system quantities such as a bus voltage of the secondary converter and independent active and reactive power flows of two lines, which are connected with the primary converters.

In the steady state operation, the main objective of the M-VSC-HVDC is to control voltage and power flow. Real power can be exchanged among these converters via the common DC link. The sum of the real power exchange should be zero if we neglect the losses of the converter circuits. For the M-VSC-HVDC shown in Fig. 4.2, if explicit controls are applied, it has total 5 degrees of control freedom, that means it can control five power system quantities such as one bus voltage, and 4 active and reactive power flows of two lines. It can be seen that with more converters included within the M-VSC-HVDC, more degrees of control freedom can be introduced and hence more control objectives can be achieved.

### 4.4.2 Operating and Control Constraints of the M-VSC-HVDC

In Fig. 4.2, the AC power flow constraints of the M-VSC-HVDC at buses $i$, $j$, $k$ can be explicitly incorporated into the power mismatch equations at these AC buses.

The active power exchange among the converters via the DC link should be balanced at any instant, which is described by:

$$PEx = Pdc_i + Pdc_j + Pdc_k + Ploss = 0 \qquad (4.111)$$

where *Ploss* represents losses in converter circuits. The handling of *Ploss* has been discussed in chapter 3.

124    4 Modeling of FACTS-Devices in Optimal Power Flow Analysis

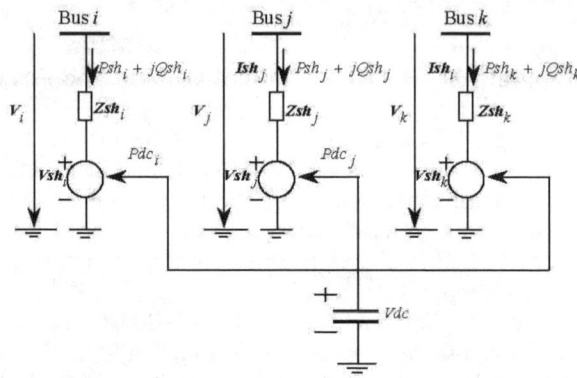

**Fig. 4.2.** The equivalent circuit of the multi-terminal VSC HVDC

$Pdc_m$ ($m = i, j, k$) as shown in Fig. 4.2 is the power exchange of the converter with the DC link and can be described by the following equalities:

$$Pdc_m - \text{Re}(-Vsh_m Ish_m^*) = 0 \qquad m = i, j, k \qquad (4.112)$$

where

$$\text{Re}(-Vsh_m Ish_m^*) = Vsh_m^2 gsh_m - V_i Vsh_m (gsh_m \cos(\theta_i - \theta sh_m) - bsh_m \sin(\theta_i - \theta sh_m)).$$

Voltage and power flow control constraints of the M-VSC-HVDC consists of explicit PQ or PV control of primary converters as given by (3.79)-(3.82) and voltage control of secondary converter. In the implementation of the voltage control, the equality is simply replaced by an inequality of bus voltage constraint since the implementation of the simple variable inequality constraint is very simple and straightforward.

In addition to the above equality constraints, voltage and current inequality constraints of the M-VSC-HVDC as shown in (3.84) and (3.85) should be considered.

### 4.4.3 Modeling of M-VSC-HVDC in the Nonlinear Interior Point OPF

Following the similar procedure in the derivation of Nonlinear Interior Point OPF with incorporation of the GUPFC, a reduced Newton equation can be obtained:

With the incorporation of the augmented Lagrangian functions above into the OPF problem in section 4.2, a reduced Newton equation can be derived:

$$\begin{bmatrix} \mathbf{A} & \mathbf{C} \\ \hline \mathbf{C}^T & \mathbf{B} \end{bmatrix} \begin{bmatrix} \Delta \mathbf{x}^{HVDC} \\ \Delta \mathbf{x}^{sys} \end{bmatrix} = \begin{bmatrix} \mathbf{a} \\ \mathbf{b} \end{bmatrix} \qquad (4.113)$$

where

$\Delta \mathbf{X}^{HVDC} = [\Delta \mathbf{X}_i^{HVDC}, \Delta \mathbf{X}_j^{HVDC}, \Delta \mathbf{X}_k^{HVDC}]^T$ - the incremental vector of the M-VSC-HVDC variables, and

$\Delta \mathbf{X}_i^{HVDC} = [\Delta m I sh_i, \Delta \theta sh_i, \Delta V sh_i, \Delta PEsh_i, \Delta \lambda PEsh_i, \Delta \lambda PEx]^T$ - the incremental vector of the variables of the M-VSC-HVDC branch $i$.

$\Delta \mathbf{X}_j^{HVDC} = [\Delta m I sh_j, \Delta \theta sh_j, \Delta V sh_j, \Delta PEsh_j, \Delta \lambda PEsh_j, \Delta \lambda Psh_j, \lambda QEsh_j]^T$ - the incremental vector of the variables of the M-VSC-HVDC branch $j$.

$\Delta \mathbf{X}_k^{HVDC} = [\Delta m I sh_k, \Delta \theta sh_k, \Delta V sh_k, \Delta PEsh_k, \Delta \lambda PEsh_k, \Delta \lambda Psh_k, \lambda QEsh_k]^T$ the incremental vector of the variables of the M-VSC-HVDC branch $k$.

$\Delta \mathbf{X}^{sys} = [\Delta \mathbf{X}_i^{sys}, \Delta \mathbf{X}_j^{sys}, \Delta \mathbf{X}_k^{sys}]^T$ - the incremental vector of the variables of the system buses.

$\mathbf{a} = [\mathbf{a}_i, \mathbf{a}_j, \mathbf{a}_k]^T$ - the right hand vector of the M-VSC-HVDC.

$\Delta \mathbf{X}_m^{sys} = [\Delta \theta_m, \Delta V_m, \Delta \lambda p_m, \Delta \lambda q_m]^T$ ($m = i, j, k$) - the incremental vector of the variables of system bus $m$.

In (4.113), all the slack and dual variables of the simple variable inequalities have been eliminated from the formulation. **B** and **b** are the system matrix and right hand vector have similar structure to the system matrix and right hand of (4.54), respectively except that in calculating the former, the contributions from the M-VSC-HVDC should be considered. $\mathbf{a}_i$ and $\mathbf{a}_m$ are given by:

$$\mathbf{a}_i = \begin{bmatrix} -\nabla_{m I sh_i} L_\mu - (m I sh_i)^{-1} \nabla_{SuIsh_i} L_\mu \\ -\nabla_{\theta sh_i} L_\mu \\ -\nabla_{Vsh_i} L_\mu + \mu(1/SlVsh_i - 1/SuVsh_i) \\ -\nabla_{PEsh_i} L_\mu + +\mu(1/SlPEsh_i - 1/SuPEsh_i) \\ -\nabla_{\lambda PEsh_i} L_\mu \\ -\nabla_{\lambda PEx} L_\mu \end{bmatrix} \qquad (4.114)$$

$$\mathbf{a}_m = \begin{bmatrix} -\nabla_{m I sh_m} L_\mu - (m I sh_m)^{-1} \nabla_{SuIsh_m} L_\mu \\ -\nabla_{\theta se_m} L_\mu \\ -\nabla_{Vsh_m} L_\mu + \mu(1/SlVsh_m - 1/SuVsh_m) \\ -\nabla_{PEsh_m} L_\mu + \mu(1/SlPEsh_m - 1/SuPEsh_m) \\ -\nabla_{\lambda PEsh_m} L_\mu \\ -\nabla_{\lambda Psh_m} L_\mu \\ -\nabla_{\lambda Qsh_m} L_\mu \end{bmatrix} \quad (m = j, k) \qquad (4.115)$$

In (4.113), **A** and **C** are given by:

$$\mathbf{A} = \left[\frac{\partial^2 L}{\partial \mathbf{X}^{HVDC} \partial \mathbf{X}^{HVDC}}\right] + \mathbf{D} \tag{4.116}$$

$$\mathbf{C} = \left[\frac{\partial^2 L}{\partial \mathbf{X}^{HVDC} \partial \mathbf{X}^{sys}}\right] \tag{4.117}$$

where **D** is given by:

$$\mathbf{D} = Diag[\mathbf{d}_i, \mathbf{d}_j, \mathbf{d}_k] \tag{4.118}$$

$$\mathbf{d}_i = Diag\begin{bmatrix} 0,0,(\pi dVsh_i / SlVsh_i - \pi uVsh_i / SuVsh_i) \\ (\pi dPEsh_i / SlPEsh_i - \pi uPEsh_i / SuPEsh_i),0,0 \end{bmatrix} \tag{4.119}$$

$$\mathbf{d}_m = Diag\begin{bmatrix} 0,0,(\pi dVsh_m / SlVsh_m - \pi uVsh_m / SuVsh_m) \\ (\pi dPEsh_m / SlPEsh_m - \pi uPEsh_m / SuPEsh_m),0,0,0 \end{bmatrix} \tag{4.120}$$

$$(m = j,k)$$

## 4.5 Comparison of FACTS-Devices with VSC-HVDC

The test cases in this section are carried out on the IEEE 30 bus system and IEEE118-bus system. The IEEE 30-bus system has 6 generators, 4 OLTC transformers and 37 transmission lines. The IEEE 118-bus system has 18 controllable active power generation, 54 controllable reactive power generation, 9 OLTC transformers, 177 transmission lines. For all cases in this section, the convergence tolerances are 5.0e-4 for complementary gap and 1.0e-4 (0.01MW/Mvar) for maximal absolute bus power mismatch, respectively.

### 4.5.1 Comparison of UPFC with BTB-VSC-HVDC

As a special case, Back-to-Back (BTB) VSC-HVDC should have similar a control capability as UPFC. Cases, which are used to show the control performance of both BTB-VSC-HVDC and UPFC, are presented. Four cases are given on the IEEE 30-bus system:

*Case 1*: A UPFC is installed for control of the voltage at bus 12 and control of active and reactive power flows in line 12-15. Suppose the FACTS bus 15' is created, and assume that the sending end of the transmission line 12-15 is now connected with the FACTS buses 15' while the series converter is

## 4.5 Comparison of FACTS-Devices with VSC-HVDC

installed between buses 12 and 15'.

*Case 2*: Similar to case 1, but no explicit UPFC control is applied.

*Case 3*: A BTB VSC-HVDC is used to replace the UPFC in case 1 while the control settings for voltage and power flows are as the same as that of case 1. This also means the two primary converters are using the PQ control mode.

*Case 4*: Similar to case 3, but no BTB-VSC-HVDC is applied.

The transformer impedance is $0+j0.025$ p.u. The active power flow settings on the transmission line 12-15 are 25 MW and 10 MVar, respectively, for active and reactive power flows while voltage control setting is 1.0 p.u.

For cases 1 and 2 and cases 3 and 4, the results have been obtained for two different situations: (a) there are explicit voltage and power flow control objectives; (b) there are no explicit voltage and power flow control objectives. For the second situation, global optimization of control settings of voltage and power flows of the FACTS-devices is expected. Table 4.2 and Table 4.3 show the test results.

From Table 4.2, it can be seen that the OPF solution with UPFC needs more iterations in comparison to that with BTB-VSC-HVDC. Furthermore, UPFC cases without explicit control will need more iterations than that with explicit control. In contrast, it has been found that the OPF algorithm with BTB-VSC-HVDC is not sensitive to the initial point and can converge more quickly.

**Table 4.2.** Optimal power flow results for the IEEE 30-bus system with UPFC or BTB VSC HVDC

| Case No. | Case 1 | Case 2 | Case 3 | Case 4 |
|---|---|---|---|---|
| Control type | With explicit FACTS control | Without explicit FACTS control | With explicit FACTS control | Without explicit FACTS control |
| Objective function | 8.0413e+002 | 8.0344e+002 | 8.0413e+002 | 8.0344e+002 |
| Active power loss | 9.97 MW | 9.76 MW | 9.97 MW | 9.76 MW |
| Total active power generation | 293.37 MW | 293.16 MW | 293.37 MW | 293.16 MW |
| Total reactive power generation | 127.84 MVar | 112.52 Mvar | 129.78 MVar | 112.08 MVar |
| Number of iterations | 11 | 17 | 9 | 9 |

**Table 4.3.** The OPF solution of UPFC or BTB VSC HVDC for the IEEE 30-bus system

| Case No. | Case 1 | Case 2 | Case 3 | Case 4 |
|---|---|---|---|---|
| Control type | With explicit FACTS control | Without explicit FACTS control | With explicit FACTS control | Without explicit FACTS control |
| FACTS solution | UPFC: $Vsh_{12,15'} = 0.0435$ p.u. $\theta se_{12,15'} = 151.94°$ $Vsh_{12} = 1.002$ p.u. $\theta sh_{12} = 11.70°$ | UPFC: $Vsh_{12,15'} = 0.00549$ p.u. $\theta se_{12,15'} = -147.35°$ $Vsh_{12} = 1.048$ p.u. $\theta sh_{12} = -11.21°$ | BTB HVDC: $Vsh_{15'} = 1.034$ p.u. $\theta sh_{15'} = -10.15°$ $Vsh_{12} = 1.000$ p.u. $\theta sh_{12} = -12.07°$ | BTB HVDC: $Vsh_{15'} = 1.046$ p.u. $\theta sh_{15'} = -11.45°$ $Vsh_{12} = 1.047$ p.u. $\theta sh_{12} = -11.01°$ |
| FACTS converter max. power | UPFC: Series converter: 1.1 MVA Shunt converter: 5.4 MVA | UPFC: Series converter: 0.1 MVA Shunt converter: 19 MVA | BTB HVDC: Primary conv.: 27 MVA Secondary conv.: 26 MVA | BTB HVDC: Primary conv..: 19 MVA Secondary conv.: 22 MVA |

From Table 4.3, it can be seen that using UPFC and BTB VSC HVDC can achieve the similar control purposes. However, ratings of the converters are quite different. In the studies, it has been found that the rating of the series converter of a UPFC is very small. This means a reduced investment in comparison to a BTB VSC HVDC.

### 4.5.2 Comparison of GUPFC with M-VSC-HVDC

Case studies are carried out on the IEEE 118-bus system with GUPFC installed. Five cases are presented as follows:

*Case 5*: The base case of the IEEE 118-bus system.

*Case 6*: There is a GUPFC installed for control of voltage at bus 45 and active and reactive power flow of line 45-44 and line 45-46. The control setting of the bus voltage is 1.0p.u. The control settings for active and reactive power flow of line 45-44 and line 45-46 are 40MW +$j$7Mvar and -50MW-$j$7Mvar, respectively. There is second GUPFC further installed for control voltage of bus 94 and power flow of line 94-95, line 94-93, line 94-100. The voltage control objective is 1.0 p.u. The control settings for active and reactive power flow of line 94-95, line 94-93, line 94-100 are 50MW +$j$5Mvar, -50MW-$j$20Mvar and -35MW –$j$10Mvar, respectively. There is third GUPFC further installed for control voltage of bus 12 and power flow of line 12-3 and line 12-11. The control settings for active and reactive power flow of line 12-3, line 12-11 are 15MW+$j$4Mvar and $-40$ MW + $j$15Mvar, respectively.

## 4.5 Comparison of FACTS-Devices with VSC-HVDC

*Case 7*: This is similar to case 6 except that there are no explicit FACTS controls applied.

The following cases are carried out on the IEEE 118-bus system with M-VSC-HVDC installed:

*Case 8*: This is similar to case 6 except that all GUPFC-devices are replaced by M-VSC-HVDCs.

*Case 9*: This is similar to case 8 except that there are no explicit M-VSC-HVDC controls applied.

Test results based on the cases 5-9 are summarized in Table 4.4. In these cases, active power flow settings are over 125% of their corresponding base case active power flows.

The power flow solutions of case 6 and case 8 for the GUPFCs and M-VSC-HVDCs are shown in Table 4.5. As expected, multi-terminal VSC HVDCs have similar steady state control capability as that of GUPFCs. However, it has been found that the actual power of a converter of GUPFC is much less than that of a converter of M-VSC-HVDC.

**Table 4.4.** Test results of the IEEE 118-bus system

|  | Case 5 | Case 6 | Case 7 | Case 8 | Case 9 |
|---|---|---|---|---|---|
| FACTS type | None | GUPFC | GUPFC | M-VSC-HVDC | M-VSC-HVDC |
| Control type | None | With explicit control | Without explicit control | With explicit control | Without explicit control |
| Number of devices | None | 3 | 3 | 3 | 3 |
| Total number of active and reactive power flow control | None | 7P Flow 7Q Flow | 0 | 7P Flow 7Q Flow | 0 |
| Total number of voltage control | None | 3 | 0 | 3 | 0 |
| Number of iterations | 12 | 14 | 21 | 12 | 12 |

**Table 4.5.** FACTS-solutions of case 6 and case 8

|  | Case 6 | Case 8 |
|---|---|---|
| FACTS type | GUPFC | M-VSC-HVDC |
| Control type | With explicit control | With explicit control |
| Number of devices | 3 | 3 |
| Maximum power of converter | GUPFC at bus 12 | M-VSC-HVDC at bus 12 |
|  | *Series converter: 0.6 MVA* | *Primary converter: 21.2 MVA* |
|  | *Series converter : 0.01 MVA* | *Primary converter: 25.1 MVA* |
|  | *Shunt converter: 19.7 MVA* | *Secondary converter: 20.9 MVA* |
|  | GUPFC at bus 45 | M-VSC-HVDC at bus 45 |
|  | *Series converter 1: 2.4 MVA* | *Primary converter 1: 40.6 MVA* |
|  | *Series converter 2: 2.6 MVA* | *Primary converter 2: 50.2 MVA* |
|  | *Shunt converter: 10 MVA* | *Secondary converter: 25.2 MVA* |
|  | GUPFC at bus 94 | M-VSC-HVDC at bus 94 |
|  | *Series converter 1: 0.4 MVA* | *Primary converter 1: 50.3 MVA* |
|  | *Series converter 2: 2.9 MVA* | *Primary converter 2: 53.6 MVA* |
|  | *Series converter 3: 2.9 MVA* | *Primary converter 3: 36.3 MVA* |
|  | *Shunt converter: 49.7 MVA* | *Secondary converter: 52.9 MVA* |

From these results on the IEEE 30-bus and 118-bus systems, it can be seen:

1. Numerical results demonstrate the feasibility as well as the effectiveness of the FACTS and VSC-HVDC models established and the OPF method proposed.
2. The GUPFC and the M-VSC-HVDC are quite flexible and powerful FACTS-device. Both of them can control bus voltage and active and reactive power flows of several lines simultaneously. They may be installed in some central substations to manage power flows of multi-lines or a group of lines and provide voltage support as well.
3. The OPF with global coordinating capability is a very useful tool to minimize (or maximize) an objective while satisfying power flow constraints, thermal constraints, as well as the operating and control constraints of the GUPFC devices.
4. The flexibility of the GUPFC and the M-VSC-HVDC with controlling bus voltage and multi-line active and reactive power flows offers a great potential in solving many of the problems facing the electric utilities in a competitive environment.

# 4.6 Appendix: Derivatives of Nonlinear Interior Point OPF with GUPFC

5. The power rating of a primary converter of the M-VSC-HVDC may be higher than that of a corresponding series converter of the GUPFC since the voltage rating of the former is higher that of the latter. Hence, the power rating of the secondary converter of the M-VSC-HVDC may be higher than that of the shunt converter of the GUPFC.
6. VSC-HVDC and UPFC, and M-VSC-HVDC and GUPFC may be used interchangeably. However, the investment of the former may be higher than that of the latter.

## 4.6 Appendix: Derivatives of Nonlinear Interior Point OPF with GUPFC

The power mismatches at bus $m$ are $\Delta P_m$, $\Delta Q_m$.

$$\Delta P_m = Pg_m - Pd_m - P_m \tag{4.121}$$

$$\Delta Q_m = Qg_m - Qd_m - Q_m \tag{4.122}$$

where $P_m$ and $Q_m$ are the sum of the active power flow and reactive power flow at bus $m$, respectively.

### 4.6.1 First Derivatives of Nonlinear Interior Point OPF

$$\frac{\partial P_i}{\partial \theta se_{ij}} = V_i Vse_{ij}(g_{ij} \sin(\theta_i - \theta se_{ij}) - b_{ij} \cos(\theta_i - \theta se_{ij})) \tag{4.123}$$

$$\frac{\partial P_i}{\partial Vse_{ij}} = -V_i(g_{ij} \cos(\theta_i - \theta se_{ij}) + b_{ij} \sin(\theta_i - \theta se_{ij})) \tag{4.124}$$

$$\frac{\partial P_i}{\partial \theta sh_i} = -V_i Vsh_i(gsh_i \sin(\theta_i - \theta sh_i) - bsh_i \cos(\theta_i - \theta sh_i)) \tag{4.125}$$

$$\frac{\partial P_i}{\partial Vsh_i} = -V_i(gsh_i \cos(\theta_i - \theta sh_i) + bsh_i \sin(\theta_i - \theta sh_i)) \tag{4.126}$$

$$\frac{\partial Q_i}{\partial \theta se_{ij}} = V_i Vse_{ij}(g_{ij} \cos(\theta_i - \theta se_{ij}) + b_{ij} \sin(\theta_i - \theta se_{ij})) \tag{4.127}$$

$$\frac{\partial Q_i}{\partial Vse_{ij}} = -V_i(g_{ij} \sin(\theta_i - \theta se_{ij}) - b_{ij} \cos(\theta_i - \theta se_{ij})) \tag{4.128}$$

$$\frac{\partial Q_i}{\partial \theta sh_i} = V_i Vsh_i(gsh_i \cos(\theta_i - \theta sh_i) + bsh_i \sin(\theta_i - \theta sh_i)) \tag{4.129}$$

$$\frac{\partial Q_i}{\partial Vsh_i} = -V_i(gsh_i\sin(\theta_i-\theta sh_i)-bsh_i\cos(\theta_i-\theta sh_i)) \tag{4.130}$$

$$\frac{\partial P_j}{\partial \theta se_{ij}} = V_j Vse_{ij}(g_{ij}\sin(\theta_j-\theta se_{ij})-b_{ij}\cos(\theta_j-\theta se_{ij})) \tag{4.131}$$

$$\frac{\partial P_j}{\partial Vse_{ij}} = -V_j(g_{ij}\cos(\theta_j-\theta se_{ij})+b_{ij}\sin(\theta_j-\theta se_{ij})) \tag{4.132}$$

$$\frac{\partial Q_j}{\partial \theta se_{ij}} = -V_j Vse_{ij}(g_{ij}\cos(\theta_j-\theta se_{ij})+b_{ij}\sin(\theta_j-\theta se_{ij})) \tag{4.133}$$

$$\frac{\partial Q_j}{\partial Vse_{ij}} = V_j(g_{ij}\sin(\theta_j-\theta se_{ij})-b_{ij}\cos(\theta_j-\theta se_{ij})) \tag{4.134}$$

$$\frac{\partial (Vse_{ij}I_{ji}^*)}{\partial \theta se_{ij}} = -V_i Vse_{ij}(g_{ij}\sin(\theta se_{ij}-\theta_i)-b_{ij}\cos(\theta se_{ij}-\theta_i)) \\ +V_j Vse_{ij}(g_{ij}\sin(\theta se_{ij}-\theta_j)-b_{ij}\cos(\theta se_{ij}-\theta_j)) \tag{4.135}$$

$$\frac{\partial (Vse_{ij}I_{ji}^*)}{\partial Vse_{ij}} = -2g_{ij}Vse_{ij}+V_i(g_{ij}\cos(\theta se_{ij}-\theta_i)+b_{ij}\sin(\theta se_{ij}-\theta_i)) \\ -V_j(g_{ij}\cos(\theta se_{ij}-\theta_j)+b_{ij}\sin(\theta se_{ij}-\theta_j)) \tag{4.136}$$

$$\frac{\partial (Vse_{ij}I_{ji}^*)}{\partial \theta_i} = V_i Vse_{ij}(g_{ij}\sin(\theta se_{ij}-\theta_i)-b_{ij}\cos(\theta se_{ij}-\theta_i)) \tag{4.137}$$

$$\frac{\partial (Vse_{ij}I_{ji}^*)}{\partial V_i} = Vse_{ij}(g_{ij}\cos(\theta se_{ij}-\theta_i)+b_{ij}\sin(\theta se_{ij}-\theta_i)) \tag{4.138}$$

$$\frac{\partial (Vse_{ij}I_{ji}^*)}{\partial \theta_j} = -V_j Vse_{ij}(g_{ij}\sin(\theta se_{ij}-\theta_j)-b_{ij}\cos(\theta se_{ij}-\theta_j)) \tag{4.139}$$

$$\frac{\partial (Vse_{ij}I_{ji}^*)}{\partial V_j} = -Vse_{ij}(g_{ij}\cos(\theta se_{ij}-\theta_j)+b_{ij}\sin(\theta se_{ij}-\theta_j)) \tag{4.140}$$

$$\frac{\partial (Vsh_i Ish_i^*)}{\partial \theta sh_i} = -V_i Vsh_i(gsh_i\sin(\theta sh_i-\theta_i)-bsh_i\cos(\theta sh_i-\theta_i)) \tag{4.141}$$

$$\frac{\partial (Vsh_i Ish_i^*)}{\partial Vsh_i} = 2gsh_i Vsh_i - V_i(gsh_i\cos(\theta sh_i-\theta_i)+bsh_i\sin(\theta sh_i-\theta_i)) \tag{4.142}$$

$$\frac{\partial (Vsh_i Ish_i^*)}{\partial \theta_i} = -V_i Vsh_i(gsh_i\sin(\theta sh_i-\theta_i)-bsh_i\cos(\theta sh_i-\theta_i)) \tag{4.143}$$

$$\frac{\partial (Vsh_i Ish_i^*)}{\partial V_i} = -Vsh_i(gsh_i\cos(\theta sh_i-\theta_i)+bsh_i\sin(\theta sh_i-\theta_i)) \tag{4.144}$$

## 4.6.2 Second Derivatives of Nonlinear Interior Point OPF

$$\frac{\partial^2 P_i}{\partial^2 \theta se_{ij}} = V_i Vse_{ij} (g_{ij} \cos(\theta_i - \theta se_{ij}) + b_{ij} \sin(\theta_i - \theta se_{ij})) \qquad (4.145)$$

$$\frac{\partial^2 P_i}{\partial \theta se_{ij} \partial Vse_{ij}} = -V_i (g_{ij} \sin(\theta_i - \theta se_{ij}) - b_{ij} \cos(\theta_i - \theta se_{ij})) \qquad (4.146)$$

$$\frac{\partial^2 P_i}{\partial^2 Vse_{ij}} = 0 \qquad (4.147)$$

$$\frac{\partial^2 P_i}{\partial^2 \theta sh_i} = V_i Vsh_i (gsh_i \cos(\theta_i - \theta sh_i) + bsh_i \sin(\theta_i - \theta sh_i)) \qquad (4.148)$$

$$\frac{\partial^2 P_i}{\partial \theta sh_i \partial Vsh_i} = -V_i (gsh_i \sin(\theta_i - \theta sh_i) - bsh_i \cos(\theta_i - \theta sh_i)) \qquad (4.149)$$

$$\frac{\partial^2 P_i}{\partial^2 Vsh_i} = 0 \qquad (4.150)$$

$$\frac{\partial^2 Q_i}{\partial^2 \theta se_{ij}} = V_i Vse_{ij} (g_{ij} \sin(\theta_i - \theta se_{ij}) - b_{ij} \cos(\theta_i - \theta se_{ij})) \qquad (4.151)$$

$$\frac{\partial^2 Q_i}{\partial \theta se_{ij} \partial Vse_{ij}} = V_i (g_{ij} \cos(\theta_i - \theta se_{ij}) + b_{ij} \sin(\theta_i - \theta se_{ij})) \qquad (4.152)$$

$$\frac{\partial^2 Q_i}{\partial^2 Vse_{ij}} = 0 \qquad (4.153)$$

$$\frac{\partial^2 Q_i}{\partial^2 \theta sh_i} = V_i Vsh_i (gsh_i \sin(\theta_i - \theta sh_i) - bsh_i \cos(\theta_i - \theta sh_i)) \qquad (4.154)$$

$$\frac{\partial^2 Q_i}{\partial \theta sh_i \partial Vsh_i} = V_i (gsh_i \cos(\theta_i - \theta sh_i) + bsh_i \sin(\theta_i - \theta sh_i)) \qquad (4.155)$$

$$\frac{\partial^2 \Delta Q_i}{\partial^2 Vsh_i} = 0 \qquad (4.156)$$

$$\frac{\partial^2 P_j}{\partial^2 \theta se_{ij}} = -V_j Vse_{ij} (g_{ij} \cos(\theta_j - \theta se_{ij}) + b_{ij} \sin(\theta_j - \theta se_{ij})) \qquad (4.157)$$

$$\frac{\partial^2 P_j}{\partial \theta se_{ij} \partial Vse_{ij}} = V_j (g_{ij} \sin(\theta_j - \theta se_{ij}) - b_{ij} \cos(\theta_j - \theta se_{ij})) \qquad (4.158)$$

$$\frac{\partial^2 P_j}{\partial^2 Vse_{ij}} = 0 \tag{4.159}$$

$$\frac{\partial^2 Q_j}{\partial^2 \theta se_{ij}} = -V_j Vse_{ij}(g_{ij}\sin(\theta_j - \theta se_{ij}) - b_{ij}\cos(\theta_j - \theta se_{ij})) \tag{4.160}$$

$$\frac{\partial^2 Q_j}{\partial \theta se_{ij} \partial Vse_{ij}} = -V_j(g_{ij}\cos(\theta_j - \theta se_{ij}) + b_{ij}\sin(\theta_j - \theta se_{ij})) \tag{4.161}$$

$$\frac{\partial^2 Q_j}{\partial^2 Vse_{ij}} = 0 \tag{4.162}$$

$$\frac{\partial^2 (Vse_{ij} I_{ji}^*)}{\partial^2 \theta se_{ij}} = -V_i Vse_{ij}(g_{ij}\cos(\theta se_{ij} - \theta_i) + b_{ij}\sin(\theta se_{ij} - \theta_i)) \\ + V_j Vse_{ij}(g_{ij}\cos(\theta se_{ij} - \theta_j) + b_{ij}\sin(\theta se_{ij} - \theta_j)) \tag{4.163}$$

$$\frac{\partial^2 (Vse_{ij} I_{ji}^*)}{\partial \theta se_{ij} \partial Vse_{ij}} = -V_i(g_{ij}\sin(\theta se_{ij} - \theta_i) - b_{ij}\cos(\theta se_{ij} - \theta_i)) \\ + V_j(g_{ij}\sin(\theta se_{ij} - \theta_j) - b_{ij}\cos(\theta se_{ij} - \theta_j)) \tag{4.164}$$

$$\frac{\partial^2 (Vse_{ij} I_{ji}^*)}{\partial \theta se_{ij} \partial \theta_i} = V_i Vse_{ij}(g_{ij}\cos(\theta se_{ij} - \theta_i) + b_{ij}\sin(\theta se_{ij} - \theta_i)) \tag{4.165}$$

$$\frac{\partial^2 (Vse_{ij} I_{ji}^*)}{\partial \theta se_{ij} \partial V_i} = -Vse_{ij}(g_{ij}\sin(\theta se_{ij} - \theta_i) - b_{ij}\cos(\theta se_{ij} - \theta_i)) \tag{4.166}$$

$$\frac{\partial^2 (Vse_{ij} I_{ji}^*)}{\partial \theta se_{ij} \partial \theta_j} = -V_j Vse_{ij}(g_{ij}\cos(\theta se_{ij} - \theta_j) + b_{ij}\sin(\theta se_{ij} - \theta_j)) \tag{4.167}$$

$$\frac{\partial^2 (Vse_{ij} Ise_{ij}^*)}{\partial \theta se_{ij} \partial V_j} = Vse_{ij}(g_{ij}\sin(\theta se_{ij} - \theta_j) - b_{ij}\cos(\theta se_{ij} - \theta_j)) \tag{4.168}$$

$$\frac{\partial^2 (Vse_{ij} Ise_{ij}^*)}{\partial Vse_{ij} \partial \theta_i} = V_i(g_{ij}\sin(\theta se_{ij} - \theta_i) - b_{ij}\cos(\theta se_{ij} - \theta_i)) \tag{4.169}$$

$$\frac{\partial^2 (Vse_{ij} Ise_{ij}^*)}{\partial Vse_{ij} \partial V_i} = -(g_{ij}\cos(\theta se_{ij} - \theta_i) + b_{ij}\sin(\theta se_{ij} - \theta_i)) \tag{4.170}$$

$$\frac{\partial^2 (Vse_{ij} Ise_{ij}^*)}{\partial Vse_{ij} \partial \theta_j} = -V_j(g_{ij}\sin(\theta se_{ij} - \theta_j) - b_{ij}\cos(\theta se_{ij} - \theta_j)) \tag{4.171}$$

$$\frac{\partial^2 (Vse_{ij} I_{ji}^*)}{\partial Vse_{ij} \partial V_j} = -(g_{ij}\cos(\theta se_{ij} - \theta_j) + b_{ij}\sin(\theta se_{ij} - \theta_j)) \tag{4.172}$$

$$\frac{\partial^2 (Vse_{ij} I_{ji}^*)}{\partial^2 \theta_i} = -V_i Vse_{ij} (g_{ij} \cos(\theta se_{ij} - \theta_i) + b_{ij} \sin(\theta se_{ij} - \theta_i)) \qquad (4.173)$$

$$\frac{\partial^2 (Vse_{ij} I_{ji}^*)}{\partial \theta_i \partial V_i} = Vse_{ij} (g_{ij} \sin(\theta se_{ij} - \theta_i) - b_{ij} \cos(\theta se_{ij} - \theta_i)) \qquad (4.174)$$

$$\frac{\partial^2 (Vse_{ij} I_{ji}^*)}{\partial^2 \theta_j} = V_j Vse_{ij} (g_{ij} \cos(\theta se_{ij} - \theta_j) + b_{ij} \sin(\theta se_{ij} - \theta_j)) \qquad (4.175)$$

$$\frac{\partial^2 (Vse_{ij} I_{ji}^*)}{\partial \theta_j \partial V_j} = -Vse_{ij} (g_{ij} \sin(\theta se_{ij} - \theta_j) - b_{ij} \cos(\theta se_{ij} - \theta_j)) \qquad (4.176)$$

$$\frac{\partial^2 (Vsh_i Ish_i^*)}{\partial^2 \theta sh_i} = -V_i Vsh_i (gsh_i \cos(\theta sh_i - \theta_i) + bsh_i \sin(\theta sh_i - \theta_i)) \qquad (4.177)$$

$$\frac{\partial^2 (Vsh_i Ish_i^*)}{\partial \theta sh_i \partial Vsh_i} = V_i (gsh_i \sin(\theta sh_i - \theta_i) - bsh_i \cos(\theta sh_i - \theta_i)) \qquad (4.178)$$

$$\frac{\partial^2 (Vsh_i Ish_i^*)}{\partial \theta sh_i \partial \theta_i} = -V_i Vsh_i (gsh_i \cos(\theta sh_i - \theta_i) + bsh_i \sin(\theta sh_i - \theta_i)) \qquad (4.179)$$

$$\frac{\partial^2 (Vsh_i Ish_i^*)}{\partial \theta sh_i \partial V_i} = Vsh_i (gsh_i \sin(\theta sh_i - \theta_i) - bsh_i \cos(\theta sh_i - \theta_i)) \qquad (4.180)$$

$$\frac{\partial^2 (Vsh_i Ish_i^*)}{\partial^2 Vsh_i} = 2.0 gsh_i \qquad (4.181)$$

$$\frac{\partial^2 (Vsh_i Ish_i^*)}{\partial Vsh_i \partial \theta_i} = -V_i (gsh_i \sin(\theta sh_i - \theta_i) - bsh_i \cos(\theta sh_i - \theta_i)) \qquad (4.182)$$

$$\frac{\partial^2 (Vsh_i Ish_i^*)}{\partial Vsh_i \partial V_i} = -(gsh_i \cos(\theta sh_i - \theta_i) + bsh_i \sin(\theta sh_i - \theta_i)) \qquad (4.183)$$

$$\frac{\partial^2 (Vsh_i Ish_i^*)}{\partial^2 \theta_i} = V_i Vsh_i (gsh_i \cos(\theta sh_i - \theta_i) + bsh_i \sin(\theta sh_i - \theta_i)) \qquad (4.184)$$

$$\frac{\partial^2 (Vsh_i Ish_i^*)}{\partial \theta_i \partial V_i} = -Vsh_i (gsh_i \sin(\theta sh_i - \theta_i) - bsh_i \cos(\theta sh_i - \theta_i)) \qquad (4.185)$$

## References

[1] Kirchmayer LK (1958) Economic operation of power systems. John Wiley & Sons, New York
[2] Kirchmayer LK (1959) Economic control of interconnected systems. John Wiley & Sons, New York
[3] Carpentier JL (1962) Contribution a. 'l'etude du dispatching economique. Bulletin de la Societe Francaise des Electriciens, vol 3, pp 431-447
[4] Dommel HW, Tinney WF (1968) Optimal power flow solutions. IEEE Trans. on PAS, vol 87, no 10, pp1866-1876
[5] Happ HH (1977), Optimal power dispatch – A comprehensive survey. IEEE Transactions on PAS, vol 96, pp 841-854
[6] Carpentier JL (1985) Optimal power flows: uses, methods and developments, Proceedings of IFAC Conference
[7] Stott B, Alsc O, Monticelli A (1987) Security and optimization, Proceedings of the IEEE, vol 75, no 12, pp 1623-1624
[8] Carpentier JL (1987) Towards a secure and optimal automatic operation of power systems. Proceedings of Power Industry Computer Applications (PICA) conference, pp 2-37
[9] Wu FF (1988) Real-time network security monitoring, assessment and optimization. International Journal of Electrical Power and Energy Systems, vol 10, no 2, pp 83-100
[10] Chowdhury BH, Rahman S (1990) A review of recent advances in economic dispatch. IEEE Transactions on Power Systems, vol 5, no 4, pp 1248-1257
[11] Huneault M, Galiana FD (1991) A survey of the optimal power flow literature. IEEE Transactions on Power Systems, vol 6, no 2, pp 762-770
[12] IEEE Tutorial Course (1996) Optimal power flow: solution techniques, requirements and challenges. IEEE Power Engineering Society
[13] Momoh JA, El-Haway ME, Adapa R (1999) A review of selected optimal power flow literature to 1993 Part 1 and Part 2. IEEE Transactions on Power Systems, vol 14 no 1, pp 96-111
[14] Carpentier JL (1973) Differential injections method: A general method for secure and optimal load flows, Proceedings of IFAC Conference
[15] Stott B, Marinho JL (1979) Linear programming for power system network security applications. IEEE Trans. on PAS, vol 98, no 3, pp 837-848
[16] Alsc O, Bright J, Praise M, Stott B (1990) Further developments in LP-based optimal power flow. IEEE Transactions on Power Systems, vol 5, no 3, pp 697-711
[17] Burchett RC, Happ HH, Wirgau KA(1982) Large scale optimal power flow. IEEE Trans. on PAS, vol 101, no 10, pp 3722-3732
[18] Burchett RC, Happ HH, Veirath DR (1984) Quadratically convergent optimal power flow. IEEE Trans. on PAS, vol 103, no 11, pp 3267-3275
[19] El-Kady MA, Bell BD, Carvalho VF, Burchett RC, Happ HH, Veirath DR (1986) Quadratically convergent optimal power flow. IEEE Trans. on Power Systems, vol. 1, no 2, pp 98-105
[20] Glavitsch H, Spoerry M (1983) Quadratic loss formula for reactive dispatch. IEEE Trans. on PAS, vol 102, no 12, pp 3850-3858
[21] Sun DI, Ashley B, Brewer B, Hughes A, Tinney WF (1984) Optimal power flow by Newton approach. IEEE Trans. on PAS, vol 103, no 10, pp 2864-2880
[22] Maria GA, Findlay JA (1987) A Newton optimal power flow program for Ontario Hydro EMS. IEEE Trans. on Power Systems, vol 2, no 3, pp 576-584

[23] Tinny WF, Bright JM, Demaree KD, Hughes BA (1988) Some deficiencies in optimal power flow. IEEE Trans. on Power Systems, vol 3, no 2, pp 676-682
[24] Chang SK, Marks GE, Kato K (1990) Optimal real-time voltage control. IEEE Trans. on Power Systems, vol 5, no 3, pp 750-756
[25] Hollenstein W, Glavitch H (1990) Linear programming as a tool for treating constraints in a Newton OPF. Proceedings of the 10$^{th}$ Power Systems Computation Conference (PSCC), Graz, Austria, August 19.-24.
[26] Karmarkar N (1984,) A new polynomial time algorithm for linear programming, *Combinatorica* 4, pp 373-395
[27] Vargas LS, Quintana VH, Vannelli A (1993) A tutorial description of an interior point method and its applications to security-constrained economic dispatch. IEEE Trans. on Power Systems, vol 8, no 3, pp 1315-1323
[28] Lu N, Unum MR (1993) Network constrained security control using an interior point algorithm. IEEE Transactions on Power Systems, vol 8, no 3, pp 1068-1076
[29] Zhang XP and Chen Z (1997) Security-constrained economic dispatch through interior point methods. Automation of Electric Power Systems, vol 21, no 6, pp 27-29
[30] Momoh JA, Guo SX, Ogbuobiri EC, Adapa R (1994) The quadratic interior point method solving power system optimization problems. IEEE Trans. on Power Systems, vol 9, no 3, pp1327-1336
[31] Granville S (1994) Optimal reactive power dispatch through interior point methods. IEEE Transactions on Power Systems, vol 9, no 1, pp 136-146
[32] Wu Y-C, Debs A, Marsten R E (1994) A direct nonlinear predictor-corrector primal-dual interior point algorithm for optimal power flows. IEEE Trans. on Power Systems, vol 9, no 2, pp 876-883
[33] Irisarri GD, Wang X, Tong J, Mokhtari S (1997) Maximum loadability of power systems using interior point nonlinear optimisation method. IEEE Trans. on Power Systems, vol 12, no 1, pp 167-172
[34] Wei H, Sasaki H, Yokoyama R (1998) An interior point nonlinear programming for optimal power flow problems within a novel data structure. IEEE Trans. on Power Systems, vol 13, no 3, pp 870-877
[35] Torres GL, Quintana VH (1998) An interior point method for non-linear optimal power flow using voltage rectangular coordinates. IEEE Transactions on Power Systems, vol 13, no 4, pp 1211-1218
[36] Zhang XP, Petoussis SG, Godfrey KR (2005) Novel nonlinear interior point optimal power flow (OPF) method based on current mismatch formulation. IEE Proceedings–Generation, Transmission & Distribution, to appear
[37] El-Bakry S, Tapia RA, Tsuchiya T, Zhang Y (1996) On the formulation and theory of the Newton interior-point method for nonlinear programming. Journal of Optimisation Theory and Applications, vol 89, no 3, pp 507-541
[38] La Scala M, Trovato M, Antonelli C (1998) On-line dynamic preventive control: An algorithm for transient security constraints. IEEE Transactions on Power Systems, vol 13, no 2, pp 601-610
[39] Gan D, Thomas RJ, Zimmermann RD (2000) Stability constrained optimal power flow. IEEE Transactions on Power Systems, vol 15, no 2, pp 535-540
[40] Chen L, Tada Y, et al (2001) Optimal operation solutions of power systems with transient stability constraints. IEEE Transactions on Circuit and Systems – I: Fundamental Theory and Applications, vol 48, no 3, pp 327-339

[41] Yue Y, Kubokawa J, Sasaki H (2003) A solution of optimal power flow with multi-contingency transient stability constraints. IEEE Transactions on Power Systems, vol 18, no 3, pp 1094-1102
[42] Rosehart W, Canizares C, Quintana VH (1999) Optimal power flow incorporating voltage collapse constraints. Proceedings of the 1999 IEEE/PES Summer Meeting, Edmonton, Alberta, July 1999, pp 820-825
[43] Hingorani NG, Gyugyi L (2000) Understanding FACTS – concepts and technology of flexible ac transmission systems. New York: IEEE Press
[44] Handschin E, Lehmkoester C (1999) Optimal power flow for deregulated systems with FACTS-Devices. 13th PSCC, Trondheim, Norway, pp 1270-1276
[45] Lehmkoster C (2002) Security constrained optimal power flow for an economical operation of FACTS-devices in liberalized energy markets. IEEE Transactions on Power Delivery, vol 17 no 2, pp 603-608
[46] Acha E, H. Ambriz-Perez H (1999) FACTS devices modelling in optimal power flow using Newton's method. 13th PSCC, Trondheim, Norway, pp 1277-1284
[47] Zhang XP, Handschin E (2001) Advanced implementation of UPFC in a nonlinear interior point OPF. IEE Proceedings– Generation, Transmission & Distribution, vol 148, no 3, pp 489-496
[48] Zhang XP, Handschin E, (2001) Optimal power flow control by converter based FACTS controllers. 7th International Conference on AC-DC Power Transmission, 28-30 Nov. 2001
[49] Fardanesh B, Henderson M, Shperling B, Zelingher S, Gyugyi L, Schauder C, Lam B, Mounford J, Adapa R, Edris A (1998) Convertible static compensator: application to the New York transmission system. CIGRE 14-103, Paris, France, September
[50] Fardanesh B, Shperling B, Uzunovic E, Zelingher S (2000) Multi-converter FACTS devices: the generalized unified power flow controller (GUPFC). Proceedings of IEEE 2000 PES Summer Meeting, Seattle, USA
[51] Zhang XP, Handschin E, Yao MM (2001) Modeling of the generalized unified power flow controller in a nonlinear interior point OPF. IEEE Trans. on Power Systems, vol 16, no 3, pp 367-373
[52] Zhang XP (2003) Modelling of the interline power flow controller and generalized unified power flow controller in Newton power flow. IEE Proc. - Generation, Transmission and Distribution, vol 150, no 5, pp 268-274
[53] Zhang XP (2004) Multiterminal voltage-sourced converter based HVDC models for power flow analysis. IEEE Transactions on Power Systems, vol 18, no 4, 2004, pp 1877-1884

# 5 Modeling of FACTS in Three-Phase Power Flow and Three-Phase OPF Analysis

Three-phase power flow calculations are important tools to compute the realistic system operation states and evaluate the control performance of various control devices such as transformer, synchronous machines and FACTS-devices, particularly because (a) there are unbalances of three-phase transmission lines in high voltage transmission networks; (b) there are unbalanced three-phase loads; (c) in addition, there are one-phase or two-phase lines in some distribution networks, etc. Under these unbalanced operating conditions, three-phase power flow studies are needed to assess the realistic operating conditions of the systems and analyze the behavior and control performance of power system components including FACTS-devices.

A number of three-phase power flow methods such as Bus-Impedance Method [1], Newton-Raphson Method [2][3], Fast-Decoupled Method [4][6], Gauss-Seidel Method [5], Hybrid Method [7], A Newton approach combining representation of linear elements using linear nodal voltage equation and representation of nonlinear elements using injected currents and associated equality constraints [8], Implicit Bus-Impedance Method [9], Decoupling-Compensation Bus-Admittance Method [9], Fast Three-phase Load Flow Methods [10], and Newton power flow in current injection form [12] etc. have been proposed since 1960s. The Newton method proposed in [8] is in particular interfaced with EMTP (**E**lectro-**M**agnetic **T**ransients **P**rogram) and can be used to initialize the simulations. The Fast Three-phase Load Flow Methods proposed in [10] have been further implemented on a parallel processor [11].

In addition to the above three-phase power flow solution methods, specialized three-phase power flow techniques [13]-[21] for distribution networks have also been proposed with various success where the special structure of distribution networks is exploited and computational efficiency is improved. Modeling of power system components can be found in [22][23][6].

An Optimal Power Flow (OPF) program can be used to determine the optimal operation state of a power system by optimizing a particular objective while satisfying specified physical and operating constraints. Because of its capability of integrating the economic and secure aspects of the system into one mathematical model, the OPF can be applied not only in three-phase power system planning, but also in real time operation optimisation of three-phase power systems. With the incorporation of FACTS-devices into power systems, a three-phase optimal power flow will be required. In contrast to the research in three-phase power flow solu-

140    5 Modeling of FACTS in Three-Phase Power Flow and Three-Phase OPF Analysis

tion techniques, the research in optimal three-phase power flow methods has been very limited.

With the increasing installation of FACTS in power systems, modeling of FACTS-devices into three-phase power flow and optimal three-phase power flow analysis will be of great interest. In recent years, three-phase FACTS models have been investigated for three-phase power flow analysis [24][25]. Positive sequence models for FACTS-devices have been discussed in chapters 2, 3 and 4. However, three-phase FACTS models are more complex than those positive sequence ones since unbalanced conditions need to be considered. This chapter introduces the following aspects:

- review of three-phase power flow solution techniques;
- three-phase Newton power flow solution methods in polar and rectangular coordinates;
- three-phase FACTS models for SSSC and UPFC and their incorporation in three-phase power flow analysis;
- formulation of optimal three-phase power flow problems.

## 5.1 Three-Phase Newton Power Flow Methods in Rectangular Coordinates

Modeling of power system components such as transmission lines, loads, etc. have been discussed in [23][6]. In the following, the formulation of three-phase Newton power flow in rectangular coordinates will be presented where the modeling of synchronous generator is discussed in detail.

### 5.1.1 Classification of Buses

In three-phase power flow calculations, all buses may be classified into the following categories:

**Slack bus.** Similar to that in single-phase positive-sequence power flow calculations, a slack bus, which is usually one of the generator terminal buses, should be selected for three-phase power flow calculations. At the slack bus, the positive-sequence voltage angle and magnitude are specified while the active and reactive power injections at the generator terminal are unknown. The voltage angle of the slack bus is taken as the reference for the angles of all other buses. Usually there is only one slack bus in a system. However, in some production grade programs, it may be possible to include more than one bus as distributed slack buses.

**PV Buses.** PV buses in three-phase power flow calculations are usually generator terminal buses. For these buses, the total active power injections and positive-sequence voltage magnitudes are specified.

**PQ Buses.** PQ buses are usually load buses in the network. For these buses, the active and reactive power injections of their three-phases are specified.

### 5.1.2 Representation of Synchronous Machines

A synchronous machine may be represented by a set of three-phase balanced voltage sources in series with a 3 by 3 impedance matrix. Such a synchronous machine model is shown in Fig. 5.1. The impedance matrix $Zg_i$ may be determined by positive-, negative-, and zero-sequence impedance parameters of a synchronous machine. $Zg_i$ is defined in Appendix A of this chapter.

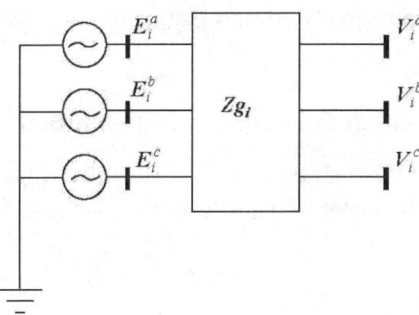

**Fig. 5.1.** A synchronous machine

It is assumed that the synchronous generator in Fig. 5.1 has a round rotor structure, and saturation of the synchronous generator is not considered in the present model. However, in principle, there is no difficulty to take into account the saturation.

In Fig. 5.1, $V_i^a = E_i^a + jF_i^a$, $V_i^b = E_i^b + jF_i^b$, $V_i^c = E_i^c + jF_i^c$, which are the three-phase voltages at the generator terminal bus, are expressed in phasors in rectangular coordinates. Similarly, the voltages at the generator internal bus may be given by $E_i^a = Eg_i^a + jFg_i^a$, $E_i^b = Eg_i^b + jFg_i^b$, $E_i^c = Eg_i^c + jEg_i^c$. In fact the voltages at the generator internal bus are balanced, that is:

$$E_i^b = E_i^a e^{-j2\pi/3} \qquad (5.1)$$

$$E_i^c = E_i^a e^{j2\pi/3} \qquad (5.2)$$

In the three-phase power flow equations of the generator, $Eg_i^a$ and $Fg_i^a$ can be considered as independent state variables of the internal generator bus while $Eg_i^b$ and $Fg_i^b$, and $Eg_i^c$ and $Eg_i^c$ are dependent state variables and can be represented by $Eg_i^a$ and $Eg_i^a$. We have:

$$Eg_i^b = -\frac{1}{2}Eg_i^a + \frac{\sqrt{3}}{2}Fg_i^a \qquad (5.3)$$

$$Fg_i^b = -\frac{\sqrt{3}}{2}Eg_i^a - \frac{1}{2}Fg_i^a \qquad (5.4)$$

$$Eg_i^c = -\frac{1}{2}Eg_i^a - \frac{\sqrt{3}}{2}Fg_i^a \qquad (5.5)$$

$$Fg_i^c = -\frac{\sqrt{3}}{2}Eg_i^a - \frac{1}{2}Fg_i^a \qquad (5.6)$$

### 5.1.3 Power and Voltage Mismatch Equations in Rectangular Coordinates

#### 5.1.3.1 Power Mismatch Equations at Network Buses

The network buses include all buses of the network except the internal buses of generators. The power mismatch equations of phase $p$ at the network bus $i$ are given by:

$$\Delta P_i^p = -Pd_i^p - \sum_{j \in i}[E_i^p(G_{ij}^{pm}E_j^m - B_{ij}^{pm}F_j^m) + F_i^p(G_{ij}^{pm}F_j^m + B_{ij}^{pm}E_j^m)] \qquad (5.7)$$

$$\Delta Q_i^p = -Qd_i^p - \sum_{j \in i}[F_i^p(G_{ij}^{pm}E_j^m - B_{ij}^{pm}F_j^m) - E_i^p(G_{ij}^{pm}F_j^m + B_{ij}^{pm}E_j^m)] \qquad (5.8)$$

where $p = a, b, c$. $Pd_i^p$ and $Qd_i^p$ are the active and reactive loads of phase $p$ at bus $i$.

#### 5.1.3.2 Power and Voltage Mismatch Equations of Synchronous Machines

**PQ Machines.** For a PQ machine, the total three-phase active and reactive powers at the terminal bus of the machine are specified:

$$\begin{aligned}\Delta Pg_i = &-Pg_i^{Spec} \\ &- \sum_{p=a,b,c}\sum_{m=a,b,c}[E_i^p(Gg_i^{pm}E_i^m - Bg_i^{pm}F_i^m) + F_i^p(Gg_i^{pm}F_i^m + Bg_i^{pm}E_i^m)] \\ &+ \sum_{p=a,b,c}\sum_{m=a,b,c}[E_i^p(Gg_i^{pm}Eg_i^m - Bg_i^{pm}Fg_i^m) + F_i^p(Gg_i^{pm}Fg_j^m + Bg_i^{pm}Eg_j^m)]\end{aligned} \qquad (5.9)$$

$$\begin{aligned}\Delta Qg_i = &-Qg_i^{Spec} \\ &- \sum_{p=a,b,c}\sum_{m=a,b,c}F_i^p(Gg_i^{pm}E_i^m - Bg_i^{pm}F_i^m) - E_i^p(Gg_i^{pm}F_i^m + Bg_i^{pm}E_i^m)] \\ &+ \sum_{p=a,b,c}\sum_{m=a,b,c}F_i^p(Gg_i^{pm}Eg_g^m - Bg_i^{pm}Fg_i^m) - E_i^p(Gg_i^{pm}Fg_i^m + Bg_i^{pm}Eg_i^m)]\end{aligned} \qquad (5.10)$$

where $Pg_i^{Spec}$ and $Qg_i^{Spec}$ are the specified active and reactive powers of the generator at bus $I$, which are in the direction of terminal bus $i$.

**PV Machines.** For a PV machine, the total three-phase active power flow and the positive sequence voltage magnitude at its terminal bus $i$ are specified. The active power flow mismatch equation is given by (5.9) while the voltage mismatch equation at bus $i$ is given by:

$$\Delta Vg_i = V_i^{Spec} - V_i^1 = V_i^{Spec} - \sqrt{(e_i^1)^2 + (f_i^1)^2} \qquad (5.11)$$

where $V_i^1$ is the positive-sequence voltage magnitude voltage at the generator terminal bus $i$. $e_i^1$ and $f_i^1$ are the real and imaginary parts of the positive-sequence voltage phasor at bus $i$ and they are given by:

$$e_i^1 = \mathrm{Re}(V_i^a + V_i^b e^{j120°} + V_i^c e^{j240°})/3 \qquad (5.12)$$

$$f_i^1 = \mathrm{Im}(V_i^a + V_i^b e^{j120°} + V_i^c e^{j240°})/3 \qquad (5.13)$$

where $V_i^a$, $V_i^b$ and $V_i^c$ are the phase $a$, phase $b$ and phase $c$ voltages at bus $i$, respectively.

**Slack Machine.** At the terminal bus of the Slack machine, the positive-sequence voltage magnitude is specified and the positive-sequence voltage angle is taken as the system reference. We have:

$$\Delta \theta g_i = f_i^1 = 0 \qquad (5.14)$$

$$\Delta Vg_i = V_i^{Spec} - V_i^1 = V_i^{Spec} - \sqrt{(e_i^1)^2 + (f_i^1)^2} \qquad (5.15)$$

where $V_i^{Spec}$ is the specified positive-sequence voltage at the terminal bus of the slack machine. $e_i^1$ and $f_i^1$ are the real and imaginary parts of the positive-sequence voltage at the terminal bus of the Slack machine, and they are defined in (5.11) and (5.12).

### 5.1.4 Formulation of Newton Equations in Rectangular Coordinates

Combining the power mismatch equations of network buses and generator active power and voltage control constraints for the case of PV machines, the following Newton equation in rectangular coordinates can be obtained:

$$\mathbf{J}\Delta\mathbf{X} = -\mathbf{F}(\mathbf{X}) \qquad (5.16)$$

where $\Delta\mathbf{X} = [\Delta\mathbf{X}_{gen}, \Delta\mathbf{X}_{sys}]^T$

$$\Delta \mathbf{X}_{gen} = [\Delta Eg_i^a, \Delta Fg_i^a]^T$$

$$\Delta \mathbf{X}_{sys} = [\Delta E_i^a, \Delta F_i^a, \Delta E_i^b, \Delta F_i^b, \Delta E_i^c, \Delta F_i^c, \Delta E_j^a, \Delta F_j^a, \Delta E_j^b, \Delta F_j^b, \Delta E_j^c, \Delta F_j^c]^T$$

$$\mathbf{F(X)} = [\mathbf{F}_{gen}, \mathbf{F}_{sys}]^T$$

$$\mathbf{F}_{gen} = [f_{gen}^1, f_{gen}^2]^T$$

$$\mathbf{F}_{sys} = [\Delta P_i^a, \Delta Q_i^a, \Delta P_i^b, \Delta Q_i^b, \Delta P_i^c, \Delta Q_i^c, \Delta P_j^a, \Delta Q_j^a, \Delta P_j^b, \Delta Q_j^b, \Delta P_j^c, \Delta Q_j^c]^T$$

$$\mathbf{J} = \frac{\partial \mathbf{F(X)}}{\partial \mathbf{X}}$$

The Jacobian elements of the network block are defined as:

$$\frac{\partial \Delta P_i^p}{\partial E_j^m} = \begin{cases} -(G_{ij}^{pm} E_i^p + B_{ij}^{pm} F_i^p) & (j \neq i, \text{ or } m \neq p) \\ -\sum_{j \in i} \sum_{m=a,b,c} (G_{ij}^{pm} E_j^m - B_{ij}^{pm} F_j^m) - G_{ii}^{pp} E_i^p - B_{ii}^{pp} F_i^p & (j = i, m = p) \end{cases} \quad (5.17)$$

$$\frac{\partial \Delta P_i^p}{\partial F_j^m} = \begin{cases} B_{ij}^{pm} E_i^p - G_{ij}^{pm} F_i^p & (j \neq i, \text{ or } m \neq p) \\ -\sum_{j \in i} \sum_{m=a,b,c} (G_{ij}^{pm} F_j^m + B_{ij}^{pm} E_j^m) + B_{ii}^{pp} E_i^p - G_{ii}^{pp} F_i^p & (j = i, m = p) \end{cases} \quad (5.18)$$

$$\frac{\partial \Delta Q_i^p}{\partial E_j^m} = \begin{cases} B_{ij}^{pm} E_i^p - G_{ij}^{pm} F_i^p & (j \neq i, \text{ or } m \neq p) \\ -\sum_{j \in i} \sum_{m=a,b,c} (G_{ij}^{pm} F_j^m + B_{ij}^{pm} E_j^m) + B_{ii}^{pp} E_i^p - G_{ii}^{pp} F_i^p & (j = i, m = p) \end{cases} \quad (5.19)$$

$$\frac{\partial \Delta Q_i^p}{\partial F_j^m} = \begin{cases} G_{ij}^{pm} E_i^p + B_{ij}^{pm} F_i^p & (j \neq i, \text{ or } m \neq p) \\ -\sum_{j \in i} \sum_{m=a,b,c} (G_{ij}^{pm} E_j^m - B_{ij}^{pm} F_j^m) + G_{ii}^{pp} E_i^p + B_{ii}^{pp} F_i^p & (j = i, m = p) \end{cases} \quad (5.20)$$

In addition, we can find the following partial differentials with respect to generator internal variables $Eg_i^m$, $Fg_i^m$ ($m = a, b, c$):

$$\frac{\partial \Delta P_i^p}{\partial Eg_i^m} = (Gg_i^{pm} E_i^p + Bg_i^{pm} F_i^p) \quad (5.21)$$

$$\frac{\partial \Delta P_i^p}{\partial Fg_i^m} = -Bg_i^{pm} E_i^p + Gg_i^{pm} F_i^p \quad (5.22)$$

## 5.1 Three-Phase Newton Power Flow Methods in Rectangular Coordinates

$$\frac{\partial \Delta Q_i^p}{\partial Eg_i^m} = -Bg_i^{pm} E_i^p + Gg_i^{pm} F_i^p \tag{5.23}$$

$$\frac{\partial \Delta Q_i^p}{\partial Fg_i^m} = -Gg_i^{pm} E_i^p - Bg_i^{pm} F_i^p \tag{5.24}$$

Assuming $Pg_i^p$ and $Qg_i^p$ are the active and reactive generator output of phase $p$ at the terminal bus $i$, we have $Pg_i = \sum_{p=a,b,c} Pg_i^p$ and $Qg_i = \sum_{p=a,b,c} Qg_i^p$. Following the above formulas, we can find $\frac{\partial \Delta Pg_i^p}{\partial E_i^m}$, $\frac{\partial \Delta Pg_i^p}{\partial E_i^m}$, $\frac{\partial \Delta Pg_i^p}{\partial Eg_i^m}$, $\frac{\partial \Delta Pg_i^p}{\partial Eg_i^m}$, $\frac{\partial \Delta Qg_i^p}{\partial E_i^m}$, $\frac{\partial \Delta Qg_i^p}{\partial E_i^m}$, $\frac{\partial \Delta Qg_i^p}{\partial Eg_i^m}$, $\frac{\partial \Delta Qg_i^p}{\partial Eg_i^m}$. The differentials of the synchronous machine power mismatches with respect to the internal voltage variables $Eg_i^m$, $Fg_i^m$ ($m = a, b, c$) are given by:

$$\frac{\partial \Delta Pg_i}{\partial Eg_i^m} = \sum_{p=a,b,c} \frac{\partial \Delta Pg_i^p}{\partial Eg_i^m} \tag{5.25}$$

$$\frac{\partial \Delta Pg_i}{\partial Fg_i^m} = \sum_{p=a,b,c} \frac{\partial \Delta Pg_i^p}{\partial Fg_i^m} \tag{5.26}$$

$$\frac{\partial \Delta Pg_i}{\partial E_i^m} = \sum_{p=a,b,c} \frac{\partial \Delta Pg_i^p}{\partial E_i^m} \tag{5.27}$$

$$\frac{\partial \Delta Pg_i}{\partial F_i^m} = \sum_{p=a,b,c} \frac{\partial \Delta Pg_i^p}{\partial F_i^m} \tag{5.28}$$

$$\frac{\partial \Delta Qg_i}{\partial Eg_i^m} = \sum_{p=a,b,c} \frac{\partial \Delta Qg_i^p}{\partial Eg_i^m} \tag{5.29}$$

$$\frac{\partial \Delta Qg_i}{\partial Fg_i^m} = \sum_{p=a,b,c} \frac{\partial \Delta Qg_i^p}{\partial Fg_i^m} \tag{5.30}$$

$$\frac{\partial \Delta Qg_i}{\partial E_i^m} = \sum_{p=a,b,c} \frac{\partial \Delta Qg_i^p}{\partial E_i^m} \tag{5.31}$$

$$\frac{\partial \Delta Qg_i}{\partial F_i^m} = \sum_{p=a,b,c} \frac{\partial \Delta Qg_i^p}{\partial F_i^m} \tag{5.32}$$

where $m = a, b, c$.

As mentioned, actually in the three-phase power flow equations of the generator, $Eg_i^a$ and $Fg_i^a$ can be considered as independent state variables of the internal generator bus while $Eg_i^b$ and $Fg_i^b$, and $Eg_i^c$ and $Eg_i^c$ are dependent state variables and can be represented by $Eg_i^a$ and $Eg_i^a$. We have:

$$\frac{\partial \Delta Pg_i}{\partial Eg_i^a} = \sum_{p=a,b,c} \frac{\partial \Delta Pg_i^p}{\partial Eg_i^a}$$
$$+ \sum_{p=a,b,c} \frac{\partial \Delta Pg_i^p}{\partial Eg_i^b} \frac{\partial Eg_i^b}{\partial Eg_i^a} + \sum_{p=a,b,c} \frac{\partial \Delta Pg_i^p}{\partial Fg_i^b} \frac{\partial Fg_i^b}{\partial Eg_i^a} \quad (5.33)$$
$$+ \sum_{p=a,b,c} \frac{\partial \Delta Pg_i^p}{\partial Eg_i^c} \frac{\partial Eg_i^c}{\partial Eg_i^a} + \sum_{p=a,b,c} \frac{\partial \Delta Pg_i^p}{\partial Fg_i^c} \frac{\partial Fg_i^c}{\partial Eg_i^a}$$

$$\frac{\partial \Delta Pg_i}{\partial Fg_i^a} = \sum_{p=a,b,c} \frac{\partial \Delta Pg_i^p}{\partial Fg_i^a}$$
$$+ \sum_{p=a,b,c} \frac{\partial \Delta Pg_i^p}{\partial Eg_i^b} \frac{\partial Eg_i^b}{\partial Fg_i^a} + \sum_{p=a,b,c} \frac{\partial \Delta Pg_i^p}{\partial Fg_i^b} \frac{\partial Fg_i^b}{\partial Fg_i^a} \quad (5.34)$$
$$+ \sum_{p=a,b,c} \frac{\partial \Delta Pg_i^p}{\partial Eg_i^c} \frac{\partial Eg_i^c}{\partial Fg_i^a} + \sum_{p=a,b,c} \frac{\partial \Delta Pg_i^p}{\partial Fg_i^c} \frac{\partial Fg_i^c}{\partial Fg_i^a}$$

$$\frac{\partial \Delta Qg_i}{\partial Eg_i^a} = \sum_{p=a,b,c} \frac{\partial \Delta Qg_i^p}{\partial Eg_i^a}$$
$$+ \sum_{p=a,b,c} \frac{\partial \Delta Qg_i^p}{\partial Eg_i^b} \frac{\partial Eg_i^b}{\partial Eg_i^a} + \sum_{p=a,b,c} \frac{\partial \Delta Qg_i^p}{\partial Fg_i^b} \frac{\partial Fg_i^b}{\partial Eg_i^a} \quad (5.35)$$
$$+ \sum_{p=a,b,c} \frac{\partial \Delta Qg_i^p}{\partial Eg_i^c} \frac{\partial Eg_i^c}{\partial Eg_i^a} + \sum_{p=a,b,c} \frac{\partial \Delta Qg_i^p}{\partial Fg_i^c} \frac{\partial Fg_i^c}{\partial Eg_i^a}$$

$$\frac{\partial \Delta Qg_i}{\partial Fg_i^a} = \sum_{p=a,b,c} \frac{\partial \Delta Qg_i^p}{\partial Fg_i^a}$$
$$+ \sum_{p=a,b,c} \frac{\partial \Delta Qg_i^p}{\partial Eg_i^b} \frac{\partial Eg_i^b}{\partial Fg_i^a} + \sum_{p=a,b,c} \frac{\partial \Delta Qg_i^p}{\partial Fg_i^b} \frac{\partial Fg_i^b}{\partial Fg_i^a} \quad (5.36)$$
$$+ \sum_{p=a,b,c} \frac{\partial \Delta Qg_i^p}{\partial Eg_i^c} \frac{\partial Eg_i^c}{\partial Fg_i^a} + \sum_{p=a,b,c} \frac{\partial \Delta Qg_i^p}{\partial Fg_i^c} \frac{\partial Fg_i^c}{\partial Fg_i^a}$$

Using the relationships in (5.3)-(5.6), (5.25)-(5.36) can be simplified as:

$$\frac{\partial \Delta Pg_i}{\partial Eg_i^a} = \sum_{p=a,b,c} \frac{\partial \Delta Pg_i^p}{\partial Eg_i^a}$$
$$- \frac{1}{2} \sum_{p=a,b,c} \frac{\partial \Delta Pg_i^p}{\partial Eg_i^b} - \frac{\sqrt{3}}{2} \sum_{p=a,b,c} \frac{\partial \Delta Pg_i^p}{\partial Fg_i^b} \quad (5.37)$$
$$- \frac{1}{2} \sum_{p=a,b,c} \frac{\partial \Delta Pg_i^p}{\partial Eg_i^c} - \frac{\sqrt{3}}{2} \sum_{p=a,b,c} \frac{\partial \Delta Pg_i^p}{\partial Fg_i^c}$$

$$\frac{\partial \Delta Pg_i}{\partial Fg_i^a} = \sum_{p=a,b,c} \frac{\partial \Delta Pg_i^p}{\partial Fg_i^a}$$
$$+ \frac{\sqrt{3}}{2} \sum_{p=a,b,c} \frac{\partial \Delta Pg_i^p}{\partial Eg_i^b} - \frac{1}{2} \sum_{p=a,b,c} \frac{\partial \Delta Pg_i^p}{\partial Fg_i^b} \quad (5.38)$$
$$- \frac{\sqrt{3}}{2} \sum_{p=a,b,c} \frac{\partial \Delta Pg_i^p}{\partial Eg_i^c} - \frac{1}{2} \sum_{p=a,b,c} \frac{\partial \Delta Pg_i^p}{\partial Fg_i^c}$$

$$\frac{\partial \Delta Qg_i}{\partial Eg_i^a} = \sum_{p=a,b,c} \frac{\partial \Delta Qg_i^p}{\partial Eg_i^a}$$
$$- \frac{1}{2} \sum_{p=a,b,c} \frac{\partial \Delta Qg_i^p}{\partial Eg_i^b} - \frac{\sqrt{3}}{2} \sum_{p=a,b,c} \frac{\partial \Delta Qg_i^p}{\partial Fg_i^b} \quad (5.39)$$
$$- \frac{1}{2} \sum_{p=a,b,c} \frac{\partial \Delta Qg_i^p}{\partial Eg_i^c} - \frac{\sqrt{3}}{2} \sum_{p=a,b,c} \frac{\partial \Delta Qg_i^p}{\partial Eg_i^c}$$

$$\frac{\partial \Delta Qg_i}{\partial Fg_i^a} = \sum_{p=a,b,c} \frac{\partial \Delta Qg_i^p}{\partial Fg_i^a}$$
$$+ \frac{\sqrt{3}}{2} \sum_{p=a,b,c} \frac{\partial \Delta Qg_i^p}{\partial Eg_i^b} - \frac{1}{2} \sum_{p=a,b,c} \frac{\partial \Delta Qg_i^p}{\partial Fg_i^b} \quad (5.40)$$
$$- \frac{\sqrt{3}}{2} \sum_{p=a,b,c} \frac{\partial \Delta Qg_i^p}{\partial Eg_i^c} - \frac{1}{2} \sum_{p=a,b,c} \frac{\partial \Delta Qg_i^p}{\partial Fg_i^c}$$

Similarly, if $Eg_i^a$ and $Fg_i^a$ can be considered as independent state variables of the internal generator bus while $Eg_i^b$ and $Fg_i^b$, and $Eg_i^c$ and $Eg_i^c$ are dependent state variables and can be represented by $Eg_i^a$ and $Eg_i^a$, then we have

$$\frac{\partial \Delta P_i^p}{\partial Eg_i^a} = (Gg_i^{pa} E_i^p + Bg_i^{pa} F_i^p)$$
$$- \frac{1}{2}(Gg_i^{pb} E_i^p + Bg_i^{pb} F_i^p) - \frac{\sqrt{3}}{2}(-Bg_i^{pb} E_i^p + Gg_i^{pb} F_i^p) \qquad (5.41)$$
$$- \frac{1}{2}(Gg_i^{pc} E_i^p + Bg_i^{pc} F_i^p) - \frac{\sqrt{3}}{2}(-Bg_i^{pc} E_i^p + Gg_i^{pc} F_i^p)$$

$$\frac{\partial \Delta P_i^p}{\partial Fg_i^a} = -Bg_i^{pa} E_i^p + Gg_i^{pa} F_i^p$$
$$\frac{\sqrt{3}}{2}(Gg_i^{pb} E_i^p + Bg_i^{pb} F_i^p) - \frac{1}{2}(-Bg_i^{pb} E_i^p + Gg_i^{pb} F_i^p) \qquad (5.42)$$
$$- \frac{\sqrt{3}}{2}(Gg_i^{pc} E_i^p + Bg_i^{pc} F_i^p) - \frac{1}{2}(-Bg_i^{pc} E_i^p + Gg_i^{pc} F_i^p)$$

$$\frac{\partial \Delta Q_i^p}{\partial Eg_i^a} = -Bg_i^{pa} E_i^p + Gg_i^{pa} F_i^p$$
$$- \frac{1}{2}(-Bg_i^{pb} E_i^p + Gg_i^{pb} F_i^p) - \frac{\sqrt{3}}{2}(-Gg_i^{pb} E_i^p - Bg_i^{pb} F_i^p) \qquad (5.43)$$
$$- \frac{1}{2}(-Bg_i^{pc} E_i^p + Gg_i^{pc} F_i^p) - \frac{\sqrt{3}}{2}(-Gg_i^{pc} E_i^p - Bg_i^{pc} F_i^p)$$

$$\frac{\partial \Delta Q_i^p}{\partial Fg_i^a} = -Gg_i^{pa} E_i^p - Bg_i^{pa} F_i^p$$
$$+ \frac{\sqrt{3}}{2}(-Bg_i^{pb} E_i^p + Gg_i^{pb} F_i^p) - \frac{1}{2}(-Gg_i^{pb} E_i^p - Bg_i^{pb} F_i^p) \qquad (5.44)$$
$$- \frac{\sqrt{3}}{2}(-Bg_i^{pc} E_i^p + Gg_i^{pc} F_i^p) - \frac{1}{2}(-Gg_i^{pc} E_i^p - Bg_i^{pc} F_i^p)$$

## 5.2 Three-Phase Newton Power Flow Methods in Polar Coordinates

### 5.2.1 Representation of Generators

In Fig. 5.1, $V_i^a = V_i^a \angle \theta_i^a$, $V_i^b = V_i^b \angle \theta_i^b$, $V_i^c = V_i^c \angle \theta_i^c$, which are the three-phase voltages at the generator terminal bus, are expressed in phasors in rectangular coordinates. Similarly, the voltages at the generator internal bus may be given by $E_i^a = E_i^a \angle \delta_i^a$, $E_i^b = E_i^b \angle \delta_i^b$, $E_i^c = E_i^c \angle \delta_i^c$. In fact the voltages at the generator internal bus are balanced, that is:

$$E_i^b = E_i^a e^{-j2\pi/3} \qquad (5.45)$$

$$E_i^c = E_i^a e^{j2\pi/3} \qquad (5.46)$$

In the three-phase power flow equations of the generator, $E_i^a$ and $\delta_i^a$ can be considered as independent state variables of the internal generator bus while $E_i^b$ and $\delta_i^b$, and $E_i^c$ and $\delta_i^c$ are dependent state variables and can be represented by $E_i^a$ and $\delta_i^a$. We have:

$$E_i^b = E_i^a \qquad (5.47)$$

$$\delta_i^b = \delta_i^a - \frac{2\pi}{3} \qquad (5.48)$$

$$E_i^c = E_i^a \qquad (5.49)$$

$$\delta_i^c = \delta_i^a + \frac{2\pi}{3} \qquad (5.50)$$

### 5.2.2 Power and Voltage Mismatch Equations in Polar Coordinates

#### 5.2.2.1 Power Mismatch Equations at Network Buses

The network buses include all buses of the network except the internal buses of generators. The power mismatch equation of phase $p$ at the network bus $i$ are given by:

$$\Delta P_i^p = -Pd_i^p - V_i^p \sum_{j \in i} \sum_{m=a,b,c} V_j^m (G_{ij}^{pm} \cos \theta_{ij}^{pm} + B_{ij}^{pm} \sin \theta_{ij}^{pm}) \qquad (5.51)$$

$$\Delta Q_i^p = -Qd_i^p - V_i^p \sum_{j \in i} \sum_{m=a,b,c} V_j^m (G_{ij}^{pm} \sin \theta_{ij}^{pm} - B_{ij}^{pm} \cos \theta_{ij}^{pm}) \quad (5.52)$$

where $p = a, b, c$. $Pd_i^p$ and $Qd_i^p$ are the active and reactive loads of phase $p$ at bus $i$.

### 5.2.2.2 Power and Voltage Mismatch Equations of Synchronous Machines

**PQ Machines.** For a PQ machine, the total three-phase active and reactive powers at the terminal bus of the machine are specified:

$$\begin{aligned}\Delta Pg_i = & -Pg_i^{Spec} \\ & - \sum_{p=a,b,c} \sum_{m=a,b,c} [V_i^p V_i^m (Gg_i^{pm} \cos \theta_i^{pm} + Bg_i^{pm} \sin \theta_i^{pm}) \\ & + \sum_{p=a,b,c} \sum_{m=a,b,c} [V_i^p E_i^p (Gg_i^{pm} \cos(\theta_i^p - \delta_i^m) + Bg_i^{pm} \sin(\theta_i^p - \delta_i^m))\end{aligned} \quad (5.53)$$

$$\begin{aligned}\Delta Qg_i = & -Qg_i^{Spec} \\ & - \sum_{p=a,b,c} \sum_{m=a,b,c} [V_i^p V_i^m (Gg_i^{pm} \sin \theta_i^{pm} - Bg_i^{pm} \cos \theta_i^{pm}) \\ & + \sum_{p=a,b,c} \sum_{m=a,b,c} [V_i^p E_i^p (Gg_i^{pm} \sin(\theta_i^p - \delta_i^m) - Bg_i^{pm} \cos(\theta_i^p - \delta_i^m))\end{aligned} \quad (5.54)$$

where $Pg_i^{Spec}$ and $Qg_i^{Spec}$ are the specified active and reactive powers of the generator at bus $I$, which are in the direction of into terminal bus $i$.

**PV Machines.** For a PV machine, the total three-phase active power flow and the positive sequence voltage magnitude at its terminal bus $i$ are specified. The active power flow mismatch equation is given by (5.9) while the voltage mismatch equation at bus $i$ is given by:

$$\Delta Vg_i = V_i^{Spec} - V_i^1 = V_i^{Spec} - \sqrt{(e_i^1)^2 + (f_i^1)^2} \quad (5.55)$$

where $V_i^1$ is the positive-sequence voltage magnitude voltage at the generator terminal bus $i$. $e_i^1$ and $f_i^1$ are the real and imaginary parts of the positive-sequence voltage phasor at bus $i$ and they are given by (5.11) and (5.12)

**Slack Machine.** At the terminal bus of the Slack machine, the positive-sequence voltage magnitude is specified and the positive-sequence voltage angle is taken as the system reference. We have:

$$\Delta \theta g_i = f_i^1 = 0 \quad (5.56)$$

$$\Delta Vg_i = V_i^{Spec} - V_i^1 = V_i^{Spec} - \sqrt{(e_i^1)^2 + (f_i^1)^2} \quad (5.57)$$

where $V_i^{Spec}$ is the specified positive-sequence voltage at the terminal bus of the slack machine. $e_i^1$ and $f_i^1$ are the real and imaginary parts of the positive-sequence voltage at the terminal bus of the Slack machine, and they are defined in (5.11) and (5.12).

### 5.2.3 Formulation of Newton Equations in Polar Coordinates

Combining the power mismatch equations of network buses and generator active power and voltage control constraints for the case of PV machines, the following Newton equation in polar coordinates can be obtained:

$$\mathbf{J}\Delta\mathbf{X} = -\mathbf{F}(\mathbf{X}) \quad (5.58)$$

where $\Delta\mathbf{X} = [\Delta\mathbf{X}_{gen}, \Delta\mathbf{X}_{sys}]^T$

$\Delta\mathbf{X}_{gen} = [\Delta\delta_i^a, \Delta E_i^a]^T$

$\Delta\mathbf{X}_{sys} = [\Delta\theta_i^a, \Delta V_i^a, \Delta\theta_i^b, \Delta V_i^b, \Delta\theta_i^c, \Delta V_i^c, \Delta\theta_j^a, \Delta V_j^a, \Delta\theta_j^b, \Delta V_j^b, \Delta\theta_j^c, \Delta V_j^c]^T$

$\mathbf{F}(\mathbf{X}) = [\mathbf{F}_{gen}, \mathbf{F}_{sys}]^T$

$\mathbf{F}_{gen} = [f_{gen}^1, f_{gen}^2]^T$

$\mathbf{F}_{sys} = [\Delta P_i^a, \Delta Q_i^a, \Delta P_i^b, \Delta Q_i^b, \Delta P_i^c, \Delta Q_i^c, \Delta P_j^a, \Delta Q_j^a, \Delta P_j^b, \Delta Q_j^b, \Delta P_j^c, \Delta Q_j^c]^T$

$\mathbf{J} = \dfrac{\partial \mathbf{F}(\mathbf{X})}{\partial \mathbf{X}}$

The Jacobian elements are defined as:

$$\dfrac{\partial \Delta P_i^p}{\partial \theta_j^m} = \begin{cases} -V_i^p V_j^m (G_{ij}^{pm} \sin\theta_{ij}^{pm} - B_{ij}^{pm} \cos\theta_{ij}^{pm}) & (j \neq i, m \neq p) \\ Q_i^p + (V_i^p)^2 B_{ii}^{pp} & (j = i, m = p) \end{cases} \quad (5.59)$$

$$\dfrac{\partial \Delta P_i^p}{\partial V_j^m} = \begin{cases} -V_i^p (G_{ij}^{pm} \cos\theta_{ij}^{pm} + B_{ij}^{pm} \sin\theta_{ij}^{pm}) & (j \neq i, m \neq p) \\ V_i^p B_{ii}^{pp} - P_i^p / V_i^p & (j = i, m = p) \end{cases} \quad (5.60)$$

$$\frac{\partial \Delta Q_i^p}{\partial \theta_j^m} = \begin{cases} V_i^p V_j^m (G_{ij}^{pm} \cos \theta_{ij}^{pm} + B_{ij}^{pm} \sin \theta_{ij}^{pm}) & (j \neq i, m \neq p) \\ -P_i^p + (V_i^p)^2 G_{ii}^{pp} & (j = i, m = p) \end{cases} \quad (5.61)$$

$$\frac{\partial \Delta Q_i^p}{\partial V_j^m} = \begin{cases} -V_i^p (G_{ij}^{pm} \sin \theta_{ij}^{pm} - B_{ij}^{pm} \cos \theta_{ij}^{pm}) & (j \neq i, m \neq p) \\ V_i^p B_{ii}^{pp} - Q_i^p / V_i^p & (j = i, m = p) \end{cases} \quad (5.62)$$

## 5.3 SSSC Modeling in Three-Phase Power Flow in Rectangular Coordinates

With the recent practical applications of converter based FACTS-devices such as the Static Synchronous Compensator (STATCOM) [26], Static Synchronous Series Compensator (SSSC) [27] and Unified Power Flow Controller (UPFC) [28], modeling and analysis of these FACTS-devices in power system operation and control is of great interest. In [24] mathematical models of the SSSC suitable for three-phase power flow analysis have been investigated.

In comparison to positive sequence model of SSSC, the three-phase SSSC models should consider:

- The differences between three-phase and positive sequence SSSC models. The three-phase SSSC models are basically different from the positive sequence SSSC models, which are able to give realistic results of power system operation with presence of unbalances of networks and loads while the positive sequence SSSC models can provide meaningful results only if both networks and loads are balanced.
- The transformer connection types. In the three-phase SSSC models, it is necessary to consider how the SSSC is connected with the transformer while, in the positive sequence SSSC models for conventional power flow calculations, such considerations are not needed.
- The similarity between three-phase models and positive sequence models. In principle, the three-phase models should be identical to the positive sequence models when both networks and loads are balanced.

## 5.3.1 Three-Phase SSSC Model with Delta/Wye Connected Transformer

### 5.3.1.1 Basic Operation Principles

Fig. 5.2 shows the basic operation principles of a three-phase SSSC. The SSSC consists of three converters, which are series connected with a three-phase transmission line via three single-phase transformers with Delta/Wye connections. The primary sides of the three single-phase transformers are delta-connected. It is assumed here that the transmission line is series connected with the SSSC bus $j$. With such an assumption, the active and reactive power flows entering the bus $j$ are equal to the sending-end active and reactive power flows of the transmission line, respectively. In principle, the SSSC can generate and insert three-phase series voltage sources, which can be regulated to change the three-phase impedances (more precisely reactance) of the transmission line. In this way, the power of the transmission line, which the SSSC is connected with, can be controlled.

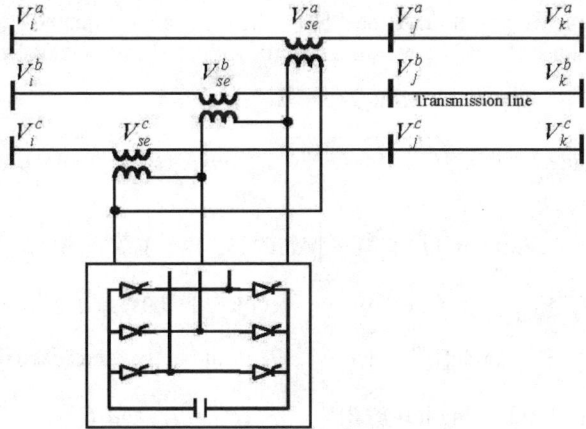

**Fig. 5.2.** Operating principles of three-phase SSSC with a Delta/Wye transformer

### 5.3.1.2 Equivalent Circuit of Three-Phase SSSC

The equivalent circuit of the three-phase SSSC is given in Fig. 5.3. The SSSC is represented by an ideal fundamental frequency three-phase voltage source vector $V_{se}^{abc}$ in series with an impedance matrix $Z_{se}^{abc}$. $Z_{se}^{abc}$ represents the impedance matrix of the three series transformers. The switching losses of SSSC may be included directly in $Z_{se}^{abc}$.

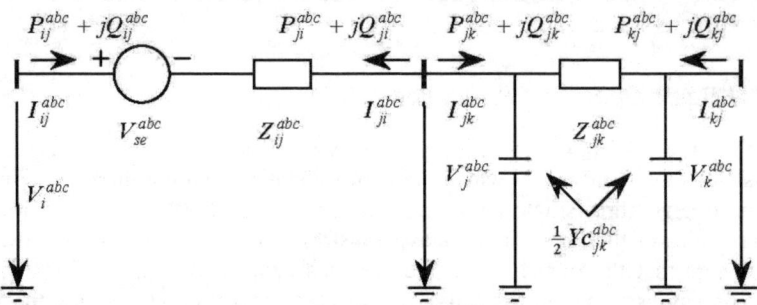

**Fig. 5.3.** Equivalent circuit of three-phase SSSC

$V_{se}^{abc} = [V_{se}^a, V_{se}^b, V_{se}^c]^T$ is the injected voltage vector. $V_{se}^p$ ($p = a, b, c$) is the voltage phasor of phase $p$, which can be further represented by real and imaginary parts $V_{se}^p = E_{se}^p + jF_{se}^p$.

In the practical operation of the SSSC, the equivalent injected voltage magnitude of each phase should be within a specific voltage limit. We define:

$$0 \leq V_{se}^p \leq Vmax_{se}^p \tag{5.63}$$

where $p = a, b, c$. $V_{se}^p = \sqrt{(E_{se}^p)^2 + (F_{se}^p)^2}$. $Vmax_{se}^p$ is the voltage limit of phase $p$.

In the equivalent circuit, $V_i^{abc} = [V_i^a, V_i^b, V_i^c]^T$, $V_j^{abc} = [V_j^a, V_j^b, V_j^c]^T$, and $V_k^{abc} = [V_k^a, V_k^b, V_k^c]^T$ are the voltage vectors at buses $i$, $j$, $k$, respectively. $P_{ij}^{abc} = [P_{ij}^a, P_{ij}^b, P_{ij}^c]^T$ and $Q_{ij}^{abc} = [Q_{ij}^a, Q_{ij}^b, Q_{ij}^c]^T$ are active and reactive power flow vectors of branch $ij$ leaving bus $i$. $P_{ji}^{abc} = [P_{ji}^a, P_{ji}^b, P_{ji}^c]^T$ and $Q_{ji}^{abc} = [Q_{ji}^a, Q_{ji}^b, Q_{ji}^c]^T$ are active and reactive power flow vectors of branch $ij$ leaving bus $j$.

### 5.3.1.3 Power Equations of the Three-Phase SSSC

Assume that $V_m^p = E_m^p + jF_m^p$ ($m=i,j,k$ and $p=a,b,c$), the following power equations of the SSSC branch are derived according to the equivalent circuit shown in Fig. 5.3:

## 5.3 SSSC Modeling in Three-Phase Power Flow in Rectangular Coordinates

$$P_{ij}^p = E_i^p \sum_{m=a,b,c} \left( G_{ii}^{pm} E_i^m - B_{ii}^{pm} F_i^m \right) + F_i^p \sum_{m=a,b,c} \left( G_{ii}^{pm} F_i^m + B_{ii}^{pm} E_i^m \right)$$
$$+ E_i^p \sum_{m=a,b,c} \left( G_{ij}^{pm} E_j^m - B_{ij}^{pm} F_j^m \right) + F_i^p \sum_{m=a,b,c} \left( G_{ij}^{pm} F_j^m + B_{ij}^{pm} E_j^m \right) \quad (5.64)$$
$$+ E_i^p \sum_{m=a,b,c} \left( G_{ij}^{pm} E_{se}^m - B_{ij}^{pm} F_{se}^m \right) + F_i^p \sum_{m=a,b,c} \left( G_{ij}^{pm} F_{se}^m + B_{ij}^{pm} E_{se}^m \right)$$

$$Q_{ij}^p = -E_i^p \sum_{m=a,b,c} \left( G_{ii}^{pm} F_i^m + B_{ii}^{pm} E_i^m \right) + F_i^p \sum_{m=a,b,c} \left( G_{ii}^{pm} E_i^m - B_{ii}^{pm} F_i^m \right)$$
$$- E_i^p \sum_{m=a,b,c} \left( G_{ij}^{pm} F_j^m + B_{ij}^{pm} E_j^m \right) + F_i^p \sum_{m=a,b,c} \left( G_{ij}^{pm} E_j^m - B_{ij}^{pm} F_j^m \right) \quad (5.65)$$
$$- E_i^p \sum_{m=a,b,c} \left( G_{ij}^{pm} F_{se}^m + B_{ij}^{pm} E_{se}^m \right) + F_i^p \sum_{m=a,b,c} \left( G_{ij}^{pm} E_{se}^m - B_{ij}^{pm} F_{se}^m \right)$$

$$P_{ji}^p = E_j^p \sum_{m=a,b,c} \left( G_{jj}^{pm} E_j^m - B_{jj}^{pm} F_j^m \right) + F_j^p \sum_{m=a,b,c} \left( G_{jj}^{pm} F_j^m + B_{jj}^{pm} E_j^m \right)$$
$$+ E_j^p \sum_{m=a,b,c} \left( G_{ji}^{pm} E_i^m - B_{ji}^{pm} F_i^m \right) + F_j^p \sum_{m=a,b,c} \left( G_{ji}^{pm} F_i^m + B_{ji}^{pm} E_i^m \right) \quad (5.66)$$
$$+ E_j^p \sum_{m=a,b,c} \left( G_{jj}^{pm} E_{se}^m - B_{jj}^{pm} F_{se}^m \right) + F_j^p \sum_{m=a,b,c} \left( G_{jj}^{pm} F_{se}^m + B_{jj}^{pm} E_{se}^m \right)$$

$$Q_{ji}^p = -E_j^p \sum_{m=a,b,c} \left( G_{jj}^{pm} F_j^m + B_{jj}^{pm} E_j^m \right) + F_j^p \sum_{m=a,b,c} \left( G_{jj}^{pm} E_j^m - B_{jj}^{pm} F_j^m \right)$$
$$- E_j^p \sum_{m=a,b,c} \left( G_{ji}^{pm} F_i^m + B_{ji}^{pm} E_i^m \right) + F_j^p \sum_{m=a,b,c} \left( G_{ji}^{pm} E_i^m - B_{ji}^{pm} F_i^m \right) \quad (5.67)$$
$$- E_j^p \sum_{m=a,b,c} \left( G_{jj}^{pm} F_{se}^m + B_{jj}^{pm} E_{se}^m \right) + F_j^p \sum_{m=a,b,c} \left( G_{jj}^{pm} E_{se}^m - B_{jj}^{pm} F_{se}^m \right)$$

where $p = a, b, c$.

The power exchange of the three converters of the SSSC with the common DC link should be zero, which is as follows:

$$\Delta P_{se}^\Sigma = \sum_{p=a,b,c} P_{se}^p = 0 \quad (5.68)$$

where $P_{se}^p$ is given by:

$$P_{se}^p = E_{se}^p \sum_{m=a,b,c} \left( G_{jj}^{pm} E_{se}^m - B_{jj}^{pm} F_{se}^m \right) + F_{se}^p \sum_{m=a,b,c} \left( G_{jj}^{pm} F_{se}^m + B_{jj}^{pm} E_{se}^m \right)$$
$$+ E_{se}^p \sum_{m=a,b,c} \left( G_{jj}^{pm} E_j^m - B_{jj}^{pm} F_j^m \right) + F_{se}^p \sum_{m=a,b,c} \left( G_{jj}^{pm} F_j^m + B_{jj}^{pm} E_j^m \right) \quad (5.69)$$
$$+ E_{se}^p \sum_{m=a,b,c} \left( G_{ji}^{pm} E_i^m - B_{ji}^{pm} F_i^m \right) + F_{se}^p \sum_{m=a,b,c} \left( G_{ji}^{pm} F_i^m + B_{ji}^{pm} E_i^m \right)$$

where $G_{ij}^{pm} + jB_{ij}^{pm} = G_{ji}^{pm} + jB_{ji}^{pm} = -y_{se}^{pm}$, $G_{ii}^{pm} + jB_{ii}^{pm} = y_{se}^{pm}$, $G_{jj}^{pm} + jB_{jj}^{pm} = y_{se}^{pm}$ ($p = a, b, c$ and $m = a, b, c$). Here $y_{se}^{pm}$ is given by:

$$Y_{se}^{abc} = \begin{bmatrix} y_{se}^{aa} & y_{se}^{ab} & y_{se}^{aa} \\ y_{se}^{ba} & y_{se}^{bb} & y_{se}^{bc} \\ y_{se}^{ca} & y_{se}^{cb} & y_{se}^{cc} \end{bmatrix} = [Z_{se}^{abc}]^{-1} = \begin{bmatrix} z_{se}^{aa} & 0 & 0 \\ 0 & z_{se}^{bb} & 0 \\ 0 & 0 & z_{se}^{cc} \end{bmatrix}^{-1} \quad (5.70)$$

where $z_{se}^{aa}$, $z_{se}^{bb}$ and $z_{se}^{cc}$ are the impedances of the three series transformers, respectively, in Fig. 5.2.

In the following, three models for the SSSC in Fig. 5.2 will be presented. They are the three-phase SSSC model with independent phase power control, three-phase SSSC model with total three-phase power control and three-phase SSSC model with symmetrical injected voltage control.

### 5.3.1.4 Three-Phase SSSC Model with Independent Phase Power Control

In the operation of the three-phase SSSC, the active power exchange of the three converters with the DC link should be zero. Such a constraint is described by (5.68). Besides, due to the fact that the SSSC is delta-connected with the three single-phase series transformers, the zero sequence component of the equivalent injected voltage vector $V_{se}^{abc}$ should be zero. In other words, the following constraints should hold,

$$\Delta E_{se}^{\Sigma} = \text{Re}(\sum_{p=a,b,c} V_{se}^{p}) = \sum_{p=a,b,c} E_{se}^{p} = 0 \quad (5.71)$$

$$\Delta F_{se}^{\Sigma} = \text{Im}(\sum_{p=a,b,c} V_{se}^{p}) = \sum_{p=a,b,c} F_{se}^{p} = 0 \quad (5.72)$$

Since the SSSC steady model has six state variables such as $E_{se}^{a}, F_{se}^{a}, E_{se}^{b}, F_{se}^{b}, E_{se}^{c}, F_{se}^{c}$, it still has three control degrees of freedom. Here assuming that the three three-phase transmission line phase power flows can be controlled, we have:

$$\Delta P_{ji}^{p} = P_{ji}^{p} - Pspec_{ji}^{p} = 0$$
$$\Delta Q_{ji}^{p} = Q_{ji}^{p} - Qspec_{ji}^{p} = 0 \quad (5.73)$$

where $p=a, b, c$. $Pspec_{ji}^{p}, Qspec_{ji}^{p}$ are the control references of the active and reactive power flows, respectively, of phase $p$.

## 5.3 SSSC Modeling in Three-Phase Power Flow in Rectangular Coordinates

Combining the six operation and control constraint equations (5.68), (5.71)-(5.73) and six power mismatch equations at buses $i, j$ together, the Newton power flow equation including the SSSC in rectangular coordinates may be given by:

$$\mathbf{J}\Delta\mathbf{X} = -\mathbf{F}(\mathbf{X}) \qquad (5.74)$$

where

$$\Delta\mathbf{X} = [\Delta\mathbf{X}_{sssc}, \Delta\mathbf{X}_{sys}]^T$$

$$\Delta\mathbf{X}_{sssc} = [\Delta E_{se}^a, \Delta F_{se}^a, \Delta E_{se}^b, \Delta F_{se}^b, \Delta E_{se}^c, \Delta F_{se}^c]^T$$

$$\Delta\mathbf{X}_{sys} = [\Delta E_i^a, \Delta F_i^a, \Delta E_i^b, \Delta F_i^b, \Delta E_i^c, \Delta F_i^c, \Delta E_j^a, \Delta F_j^a, \Delta E_j^b, \Delta F_j^b, \Delta E_j^c, \Delta F_j^c]^T$$

$$\mathbf{F}(\mathbf{X}) = [\mathbf{F}_{sssc}, \mathbf{F}_{sys}]^T$$

$$\mathbf{F}_{sssc} = [\Delta P_{se}^\Sigma, \Delta E_{se}^\Sigma, \Delta F_{se}^\Sigma, \Delta P_{ji}^a, \Delta P_{ji}^b, \Delta P_{ji}^c]^T$$

$$\mathbf{F}_{sys} = [\Delta P_i^a, \Delta Q_i^a, \Delta P_i^b, \Delta Q_i^b, \Delta P_i^c, \Delta Q_i^c, \Delta P_j^a, \Delta Q_j^a, \Delta P_j^b, \Delta Q_j^b, \Delta P_j^c, \Delta Q_j^c]^T$$

$$\mathbf{J} = \frac{\partial \mathbf{F}(\mathbf{X})}{\partial \mathbf{X}}$$

### 5.3.1.5 Three-Phase SSSC Model with Total Three-Phase Power Control

Assume, for the SSSC in Fig. 5.2, that (a) the three equivalent injected voltages $V_{se}^a, V_{se}^b, V_{se}^c$ are perpendicular to the line currents of phase a, phase b, and phase c, respectively; (b) the total three-phase power is controlled, then:

$$P_{se}^p = 0 \qquad (5.75)$$

and

$$\Delta P_{ji}^\Sigma = \sum_{p=a,b,c} P_{ji}^p - Pspec_{ji}^\Sigma = 0$$

$$\text{or } \Delta Q_{ji}^\Sigma = \sum_{p=a,b,c} Q_{ji}^p - Qspec_{ji}^\Sigma = 0 \qquad (5.76)$$

where $p = a, b, c$. $PSpec_{ji}^\Sigma$ and $QSpec_{ji}^\Sigma$ are the specified total three-phase active and reactive power flow control references, respectively.

Combining the six operation and control constraint equations (5.71), (5.72), (5.75) and (5.76) and six power mismatch equations at buses $i, j$ together, the Newton power flow equation including the SSSC in rectangular coordinates may be given by:

$$\mathbf{J}\Delta\mathbf{X} = -\mathbf{F}(\mathbf{X}) \qquad (5.77)$$

where

$$\Delta\mathbf{X} = [\Delta\mathbf{X}_{sssc}, \Delta\mathbf{X}_{sys}]^T$$

$$\Delta \mathbf{X}_{sssc} = [\Delta E^a_{se}, \Delta F^a_{se}, \Delta E^b_{se}, \Delta F^b_{se}, \Delta E^c_{se}, \Delta F^c_{se}]^T$$

$$\Delta \mathbf{X}_{sys} = [\Delta E^a_i, \Delta F^a_i, \Delta E^b_i, \Delta F^b_i, \Delta E^c_i, \Delta F^c_i, \Delta E^a_j, \Delta F^a_j, \Delta E^b_j, \Delta F^b_j, \Delta E^c_j, \Delta F^c_j]^T$$

$$\mathbf{F}(\mathbf{X}) = [\mathbf{F}_{sssc}, \mathbf{F}_{sys}]^T$$

$$\mathbf{F}_{sssc} = [P^a_{se}, P^b_{se}, P^c_{se}, \Delta E^{\Sigma}_{se}, \Delta F^{\Sigma}_{se}, \Delta P^{\Sigma}_{ji}]^T$$

$$\mathbf{F}_{sys} = [\Delta P^a_i, \Delta Q^a_i, \Delta P^b_i, \Delta Q^b_i, \Delta P^c_i, \Delta Q^c_i, \Delta P^a_j, \Delta Q^a_j, \Delta P^b_j, \Delta Q^b_j, \Delta P^c_j, \Delta Q^c_j]^T$$

$$\mathbf{J} = \frac{\partial \mathbf{F}(\mathbf{X})}{\partial \mathbf{X}}$$

### 5.3.1.6 Three-Phase SSSC Model with Symmetrical Injected Voltage Control

If we assume the series injected three-phase voltage sources of the three-phase SSSC are balanced or symmetrical, then we have the following control constraint equations:

$$V^a_{se} = V^b_{se} e^{j120°} = V^c_{se} e^{j240°} \qquad (5.78)$$

A set of symmetrical or balanced three-phase voltage phasors are equal in magnitude while their phase angles have 120° displacement among them. For the sake of computation, equation (5.78) may be replaced by the following four equations in real and imaginary parts:

$$\Delta V^1_{Re} = \text{Re}(V^a_{se} - V^b_{se} e^{j120°}) = E^a_{se} + \frac{1}{2}E^b_{se} + \frac{\sqrt{3}}{2}F^b_{se} = 0 \qquad (5.79)$$

$$\Delta V^1_{Im} = \text{Im}(V^a_{se} - V^b_{se} e^{j120°}) = E^a_{se} - \frac{\sqrt{3}}{2}E^b_{se} + \frac{1}{2}F^b_{se} = 0 \qquad (5.80)$$

$$\Delta V^2_{Re} = \text{Re}(V^a_{se} - V^c_{se} e^{j240°}) = E^a_{se} + \frac{1}{2}E^c_{se} - \frac{\sqrt{3}}{2}F^c_{se} = 0 \qquad (5.81)$$

$$\Delta V^2_{Im} = \text{Im}(V^a_{se} - V^c_{se} e^{j240°}) = F^a_{se} + \frac{\sqrt{3}}{2}E^c_{se} + \frac{1}{2}F^c_{se} = 0 \qquad (5.82)$$

When (5.79)-(5.82) hold, equations (5.71) and (5.72) will be satisfied. Furthermore, the active power exchange constraint (5.68) should be balanced at any instant. So the SSSC with symmetrical control has only one control degree of freedom. Assuming that the total active power or reactive power of the three-phase transmission line is controlled, we have the following power flow control constraint:

$$\Delta P^{\Sigma}_{ji} = \sum_{p=a,b,c} P^p_{ji} - P\text{spec}^{\Sigma}_{ji} = 0$$

$$\text{or } \Delta Q^{\Sigma}_{ji} = \sum_{p=a,b,c} Q^p_{ji} - Q\text{spec}^{\Sigma}_{ji} = 0 \qquad (5.83)$$

where $p = a, b, c$. $PSpec_{ji}^{\Sigma}$ and $QSpec_{ji}^{\Sigma}$ are the specified total three-phase active and reactive power flow control references, respectively.

The SSSC model with symmetrical voltage control has six operation and control constraint equations (5.68), (5.79)-(5.82). Combining the six operation and control constraint equations (5.68), (5.79)-(5.82) and six power mismatch equations at buses $i, j$ together, the Newton power flow equation including the SSSC in rectangular coordinates may be given by:

$$\mathbf{J}\Delta\mathbf{X} = -\mathbf{F}(\mathbf{X}) \tag{5.84}$$

where $\Delta\mathbf{X} = [\Delta\mathbf{X}_{sssc}, \Delta\mathbf{X}_{sys}]^T$

$\Delta\mathbf{X}_{sssc} = [\Delta E_{se}^a, \Delta F_{se}^a, \Delta E_{se}^b, \Delta F_{se}^b, \Delta E_{se}^c, \Delta F_{se}^c]^T$

$\Delta\mathbf{X}_{sys} = [\Delta E_i^a, \Delta F_i^a, \Delta E_i^b, \Delta F_i^b, \Delta E_i^c, \Delta F_i^c, \Delta E_j^a, \Delta F_j^a, \Delta E_j^b, \Delta F_j^b, \Delta E_j^c, \Delta F_j^c]^T$

$\mathbf{F}(\mathbf{X}) = [\mathbf{F}_{sssc}, \mathbf{F}_{sys}]^T$

$\mathbf{F}_{sssc} = [\Delta P_{se}^{\Sigma}, \Delta V_{Re}^1, \Delta V_{Im}^1, \Delta V_{Re}^2, \Delta V_{Im}^2, \Delta P_{ji}^{\Sigma}]^T$

$\mathbf{F}_{sys} = [\Delta P_i^a, \Delta Q_i^a, \Delta P_i^b, \Delta Q_i^b, \Delta P_i^c, \Delta Q_i^c, \Delta P_j^a, \Delta Q_j^a, \Delta P_j^b, \Delta Q_j^b, \Delta P_j^c, \Delta Q_j^c]^T$

$\mathbf{J} = \dfrac{\partial \mathbf{F}(\mathbf{X})}{\partial \mathbf{X}}$

### 5.3.2 Single-Phase/Three-Phase SSSC Models with Separate Single Phase Transformers

#### 5.3.2.1 Basic Operating Principles of Single Phase SSSC

In Fig 5.4, three single-phase SSSCs are series connected with phase $a$, $b$, $c$ of a transmission line, respectively. The three SSSCs have neither electrical nor magnetic connections between them. Each single phase SSSC is series-connected with the transmission line via a single-phase transformer. Each can independently control the phase power flow of the transmission line. The single-phase SSSC is attractive and practical when there are unbalanced loads and one or two phase lines existing in the systems.

#### 5.3.2.2 Equivalent Circuit of Single Phase SSSC

Due to the fact that there are no electrical and magnetic couplings between the three single-phase SSSCs, each SSSC branch can be represented by an equivalent circuit shown in Fig. 5.5. Such an equivalent circuit is exactly the same to that of the SSSC for the positive sequence power flow calculations. However, the physical meaning of the single-phase equivalent circuit here is quite different from that of the positive sequence SSSC in the positive sequence power flow calculations.

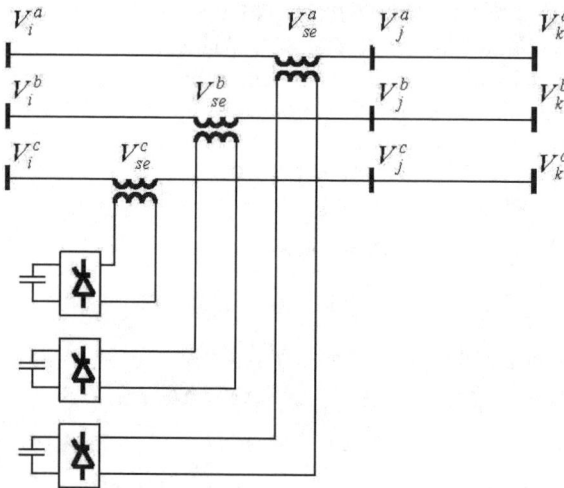

**Fig. 5.4.** Operation principle of single phase SSSC for three-phase power flow analysis

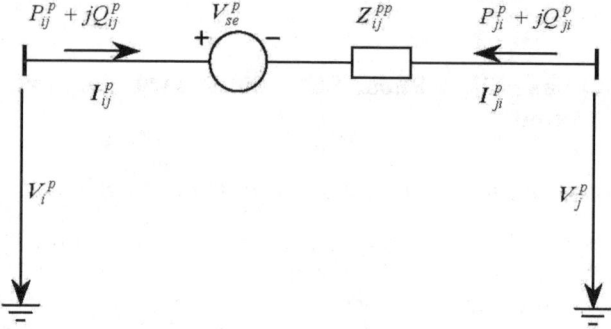

**Fig. 5.5.** Equivalent circuit of single phase SSSC for three-phase power flow analysis

### 5.3.2.3 Single-Phase SSSC

The power flow equations (5.64)-(5.67) for the three-phase SSSC are still applicable to the system with the three separate single-phase SSSCs installed on phase $a$, $b$, $c$ of the transmission line in Fig. 5.4, respectively.

Since each single phase SSSC can neither generate nor absorb active power, the power exchange of each SSSC with the system should be zero. Mathematically, such a constraint may be represented by:

## 5.3 SSSC Modeling in Three-Phase Power Flow in Rectangular Coordinates

$$P_{se}^{p} = 0 \tag{5.85}$$

where $p=a, b, c$. $P_{se}^{p}$, which is given by (5.69), is the active power exchange of SSSC with the DC link or the system.

Assuming that each SSSC independently controls the phase active or reactive power flow of the transmission line, the power flow control constraint may be represented by:

$$P_{ji}^{p} - Pspec_{ji}^{p} = 0 \quad \text{or} \quad Q_{ji}^{p} - Qspec_{ji}^{p} = 0 \tag{5.86}$$

where $p=a, b, c$. $Pspec_{ji}^{p}$ and $Qspec_{ji}^{p}$ are the specified active and reactive power flow control references of phase $p$, respectively.

Combining the six operation and control equations (5.85) and (5.86) of the three single phase SSSCs and six power mismatch equations of buses $i$ and $j$, the three-phase Newton equation may be given by:

$$\mathbf{J}\Delta\mathbf{X} = -\mathbf{F}(\mathbf{X}) \tag{5.87}$$

where $\Delta\mathbf{X} = [\Delta\mathbf{X}_{sssc}, \Delta\mathbf{X}_{sys}]^T$

$\Delta\mathbf{X}_{sssc} = [\Delta E_{se}^{a}, \Delta F_{se}^{a}, \Delta E_{se}^{b}, \Delta F_{se}^{b}, \Delta E_{se}^{c}, \Delta F_{se}^{c}]^T$

$\Delta\mathbf{X}_{sys} = [\Delta E_{i}^{a}, \Delta F_{i}^{a}, \Delta E_{i}^{b}, \Delta F_{i}^{b}, \Delta E_{i}^{c}, \Delta F_{i}^{c}, \Delta E_{j}^{a}, \Delta F_{j}^{a}, \Delta E_{j}^{b}, \Delta F_{j}^{b}, \Delta E_{j}^{c}, \Delta F_{j}^{c}]^T$

$\mathbf{F}(\mathbf{X}) = [\mathbf{F}_{sssc}, \mathbf{F}_{sys}]^T$

$\mathbf{F}_{sssc} = [P_{se}^{a}, P_{se}^{b}, P_{se}^{c}, \Delta P_{ji}^{a}, \Delta P_{ji}^{b}, \Delta P_{ji}^{c}]^T$

$\mathbf{F}_{sys} = [\Delta P_{i}^{a}, \Delta Q_{i}^{a}, \Delta P_{i}^{b}, \Delta Q_{i}^{b}, \Delta P_{i}^{c}, \Delta Q_{i}^{c}, \Delta P_{j}^{a}, \Delta Q_{j}^{a}, \Delta P_{j}^{b}, \Delta Q_{j}^{b}, \Delta P_{j}^{c}, \Delta Q_{j}^{c}]^T$

$\mathbf{J} = \dfrac{\partial \mathbf{F}(\mathbf{X})}{\partial \mathbf{X}}$

### 5.3.2.4 Three-Phase SSSC Model with Three Separate Single Phase Transformers

If we assume (a) a three-phase SSSC is connected with a three-phase transmission line via three separate single phase transformers; (b) the three injected voltages $V_{se}^{a}, V_{se}^{b}, V_{se}^{c}$ of the SSSC are perpendicular to the line currents of phase a, phase b, and phase c of the transmission line, respectively; (c) the three single phase power flows can be controlled, then the three-phase SSSC will have similar constraint equations of (5.85) and (5.86). Subsequently, for the three-phase SSSC, we have the similar Newton equation as given by (5.87).

### 5.3.3 Numerical Examples

A 5-bus system and the IEEE 118 bus system have been used to test the three-phase Newton power flow algorithm with modeling of the SSSC. The 5 bus three-phase system is shown in Fig. 5.8 in the Appendix of this chapter while the system parameters are listed in Table 5.11 - Table 5.14. In the tests, a convergence tolerance of 1.0e-12 p.u. is used. For the sake of convenience, the three-phase SSSC model with independent phase power control, three-phase SSSC model with total three-phase power control and three-phase SSSC model with symmetrical injected voltage control in Section 2 are denoted as Model 1, Model 2 and Model 3, respectively, while the three-phase and single phase SSSC models in Section 3 are referred to Model 4 and Model 5, respectively.

#### 5.3.3.1 Test Results for the 5-Bus System

Based on the 5-bus system, tests under the following conditions have been carried out:

*Case 1:* Well transposed transmission lines and the whole system with balanced load.

*Case 2:* Non-transposed transmission lines and unbalanced load at bus 3 as given by Table 5.13 and Table 5.14.

*Case 3*: As for case 1, but a SSSC is installed at the sending-end of the transmission line 1-3.

*Case 4*: As for case 2, but a SSSC is installed at the sending-end of the transmission line 1-3.

*Case 5*: As for case 3, but the whole system is represented by the positive sequence network only.

The number of iterations of the three-phase power flow algorithm on cases 1-4 are summarized in Table 5.1. For cases 3 and 4, the control references of the SSSC Models 1, 4 and 5 are $Pspec_{ji}^a = Pspec_{ji}^b = Pspec_{ji}^c = 7.0 p.u.$ while the control reference for the SSSC Models 2 and 3 is $Pspec_{ji}^\Sigma = 21.0 p.u.$ In order to verify the validity of the three-phase power flow algorithm and the SSSC models, case 5 has been carried out, in which the whole system is represented only by the positive sequence network since the system is balanced. The power flow solution of case 5 is obtained by a positive sequence power flow program. The detailed power flow solutions of case 3 and case 5 are given by Table 5.2.

5.3 SSSC Modeling in Three-Phase Power Flow in Rectangular Coordinates 163

**Table 5.1.** Number of iterations of three-phase power flow algorithm for the 5-bus system

| Case No. | Base case power flows / SSSC power flow solutions | Total power flow increase (%) | Number of iterations |
|---|---|---|---|
| 1 | $P_{13}^a = P_{13}^b = P_{13}^c = 4.94 p.u.$  $Q_{13}^a = Q_{13}^b = Q_{13}^c = 2.01 p.u.$ <br> $P_{13}^\Sigma = 14.82 p.u.$  $Q_{13}^\Sigma = 6.03 p.u.$ | – | 6 |
| 2 | $P_{13}^a = 4.96 p.u.$  $P_{13}^b = 5.17 p.u.$  $P_{13}^c = 4.69 p.u.$ <br> $Q_{13}^a = 1.96 p.u.$  $Q_{13}^b = 1.46 p.u.$  $Q_{13}^c = 2.33 p.u.$ <br> $P_{13}^\Sigma = 14.82 p.u.$  $Q_{13}^\Sigma = 5.75 p.u.$ | – | 6 |
| 3 | SSSC Model 1: <br> $V_{se}^a = V_{se}^b = V_{se}^c = 0.1933 p.u.$ <br> $\theta_{se}^a = 283.83°$  $\theta_{se}^b = 163.83°$  $\theta_{se}^c = 43.83°$ | 42% | 6 |
|   | SSSC Model 2: <br> $V_{se}^a = V_{se}^b = V_{se}^c = 0.1933 p.u.$ <br> $\theta_{se}^a = 283.83°$  $\theta_{se}^b = 163.83°$  $\theta_{se}^c = 43.83°$ | 42% | 6 |
|   | SSSC Model 3: $V_{se}^a = V_{se}^b = V_{se}^c = 0.1933 p.u.$ <br> $\theta_{se}^a = 283.83°$  $\theta_{se}^b = 163.83°$  $\theta_{se}^c = 43.83°$ | 42% | 6 |
|   | SSSC Models 4 and 5: <br> $V_{se}^a = V_{se}^b = V_{se}^c = 0.1933 p.u.$ <br> $\theta_{se}^a = 283.83°$  $\theta_{se}^b = 163.83°$  $\theta_{se}^c = 43.83°$ | 42% | 6 |
| 4 | SSSC Model 1: <br> $V_{se}^a = 0.1781 p.u.$  $V_{se}^b = 0.1400 p.u.$  $V_{se}^c = 0.2301 p.u.$ <br> $\theta_{se}^a = 265.68°$  $\theta_{se}^b = 177.55°$  $\theta_{se}^c = 48.22°$ | 42% | 6 |
|   | SSSC Model 2: <br> $V_{se}^a = 0.1437 p.u.$  $V_{se}^b = 0.1836 p.u.$  $V_{se}^c = 0.1663 p.u.$ <br> $\theta_{se}^a = 288.49°$  $\theta_{se}^b = 168.08°$  $\theta_{se}^c = 36.25°$ | 42% | 6 |
|   | SSSC Model 3: <br> $V_{se}^a = V_{se}^b = V_{se}^c = 0.1661 p.u.$ <br> $\theta_{se}^a = 283.96°$  $\theta_{se}^b = 163.95°$  $\theta_{se}^c = 43.95°$ | 42% | 6 |
|   | SSSC Models 4 and 5: <br> $V_{se}^a = 0.1730 p.u.$  $V_{se}^b = 0.1100 p.u.$  $V_{se}^c = 0.2443 p.u.$ <br> $\theta_{se}^a = 284.64°$  $\theta_{se}^b = 165.36°$  $\theta_{se}^c = 41.53°$ | 42% | 6 |

**Table 5.2.** Power flow solutions for the 5 bus system by three-phase and single-phase power flow algorithms

| Bus No. | Case 3 | | Bus No. | Case 5 | |
|---|---|---|---|---|---|
| | $V^a$ | $\theta^a$ | | $V$ | $\theta$ |
| | (p.u.) | (deg.) | | (p.u.) | (deg.) |
| 1 | 1.0183 | 27.35 | 1 | 1.0183 | -2.65 |
| 2 | 1.0238 | 28.64 | 2 | 1.0238 | -1.36 |
| 3 | 1.0101 | 30.47 | 3 | 1.0101 | 0.47 |
| 4 | 1.0450 | 0.00 | 4 | 1.0450 | 0.00 |
| 5 | 1.0610 | 2.70 | 5 | 1.0610 | 2.70 |
| SSSC | $V^a_{se}=0.1933$ | $\theta^a_{se}=283.83$ | SSSC | $V_{se}=0.1933$ | $\theta_{se}=253.83$ |

From Table 5.1 and Table 5.2, it can be seen,

1. The three-phase power flow algorithm with incorporation of the SSSC models can converge in only 6 iterations.
2. In case 3, the power flow solutions with the different SSSC models are the same when the system is balanced. Such a coincidence of computation results implies the validity of the SSSC models proposed.
3. Comparison of power flow solutions of case 3 and case 5 in Table 5.2 indicates that two sets of solutions by the conventional positive sequence power flow algorithm and three-phase power flow algorithm proposed are almost identical. The only difference is that there is 30 degree angle shift in the three-phase power flow results which is caused by the Delta/Wye transformers. The comparison of the two power flow solutions again illustrates the validity of the three-phase power flow algorithm and the SSSC models. It can be anticipated that if Wye/Wye transformers rather than Delta/Wye transformers are used, the power flow solutions by the conventional positive sequence power flow and the three-phase power flow computations should be the same, and the 30 degree shift should disappear. This theoretical analysis has been confirmed by power flow calculations. Due to the limitation of space, the power flow calculation results are not presented herein.
4. Case 4 in Table 5.1 shows that the different SSSC models will have different power flow solutions when the system is unbalanced. This implies that the appropriate modeling of the SSSC in three-phase power flow calculations is very important.

### 5.3.3.2 Test Results for the IEEE 118-Bus System

Further tests have been carried out on the IEEE 118-bus system, which are as follows:

*Case 6:* Well-transposed transmission lines and the whole system with balanced load.

## 5.3 SSSC Modeling in Three-Phase Power Flow in Rectangular Coordinates

*Case 7*: Well-transposed transmission lines and the system with unbalanced load at bus 78 with 0.51+$j$0.26 p.u., 0.71+$j$0.26 p.u., 0.91+$j$0.26 p.u. for phase a, b, c loading, respectively.

*Case 8*: As for case 6, there are two SSSCs installed on the transmission lines 30-38 and 68-81.

*Case 9*: As for case 7, there are two SSSCs installed on the transmission lines 30-38 and 68-81.

In cases 8 and 9, the control references of the SSSCs are 140% of the base case power flows, respectively. The test results are given by Table 5.3. The convergence characteristics of case 7 and case 9 are shown in Fig. 5.6, from which it can be seen that the power flow algorithm exhibits excellent quadratic convergence characteristics.

**Table 5.3.** Test results on the IEEE 118-bus system

| Case No. | SSSC models | Number of iterations |
| --- | --- | --- |
| 6 | None | 6 |
| 7 | None | 6 |
| 8 | The SSSC on line 30-38: Model 1 | 6 |
|   | The SSSC on line 68-81: Model 3 | 6 |
| 9 | The SSSC on line 30-38: Model 3 | 6 |
|   | The SSSC on line 68-81: Model 4 | 6 |

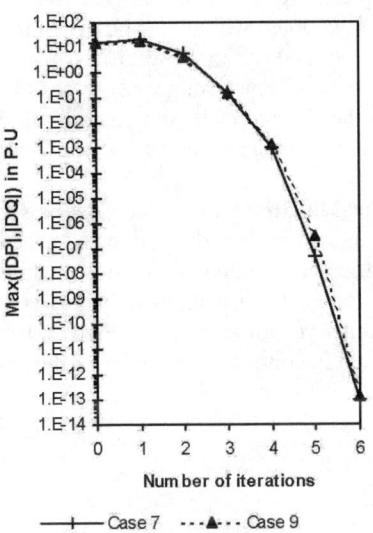

**Fig. 5.6.** Power mismatches as function of number of iterations on the IEEE 118-bus three-phase system

## 5.4 UPFC Modeling in Three-Phase Newton Power Flow in Polar Coordinates

In section 5.3, the mathematical models of SSSC for three-phase power flow analysis have been proposed. The UPFC combining two three-phase converters has been considered as one of the most powerful converter based FACTS-devices and can be used to control power flows and bus voltages. It has been recognized that due to the relative simplicity of the SSSC transformers, the transformer connection types are just implicitly represented.

However, due to the complicated combinations of the converter topologies and transformer connection types, for the modeling of the UPFC in three-phase power flow analysis, the implicit representation of the converter transformers cannot be considered adequate and may have modeling limitations. Hence, the representation of transformer connection types and UPFC control constraints becomes essential [25]. In this section, the mathematical models of UPFC, in polar coordinates, for three-phase power flow analysis are discussed. With the UPFC models derived, three-phase STATCOM models can be easily derived by eliminating the series part constraints from the equations.

### 5.4.1 Operation Principles of the Three-Phase UPFC

Fig. 5.7 shows the basic operating principles of a three-phase UPFC. The UPFC consists of series converter and shunt converter. The series converter is series connected with a three-phase transmission line via three single-phase transformers. The shunt-converter is coupled with the ac bus via a three-phase Wye-G/Delta transformer. It should be pointed out that Fig. 5.7 is just used to show one of the topologies and the related transformer connection types of the three-phase UPFC. However, in addition to the Wye-G/Delta connection, the UPFC may have other kinds of seires and shunt transformer connection types, which will be considered in the following derivation.

In Fig. 5.7, it is assumed that the transmission line is series connected with the UPFC bus $j$. With such an assumption, the active and reactive power flows entering bus $j$ are equal to the sending-end active and reactive power flows of the transmission line, respectively. In principle, the series converter may be used to control the active and reactive power flows of the transmission line while the shunt converter can be used to control the voltages of the shunt bus.

## 5.4 UPFC Modeling in Three-Phase Newton Power Flow in Polar Coordinates

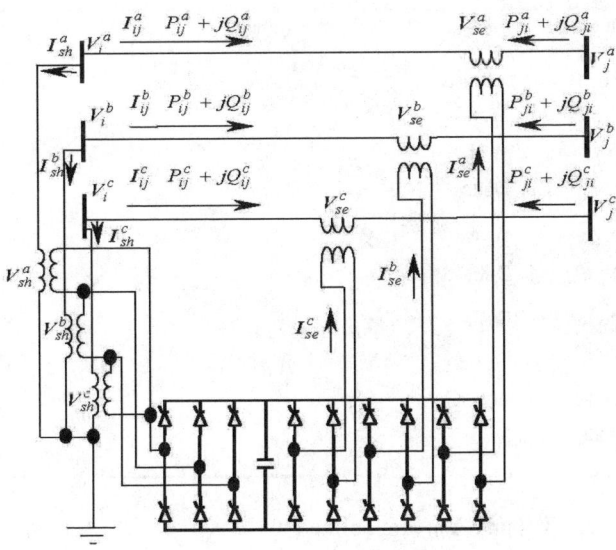

**Fig. 5.7.** Schematic description of a three-phase UPFC

### 5.4.2 Three-Phase Converter Transformer Models

In single-phase positive-sequence power flow analysis, it is usually sufficient to represent a three-phase transformer as a positive-sequence impedance in series with an ideal transformer. However, in three-phase power flow analysis where unbalanced operating conditions of network and load are considered, such a transformer model can no longer be considered appropriate. In principle, the three-phase transformer should be described in three-phase coordinates and the connection type of the transformer should also be fully represented. The admittance transformer models of various connections in three-phase coordinates were derived [23]. A two winding three-phase transformer may be represented by:

$$\begin{bmatrix} \mathbf{I}_P^{abc} \\ \mathbf{I}_S^{abc} \end{bmatrix} = \begin{bmatrix} \mathbf{Y}_{PP}^{abc} & \mathbf{Y}_{PS}^{abc} \\ \mathbf{Y}_{SP}^{abc} & \mathbf{Y}_{SS}^{abc} \end{bmatrix} \begin{bmatrix} \mathbf{V}_P^{abc} \\ \mathbf{V}_S^{abc} \end{bmatrix} \tag{5.88}$$

where $\mathbf{I}_P^{abc}$ and $\mathbf{I}_S^{abc}$ are the current vectors of the primary and secondary windings, respectively while $\mathbf{V}_P^{abc}$ and $\mathbf{V}_S^{abc}$ are the voltage vectors of the primary and secondary windings, respectively. $\mathbf{Y}_{PP}^{abc}$, $\mathbf{Y}_{SS}^{abc}$, $\mathbf{Y}_{PS}^{abc}$ and $\mathbf{Y}_{SP}^{abc}$ are 3 by 3 submatrices and they are given by Table 5.4 based on transformer connection types.

**Table 5.4.** Submatrices of three-phase transformers with different connection types

| Type No. | Transformer connection Bus P | Bus S | Self admittance $\mathbf{Y}_{PP}^{abc}$ | $\mathbf{Y}_{SS}^{abc}$ | Mutual admittance $\mathbf{Y}_{PS}^{abc}$ | $\mathbf{Y}_{SP}^{abc}$ |
|---|---|---|---|---|---|---|
| 1 | Wye-G | Wye-G | $\mathbf{Y}_I$ | $\mathbf{Y}_I$ | $-\mathbf{Y}_I$ | $-\mathbf{Y}_I$ |
| 2 | Wye-G | Wye | $\mathbf{Y}_{II}$ | $\mathbf{Y}_{II}$ | $-\mathbf{Y}_{II}$ | $-\mathbf{Y}_{II}$ |
| 3 | Wye-G | Delta | $\mathbf{Y}_I$ | $\mathbf{Y}_{II}$ | $\mathbf{Y}_{III}$ | $\mathbf{Y}_{III}^T$ |
| 4 | Wye | Wye | $\mathbf{Y}_{II}$ | $\mathbf{Y}_{II}$ | $-\mathbf{Y}_{II}$ | $-\mathbf{Y}_{II}$ |
| 5 | Wye | Delta | $\mathbf{Y}_{II}$ | $\mathbf{Y}_{II}$ | $\mathbf{Y}_{III}$ | $\mathbf{Y}_{III}^T$ |
| 6 | Delta | Delta | $\mathbf{Y}_{II}$ | $\mathbf{Y}_{II}$ | $-\mathbf{Y}_{II}$ | $-\mathbf{Y}_{II}$ |
| 7 | Wye | Wye-G | $\mathbf{Y}_{II}$ | $\mathbf{Y}_{II}$ | $-\mathbf{Y}_{II}$ | $-\mathbf{Y}_{II}$ |
| 8 | Delta | Wye-G | $\mathbf{Y}_{II}$ | $\mathbf{Y}_I$ | $\mathbf{Y}_{III}^T$ | $\mathbf{Y}_{III}$ |
| 9 | Delta | Wye | $\mathbf{Y}_{II}$ | $\mathbf{Y}_{II}$ | $\mathbf{Y}_{III}^T$ | $\mathbf{Y}_{III}$ |

In Table 5.4, $\mathbf{Y}_I$, $\mathbf{Y}_{II}$ and $\mathbf{Y}_{III}$ are defined by:

$$\mathbf{Y}_I = \begin{bmatrix} y_t & 0 & 0 \\ 0 & y_t & 0 \\ 0 & 0 & y_t \end{bmatrix} \quad (5.89)$$

$$\mathbf{Y}_{II} = \frac{1}{3}\begin{bmatrix} 2y_t & -y_t & -y_t \\ -y_t & 2y_t & -y_t \\ -y_t & -y_t & 2y_t \end{bmatrix} \quad (5.90)$$

$$\mathbf{Y}_{III} = \frac{1}{\sqrt{3}}\begin{bmatrix} -y_t & y_t & 0 \\ 0 & -y_t & y_t \\ y_t & 0 & -y_t \end{bmatrix} \quad (5.91)$$

where $y_t$ is the per unit leakage admittance.

If the transformer has off-nominal tap ratios $\alpha$ and $\beta$ of the primary and secondary windings, respectively, then the self and mutual matrices need to be modified by:

1. dividing the primary self admittance matrix $\mathbf{Y}_{PP}^{abc}$ by $\alpha^2$;
2. dividing the secondary self admittance matrix $\mathbf{Y}_{SS}^{abc}$ by $\beta^2$;
3. dividing the mutual admittance matrices $\mathbf{Y}_{PS}^{abc}$ and $\mathbf{Y}_{SP}^{abc}$ by $\alpha\beta$.

## 5.4.3 Power Flow Constraints of the Three-Phase UPFC

### 5.4.3.1 Power Flow Constraints of the Shunt Converter

Based on the operating principles shown in Fig. 5.6, the bus voltage equation of the three-phase shunt converter transformer of the UPFC may be given by:

$$\begin{bmatrix} \mathbf{I}_{ii}^{abc} \\ \mathbf{I}_{sh}^{abc} \end{bmatrix} = \begin{bmatrix} \mathbf{Y}sh_{PP}^{abc} & \mathbf{Y}sh_{PS}^{abc} \\ \mathbf{Y}sh_{SP}^{abc} & \mathbf{Y}sh_{SS}^{abc} \end{bmatrix} \begin{bmatrix} \mathbf{V}_{i}^{abc} \\ \mathbf{V}_{sh}^{abc} \end{bmatrix} \quad (5.92)$$

where $\mathbf{I}_{ii}^{abc}$ and $\mathbf{I}_{sh}^{abc}$ are the current vectors of the primary and secondary windings of the shunt converter transformer and given by:

$$\mathbf{I}_{ii}^{abc} = [I_{ii}^{a}, I_{ii}^{b}, I_{ii}^{c}]^{T} \quad (5.93)$$

$$\mathbf{I}_{sh}^{abc} = [I_{sh}^{a}, I_{sh}^{b}, I_{sh}^{c}]^{T} \quad (5.94)$$

The voltage vectors are:

$$\mathbf{V}_{i}^{abc} = [V_{i}^{a}, V_{i}^{b}, V_{i}^{c}]^{T} \quad (5.95)$$

$$\mathbf{V}_{sh}^{abc} = [V_{sh}^{a}, V_{sh}^{b}, V_{sh}^{c}]^{T} \quad (5.96)$$

The active and reactive power flows of the primary side of the shunt converter transformer are:

$$\begin{aligned} P_{ii}^{p} &= \text{Re}[V_{i}^{p}(I_{ii}^{p})^{*}] \\ &= \sum_{m=a,b,c} V_{i}^{p} V_{i}^{m}(Gsh_{PP}^{pm}\cos(\theta_{i}^{p}-\theta_{i}^{m}) + Bsh_{PP}^{pm}\sin(\theta_{i}^{p}-\theta_{i}^{m})) \\ &+ \sum_{m=a,b,c} V_{i}^{p} V_{sh}^{m}(Gsh_{PS}^{pm}\cos(\theta_{i}^{p}-\theta_{sh}^{m}) + Bsh_{PS}^{pm}\sin(\theta_{i}^{p}-\theta_{sh}^{m})) \end{aligned} \quad (5.97)$$

$$\begin{aligned} Q_{ii}^{p} &= \text{Im}[V_{i}^{p}(I_{ii}^{p})^{*}] \\ &= \sum_{m=a,b,c} V_{i}^{p} V_{i}^{m}(Gsh_{PP}^{pm}\sin(\theta_{i}^{p}-\theta_{i}^{m}) - Bsh_{PP}^{pm}\cos(\theta_{i}^{p}-\theta_{i}^{m})) \\ &+ \sum_{m=a,b,c} V_{i}^{p} V_{sh}^{m}(Gsh_{PS}^{pm}\sin(\theta_{i}^{p}-\theta_{se}^{m}) - Bsh_{PS}^{pm}\cos(\theta_{i}^{p}-\theta_{sh}^{m})) \end{aligned} \quad (5.98)$$

where $p, m = a, b, c$. $Gsh_{PP}^{pm} + jBsh_{PP}^{pm} \in \mathbf{Y}sh_{PP}^{abc}$, and $Gsh_{PS}^{pm} + jBsh_{PS}^{pm} \in \mathbf{Y}se_{Ph}^{abc}$.

Similarly, the active and reactive power flows at the secondary side of the shunt converter transformer are given by:

$$P_{sh}^p = \text{Re}[V_{sh}^p (I_{sh}^p)^*]$$

$$= \sum_{m=a,b,c} V_{sh}^p V_{sh}^m (Gsh_{SS}^{pm} \cos(\theta_{sh}^p - \theta_{sh}^m) + Bsh_{SS}^{pm} \sin(\theta_{sh}^p - \theta_{sh}^m))$$

$$+ \sum_{m=a,b,c} V_{sh}^p V_i^m (Gsh_{SP}^{pm} \cos(\theta_{sh}^p - \theta_i^m) + Bsh_{SP}^{pm} \sin(\theta_{sh}^p - \theta_i^m)) \quad (5.99)$$

$$Q_{sh}^p = \text{Im}[V_{sh}^p (I_{sh}^p)^*]$$

$$= \sum_{m=a,b,c} V_{sh}^p V_{sh}^m (Gsh_{SS}^{pm} \sin(\theta_{sh}^p - \theta_{sh}^m) - Bsh_{SS}^{pm} \cos(\theta_{sh}^p - \theta_{sh}^m))$$

$$+ \sum_{m=a,b,c} V_{sh}^p V_i^m (Gsh_{SP}^{pm} \sin(\theta_{sh}^p - \theta_i^m) - Bsh_{SP}^{pm} \cos(\theta_{sh}^p - \theta_i^m)) \quad (5.100)$$

### 5.4.3.2 Power Flow Constraints of the Series Converter

Based on the operating principles shown in Fig. 5.7, the bus voltage equation of the series converter transformer may be generally given by:

$$\begin{bmatrix} \mathbf{I}_{ij}^{abc} \\ \mathbf{I}_{se}^{abc} \end{bmatrix} = \begin{bmatrix} \mathbf{Y}se_{PP}^{abc} & \mathbf{Y}se_{PS}^{abc} \\ \mathbf{Y}se_{SP}^{abc} & \mathbf{Y}se_{SS}^{abc} \end{bmatrix} \begin{bmatrix} \mathbf{V}_i^{abc} - \mathbf{V}_j^{abc} \\ \mathbf{V}_{se}^{abc} \end{bmatrix} \quad (5.101)$$

where the series transformer may consist of three separate single-phase units or three single-phase units with the secondary sides being delta-connected. For the former, the submatrices are similar to type 1 transformer in Table 5.4 while for the later, the submatrices are similar to type 3 transformer in Table 5.4. $\mathbf{I}_{ij}^{abc}$ and $\mathbf{I}_{se}^{abc}$ are the current vectors of the primary and secondary windings of the series converter transformer, respectively and given by:

$$\mathbf{I}_{ij}^{abc} = [I_{ij}^a, I_{ij}^b, I_{ij}^c]^T \quad (5.102)$$

$$\mathbf{I}_{se}^{abc} = [I_{se}^a, I_{se}^b, I_{se}^c]^T \quad (5.103)$$

The voltage vectors $\mathbf{V}_j^{abc}$ and $\mathbf{V}_{se}^{abc}$ are:

$$\mathbf{V}_j^{abc} = [V_j^a, V_j^b, V_j^c]^T \quad (5.104)$$

$$\mathbf{V}_{se}^{abc} = [V_{se}^a, V_{se}^b, V_{se}^c]^T \quad (5.105)$$

The active and reactive power flows of the primary side of the series converter transformer leaving bus $i$ are:

## 5.4 UPFC Modeling in Three-Phase Newton Power Flow in Polar Coordinates

$$P_{ij}^p = \text{Re}[V_i^p (I_{ij}^p)^*]$$
$$= \sum_{m=a,b,c} V_i^p V_i^m (Gse_{PP}^{pm} \cos(\theta_i^p - \theta_i^m) + Bse_{PP}^{pm} \sin(\theta_i^p - \theta_i^m))$$
$$- \sum_{m=a,b,c} V_i^p V_j^m (Gse_{PP}^{pm} \cos(\theta_i^p - \theta_j^m) + Bse_{PP}^{pm} \sin(\theta_i^p - \theta_j^m)) \quad (5.106)$$
$$+ \sum_{m=a,b,c} V_i^p V_{se}^m (Gse_{PS}^{pm} \cos(\theta_i^p - \theta_{se}^m) + Bse_{PS}^{pm} \sin(\theta_i^p - \theta_{se}^m))$$

$$Q_{ij}^p = \text{Im}[V_i^p (I_{ij}^p)^*]$$
$$= \sum_{m=a,b,c} V_i^p V_i^m (Gse_{PP}^{pm} \sin(\theta_i^p - \theta_i^m) - Bse_{PP}^{pm} \cos(\theta_i^p - \theta_i^m))$$
$$- \sum_{m=a,b,c} V_i^p V_j^m (Gse_{PP}^{pm} \sin(\theta_i^p - \theta_j^m) - Bse_{PP}^{pm} \cos(\theta_i^p - \theta_j^m)) \quad (5.107)$$
$$+ \sum_{m=a,b,c} V_i^p V_{se}^m (Gsh_{PS}^{pm} \sin(\theta_i^p - \theta_{se}^m) - Bsh_{PS}^{pm} \cos(\theta_i^p - \theta_{se}^m))$$

where $p, m = a, b, c$. $Gse_{PP}^{pm} + jBse_{PP}^{pm} \in \mathbf{Y}se_{PP}^{abc}$, and $Gse_{PS}^{pm} + jBse_{PS}^{pm} \in \mathbf{Y}se_{PS}^{abc}$.

The active and reactive power flows of the primary side of the series converter transformer leaving bus $j$ are:

$$P_{ji}^p = \text{Re}[V_j^p (I_{ji}^p)^*]$$
$$= \sum_{m=a,b,c} V_j^p V_j^m (Gse_{PP}^{pm} \cos(\theta_j^p - \theta_j^m) + Bse_{PP}^{pm} \sin(\theta_j^p - \theta_j^m))$$
$$- \sum_{m=a,b,c} V_j^p V_i^m (Gse_{PP}^{pm} \cos(\theta_j^p - \theta_i^m) + Bse_{PP}^{pm} \sin(\theta_j^p - \theta_i^m)) \quad (5.108)$$
$$- \sum_{m=a,b,c} V_j^p V_{se}^m (Gse_{PS}^{pm} \cos(\theta_j^p - \theta_{se}^m) + Bse_{PS}^{pm} \sin(\theta_j^p - \theta_{se}^m))$$

$$Q_{ji}^p = \text{Im}[V_j^p (I_{ji}^p)^*]$$
$$= \sum_{m=a,b,c} V_j^p V_j^m (Gse_{PP}^{pm} \sin(\theta_j^p - \theta_j^m) - Bse_{PP}^{pm} \cos(\theta_j^p - \theta_j^m))$$
$$- \sum_{m=a,b,c} V_j^p V_i^m (Gse_{PP}^{pm} \sin(\theta_j^p - \theta_i^m) - Bse_{PP}^{pm} \cos(\theta_j^p - \theta_i^m)) \quad (5.109)$$
$$- \sum_{m=a,b,c} V_j^p V_{se}^m (Gsh_{PS}^{pm} \sin(\theta_j^p - \theta_{se}^m) - Bsh_{PS}^{pm} \cos(\theta_j^p - \theta_{se}^m))$$

Similarly, the active and reactive power flows at the secondary side of the series converter transformer are given by:

$$P_{se}^p = \text{Re}[V_{se}^p (I_{se}^p)^*]$$

$$\sum_{m=a,b,c} V_{se}^p V_{se}^m (Gse_{SS}^{pm} \cos(\theta_{se}^p - \theta_{se}^m) + Bse_{SS}^{pm} \sin(\theta_{se}^p - \theta_{se}^m))$$

$$+ \sum_{m=a,b,c} V_{se}^p V_i^m (Gse_{SP}^{pm} \cos(\theta_{se}^p - \theta_i^m) + Bse_{SP}^{pm} \sin(\theta_{se}^p - \theta_i^m)) \quad (5.110)$$

$$- \sum_{m=a,b,c} V_{se}^p V_j^m (Gse_{SP}^{pm} \cos(\theta_{se}^p - \theta_j^m) + Bse_{SP}^{pm} \sin(\theta_{se}^p - \theta_j^m))$$

$$Q_{se}^p = \text{Im}[V_{sh}^p (I_{sh}^p)^*]$$

$$\sum_{m=a,b,c} V_{se}^p V_{se}^m (Gse_{SS}^{pm} \sin(\theta_{se}^p - \theta_{se}^m) - Bse_{SS}^{pm} \cos(\theta_{se}^p - \theta_{se}^m))$$

$$+ \sum_{m=a,b,c} V_{se}^p V_i^m (Gse_{SP}^{pm} \sin(\theta_{se}^p - \theta_i^m) - Bse_{SP}^{pm} \cos(\theta_{se}^p - \theta_i^m)) \quad (5.111)$$

$$- \sum_{m=a,b,c} V_{se}^p V_j^m (Gse_{SP}^{pm} \sin(\theta_{se}^p - \theta_j^m) - Bse_{SP}^{pm} \cos(\theta_{se}^p - \theta_j^m))$$

### 5.4.3.3 Active Power Balance of the UPFC

The active power exchange among the converters via the DC link should be balanced at any instant, which is described by:

$$P_\Sigma = \sum_{p=a,b,c} P_{sh}^p + \sum_{p=a,b,c} P_{se}^p + Ploss = 0 \quad (5.112)$$

where $P_{sh}^p$ and $P_{se}^p$ are defined by (5.99) and (5.110), respectively. *Ploss* represents losses in converter circuits. Each converter losses consist of two terms. The first term is proportional to its ac terminal current squared, and the second term is a constant. The former may be represented by an equivalent resistance, and can be included into its coupling transformer impedance. The latter of all the converters can be combined and represented by *Ploss*.

### 5.4.4 Symmetrical Components Control Model for Three-Phase UPFC

The symmetrical components control assumes that both the three-phase shunt and series converter injects three-phase balanced voltages. Basically such a control is applicable to a three-phase UPFC with any series and shunt transformer connection types. In principle, the control is identical to that of the positive-sequence control of the UPFC in conventional positive-sequence power flow analysis when the three-phase network and loads are balanced.

## 5.4 UPFC Modeling in Three-Phase Newton Power Flow in Polar Coordinates

### 5.4.4.1 PQ Flow Control by the Series Converter

The injected three-phase series voltages $V_{se}^p$ ($p=a, b, c$) should be balanced, this means that the three-phase voltages are identical in magnitude while their phase angles have 120° displacement between them. The balanced three-phase voltages may be represented by the constraints as follows:

$$\Delta Vse_{Re}^1 = \text{Re}(V_{se}^a - V_{se}^b e^{j120°}) = 0 \tag{5.113}$$

$$\Delta Vse_{Im}^1 = \text{Im}(V_{se}^a - V_{se}^b e^{j120°}) = 0 \tag{5.114}$$

$$\Delta Vse_{Re}^2 = \text{Re}(V_{se}^a - V_{se}^c e^{j240°}) = 0 \tag{5.115}$$

$$\Delta Vse_{Im}^2 = \text{Im}(V_{se}^a - V_{se}^c e^{j240°}) = 0 \tag{5.116}$$

For the three-phase UPFC, the series converter can be used to control the total three-phase active and reactive power flows of the transmission line. The control constraints are given by:

$$P_{ji}^\Sigma - Pspec_{ji}^\Sigma = 0 \tag{5.117}$$

$$Q_{ji}^\Sigma - Qspec_{ji}^\Sigma = 0 \tag{5.118}$$

where $Pspec_{ji}^\Sigma$ and $Qspec_{ji}^\Sigma$ are the specified total three-phase active and reactive power flow control references, respectively. $P_{ji}^\Sigma$ and $Q_{ji}^\Sigma$ are the actual total three-phase active and reactive power flows, respectively and given by:

$$P_{ji}^\Sigma = \sum_{p=a,b,c} P_{ji}^p \tag{5.119}$$

$$Q_{ji}^\Sigma = \sum_{p=a,b,c} Q_{ji}^p \tag{5.120}$$

where $P_{ji}^p$ and $Q_{ji}^p$ are defined by (5.108) and (5.109), respectively.

### 5.4.4.2 Voltage Control by the Shunt Converter

For the symmetrical components control model, it is assumed that the injected three-phase shunt voltages $V_{sh}^p$ ($p=a, b, c$) should be balanced. The balanced three-phase voltages may be represented by the following constraints:

$$\Delta Vsh_{Re}^1 = \text{Re}(V_{sh}^a - V_{sh}^b e^{j120°}) = 0 \tag{5.121}$$

$$\Delta Vsh_{Im}^1 = \text{Im}(V_{sh}^a - V_{sh}^b e^{j120°}) = 0 \tag{5.122}$$

$$\Delta Vsh_{Re}^2 = \text{Re}(V_{sh}^a - V_{sh}^c e^{j240°}) = 0 \tag{5.123}$$

$$\Delta Vsh_{Im}^2 = \text{Im}(V_{sh}^a - V_{sh}^c e^{j240°}) = 0 \quad (5.124)$$

For the three-phase UPFC, it may be used to control the positive-sequence voltage at bus $i$:

$$V_i^1 - Vspec_i^1 = 0 \quad (5.125)$$

where $V_i^1$ is the actual positive-sequence voltage at bus $i$ and can be represented by phase voltages $V_i^a$, $V_i^b$ and $V_i^c$ while $Vspec_i^1$ is the positive-sequence voltage control reference.

### 5.4.4.3 Transformer Models

For this control model of the UPFC, the shunt converter transformer may be of any of the connection types shown in Table 5.4 while the secondary sides of the three single-phase transformers may be delta-connected or may be separated as shown in Fig. 5.7.

### 5.4.4.4 Modeling of Three-Phase UPFC in Newton Power Flow

Basically, the three-phase UPFC has twelve operating and control constraints (5.112) – (5.125). In addition, the state variables such as $V_{se}^p$ and $V_{sh}^p$ may be constrained by the converter voltage ratings, and the currents through the converter should be within its current ratings.

For the symmetrical components control model of the UPFC, the Newton equation including six power mismatches at buses $i, j$ and twelve operating and control mismatches may be written as:

$$\mathbf{J}\Delta\mathbf{X} = -\mathbf{F}(\mathbf{X}) \quad (5.126)$$

where

$\Delta\mathbf{X}$ - the incremental vector of state variables, and $\Delta\mathbf{X} = [\Delta\mathbf{X}_{upfc}, \Delta\mathbf{X}_{sys}]^T$

$\Delta\mathbf{X}_{sys} = [\Delta\theta_i^p, \Delta V_i^p, \Delta\theta_j^p, \Delta V_j^p]^T$ - the incremental vector of bus voltage angles and magnitudes.

$\Delta\mathbf{X}_{upfc} = [\Delta\theta_{se}^p, \Delta V_{se}^p, \Delta\theta_{sh}^p, \Delta V_{sh}^p]^T$ - the incremental vector of the UPFC state variables.

$\mathbf{F}(\mathbf{X}) = [\mathbf{F}_{upfc}, \mathbf{F}_{sys}]^T$ - bus power and the UPFC operating and control mismatch vector.

$\mathbf{F}_{sys} = [\Delta P_i^p, \Delta Q_i^p, \Delta P_j^p, \Delta Q_j^p]^T$ - power mismatch vector.

$\mathbf{F}_{upfc} = [P_\Sigma, V_i^1 - Vspec_i^1, \Delta Vsh_{Re}^1, \Delta Vsh_{Im}^1, \Delta Vsh_{Re}^2, \Delta Vsh_{Im}^2,$
$P_{ji}^\Sigma - Pspec_{ji}^\Sigma, Q_{ji}^\Sigma - Qspec_{ji}^\Sigma, \Delta Vse_{Re}^1, \Delta Vse_{Im}^1, \Delta Vse_{Re}^2, \Delta Vse_{Im}^2]^T$ - the UPFC operating and control mismatches

$$J = \frac{\partial F(X)}{\partial X}$$ - System Jacobian matrix.

### 5.4.5 General Three-Phase Control Model for Three-Phase UPFC

For the general control model of the three-phase UPFC, the series converter can be used to control the six independent active and reactive power flows of the transmission line while the shunt converter can be used to control the three-phase voltages at the shunt bus.

#### 5.4.5.1 PQ Flow Control by the Series Converter

The six independent active and reactive power control constraints of the series control of the UPFC are:

$$P_{ji}^p - Pspec_{ji}^p = 0 \quad (p = a, b, c) \tag{5.127}$$

$$Q_{ji}^p - Qspec_{ji}^p = 0 \quad (p = a, b, c) \tag{5.128}$$

where $Pspec_{ji}^p$, $Qspec_{ji}^p$ are the specified active and reactive power flow control references of phase $p$.

#### 5.4.5.2 Voltage Control by the Shunt Converter

For the general control model of the three-phase UPFC, it may be used to control three-phase voltages at bus $i$. The control constraints are given by:

$$V_i^p - Vspec_i^p = 0 \quad (p = a, b, c) \tag{5.129}$$

where $V_i^p$ is the actual phase voltage at bus $i$ while $Vspec_i^p$ is the phase voltage control reference.

#### 5.4.5.3 Operating Constraints of the Shunt Transformer

In this control model, it is assumed that the zero-sequence voltage component at the secondary side of the shunt transformer is zero:

$$\text{Re}(\sum_{p=a,b,c} V_{sh}^p) = \sum_{p=a,b,c} V_{sh}^p \cos \theta_{sh}^p = 0 \tag{5.130}$$

$$\text{Im}(\sum_{p=a,b,c} V_{sh}^p) = \sum_{p=a,b,c} V_{sh}^p \sin \theta_{sh}^p = 0 \tag{5.131}$$

### 5.4.5.4 Transformer Models

For this control model of the UPFC, the shunt converter transformer may be of any of the connection types as shown in Table 5.4 while the series converter is connected with the system via three separate single-phase transformers where the secondary sides of the transformers are not connected.

### 5.4.5.5 Modeling of Three-Phase UPFC in Newton Power Flow

For the general three-phase control model of the UPFC, the Newton equation including six power mismatches at buses $i, j$ and twelve operating and control mismatches (5.112), (5.127)–(5.131) may be written as:

$$\mathbf{J}\Delta \mathbf{X} = -\mathbf{F}(\mathbf{X}) \tag{5.132}$$

where

$\Delta \mathbf{X}$ - the incremental vector of state variables, and $\Delta \mathbf{X} = [\Delta \mathbf{X}_{upfc}, \Delta \mathbf{X}_{sys}]^T$

$\Delta \mathbf{X}_{sys} = [\Delta \theta_i^p, \Delta V_i^p, \Delta \theta_j^p, \Delta V_j^p]^\mathbf{T}$ - the incremental vector of bus voltage angles and magnitudes.

$\Delta \mathbf{X}_{upfc} = [\Delta \theta_{se}^p, \Delta V_{se}^p, \Delta \theta_{sh}^p, \Delta V_{sh}^p]^\mathbf{T}$ - the incremental vector of the UPFC state variables .

$\mathbf{F}(\mathbf{X}) = [\mathbf{F}_{upfc}, \mathbf{F}_{sys}]^T$ - bus power and the UPFC operating and control mismatch vector.

$\mathbf{F}_{sys} = [\Delta P_i^p, \Delta Q_i^p, \Delta P_j^p, \Delta Q_j^p]^T$ - power mismatch vector.

$\mathbf{F}_{upfc} = [P_\Sigma, V_i^p - Vspec_i^p, P_{ji}^p - Pspec_{ji}^p, Q_{ji}^p - Qspec_{ji}^p, \text{Re}(\sum_{p=a,b,c} V_{sh}^p), \text{Im}(\sum_{p=a,b,c} V_{sh}^p)]^T$

- the UPFC operating and control mismatches

$\mathbf{J} = \dfrac{\partial \mathbf{F}(\mathbf{X})}{\partial \mathbf{X}}$ - System Jacobian matrix.

### 5.4.6 Hybrid Control Model for Three-Phase UPFC

In contrast to the general control model presented in the previous section, the hybrid control model assumes:

- the positive-sequence voltage at bus $i$ and the active and reactive power flows of each phase of the transmission line are controlled;
- the shunt converter injects three-phase balanced voltages only.

## 5.4 UPFC Modeling in Three-Phase Newton Power Flow in Polar Coordinates

### 5.4.6.1 PQ Flow Control by the Series Converter

For the hybrid control model, the phase series voltages $V_{se}^p$ ($p=a, b, c$) is injected to control the active and reactive power flows of that phase. The control constraints are given by:

$$P_{ji}^p - Pspec_{ji}^p = 0 \tag{5.133}$$

$$Q_{ji}^p - Qspec_{ji}^p = 0 \tag{5.134}$$

where $Pspec_{ji}^p$ and $Qspec_{ji}^p$ ($p=a, b, c$) are the specified phase active and reactive power flow control references, respectively. $P_{ji}^p$ and $Q_{ji}^p$ ($p=a, b, c$) are the actual phase active and reactive power flows, respectively.

### 5.4.6.2 Voltage Control by the Shunt Converter

For the hybrid control model, the injected three-phase shunt voltages $V_{sh}^p$ ($p=a, b, c$) should be balanced. We have:

$$\Delta Vsh_{Re}^1 = \text{Re}(V_{sh}^a - V_{sh}^b e^{j120°}) = 0 \tag{5.135}$$

$$\Delta Vsh_{Im}^1 = \text{Im}(V_{sh}^a - V_{sh}^b e^{j120°}) = 0 \tag{5.136}$$

$$\Delta Vsh_{Re}^2 = \text{Re}(V_{sh}^a - V_{sh}^c e^{j240°}) = 0 \tag{5.137}$$

$$\Delta Vsh_{Im}^2 = \text{Im}(V_{sh}^a - V_{sh}^c e^{j240°}) = 0 \tag{5.138}$$

Assuming that the shunt converter is used to control the positive-sequence voltage at bus $i$, the control constraint is given by:

$$V_i^1 - Vspec_i^1 = 0 \tag{5.139}$$

where $V_i^1$ is the actual positive-sequence voltage at bus $i$ while $Vspec_i^1$ is the positive-sequence voltage control reference.

### 5.4.6.3 Transformer Models

For this control model of the UPFC, the shunt converter transformer may be of any of the connection types as shown in Table 5.4 while the series converter is connected with the system via three separate single-phase transformers where the secondary sides of the transformers are not connected.

### 5.4.6.4 Modeling of Three-Phase UPFC in the Newton Power Flow

Basically, the hybrid UPFC control model has eleven control constraints given by (5.133)-(5.139), and the power balance constraint given by (5.112).

For the hybrid UPFC control model, the Newton equation including six power mismatches at buses $i, j$ and twelve operating and control mismatches may be written as:

$$\mathbf{J}\Delta \mathbf{X} = -\mathbf{F}(\mathbf{X}) \qquad (5.140)$$

where

$\Delta \mathbf{X}$ - the incremental vector of state variables, and $\Delta \mathbf{X} = [\Delta \mathbf{X}_{upfc}, \Delta \mathbf{X}_{sys}]^T$

$\Delta \mathbf{X}_{sys} = [\Delta \theta_i^p, \Delta V_i^p, \Delta \theta_j^p, \Delta V_j^p]^T$ - the incremental vector of bus voltage angles and magnitudes.

$\Delta \mathbf{X}_{upfc} = [\Delta \theta_{se}^p, \Delta V_{se}^p, \Delta \theta_{sh}^p, \Delta V_{sh}^p]^T$ - the incremental vector of the UPFC state variables .

$\mathbf{F}(\mathbf{X}) = [\mathbf{F}_{upfc}, \mathbf{F}_{sys}]^T$ - bus power and the UPFC operating and control mismatch vector.

$\mathbf{F}_{sys} = [\Delta P_i^p, \Delta Q_i^p, \Delta P_j^p, \Delta Q_j^p]^T$ - power mismatch vector.

$\mathbf{F}_{upfc} = [P_\Sigma, V_i^1 - Vspec_i^1, \Delta Vsh_{Re}^1, \Delta Vsh_{Im}^1, \Delta Vsh_{Re}^2, \Delta Vsh_{Im}^2, P_{ji}^p - Pspec_{ji}^p, Q_{ji}^p - Qspec_{ji}^p,]^T$ - the UPFC operating and control mismatches

$\mathbf{J} = \dfrac{\partial \mathbf{F}(\mathbf{X})}{\partial \mathbf{X}}$ - System Jacobian matrix

### 5.4.7 Numerical Examples

In this section, numerical results are presented for a 5-bus system and the IEEE 118-bus system. The 5 bus three-phase system is shown in Fig. 5.8 in the Appendix of this chapter, while the system parameters are listed in Table 5.11 - Table 5.14. In order to make simulations on the IEEE 118-bus system realistic, a Delta/Wye-G transformer is inserted between each generator and its terminal bus.

In the following tests, a convergence tolerance of 1.0e-12 p.u. (or 1.0e-10 MW/MVAr) for maximal absolute bus power mismatches and power flow control mismatches is used. In order to simplify the following presentation, the Symmetrical Components Control Model proposed in section 5.4.4 is referred to Model I, the General Control Model in section 5.4.5 is referred to Model II while the Hybrid Control Model proposed in section 5.4.6 is referred to Model III.

### 5.4.7.1 Results for the 5-Bus System

In order to validate the three-phase control models of the UPFC, two cases are carried out under the balanced network and load condition:

*Case 1:* Well transposed transmission lines and the whole system with balanced load. A UPFC is inserted between the receiving end of line 1-3 and bus 3. Suppose the receiving end bus of line 1-3 is now referred to bus 3'. The

## 5.4 UPFC Modeling in Three-Phase Newton Power Flow in Polar Coordinates

whole system is represented only by the positive-sequence network and load. The power flow is solved by the single-phase positive-sequence power flow.

*Case 2*: Well transposed transmission lines and the whole system with balanced load. A UPFC is inserted between the receiving end of line 1-3 and bus 3. The power flow is solved by the three-phase power flow.

The single-phase power flow control reference of the UPFC is 7.0+j1.6 p.u. while the total three-phase power flow control reference is 21.0+j4.8 p.u. The voltage control reference is 1.0 p.u. The power flow solutions of case 1 and case 2 are shown in Table 5.5 and Table 5.6

**Table 5.5.** Power flow solutions for the balanced 5 bus system by single-phase and three-phase power flow algorithms

| | Case 1 | | | Case 2 | |
|---|---|---|---|---|---|
| Bus No. | $V_i$ (p.u.) | $\theta_i$ (deg) | Bus No. | $V_i^a$ (p.u.) | $\theta_i^a$ (deg) |
| 1 | 1.0107 | -3.02 | 1 | 1.0107 | 26.98 |
| 2 | 1.0196 | -1.43 | 2 | 1.0196 | 28.57 |
| 3 | 1.0000 | 0.56 | 3 | 1.0000 | 30.56 |
| 4 | 1.0450 | 0.00 | 4 | 1.0450 | 0.00 |
| 5 | 1.0610 | 2.33 | 5 | 1.0610 | 2.33 |

**Table 5.6.** UPFC solutions on the 5 bus system by single-phase and three-phase power flow algorithms

| Case 1 | | Case 2 | | | |
|---|---|---|---|---|---|
| Shunt converter | Series converter | Control models | Shunt converter transformer types | Shunt converter | Series converter |
| $\theta_{sh} = 0.66°$ $V_{sh} = 0.9886$ p.u. | $\theta_{se} = 67.78°$ $V_{se} = 0.2982$ p.u. | I, II, III | 1, 2, 4, 6, 7 | $\theta_{sh}^a = 30.66°$ $V_{sh}^a = 0.9886$ p.u. | $\theta_{se}^a = 97.78°$ $V_{se}^a = 0.2982$ p.u. |
| | | I, II, III | 3, 5 | $\theta_{sh}^a = 0.66°$ $V_{sh}^a = 0.9886$ p.u. | $\theta_{se}^a = 97.78°$ $V_{se}^a = 0.2982$ p.u. |
| | | I, II, III | 8, 9 | $\theta_{sh}^a = 60.66°$ $V_{sh}^a = 0.9886$ p.u. | $\theta_{se}^a = 97.78°$ $V_{se}^a = 0.2982$ p.u. |

180    5 Modeling of FACTS in Three-Phase Power Flow and Three-Phase OPF Analysis

From Table 5.5, it can be found that the bus voltages of the two cases are identical except the 30 degree angle shifting of the voltage angles from the three-phase power flow solution caused by the Wye-G/Delta transformers. For case 2, with the different UPFC models and the different UPFC shunt transformer types, the power flow solutions shown in Table 5.5 are the same except that some of the UPFC injected voltages in Table 5.6 have 30 or 60 degree shifting caused by the Wye/Delta and Delta/Wye transformers. The computation results indicate the validity of the UPFC models proposed. The test results shown in Table 5.5 and Table 5.6 imply that positive-sequence representation of a power system is normally sufficient when the system is balanced.

In order to investigate the behavior and control performance of the three UPFC control models proposed, case studies are carried out for the 5-bus system when the network is unbalanced and there is unbalanced load at bus 3. The power flow and voltage control references of the UPFC are the same to those of the balanced case. The system data are given by Appendix while the test results are given by Table 5.7 to Table 5.9. From these tables it can be found:

1. The power flow solutions with the different UPFC control models are not the same when the system is unbalanced. This implies that under unbalanced conditions, three-phase modeling of the system is needed and proper modeling of three-phase UPFC and its controls should be considered.
2. The power flow solutions with the same UPFC control model and the different shunt converter transformer connection types are not the same when the system is unbalanced. This indicates that appropriate modeling of UPFC transformers is needed when the system is unbalanced.

**Table 5.7.** Power flow solutions for the unbalanced 5 bus system with UPFC Model I

| Case No. | 3 | 4 |
|---|---|---|
| Shunt converter transformer type | 3 | 8 |
| Shunt bus | $V_3^a = 0.9961$  $\theta_3^a = 30.35°$ | $V_3^a = 0.9994$  $\theta_3^a = 29.59°$ |
|  | $V_3^b = 1.0106$  $\theta_3^b = -89.82°$ | $V_3^b = 1.0253$  $\theta_3^b = -89.71°$ |
|  | $V_3^c = 0.9933$  $\theta_3^c = 150.17°$ | $V_3^c = 0.9854$  $\theta_3^c = 150.78°$ |
| Shunt converter | $V_{sh}^a = 0.9914$  $\theta_{sh}^a = 0.32°$ | $V_{sh}^a = 0.9915$  $\theta_{sh}^a = 60.30°$ |
|  | $V_{sh}^b = 0.9914$  $\theta_{sh}^b = -119.68°$ | $V_{sh}^b = 0.9915$  $\theta_{sh}^b = -59.70°$ |
|  | $V_{sh}^c = 0.9914$  $\theta_{sh}^c = 120.32°$ | $V_{sh}^c = 0.9915$  $\theta_{sh}^c = 180.30°$ |
| Series converter | $V_{se}^a = 0.2667$  $\theta_{se}^a = 99.45°$ | $V_{se}^a = 0.2667$  $\theta_{se}^a = 99.37°$ |
|  | $V_{se}^b = 0.2667$  $\theta_{se}^b = -20.55°$ | $V_{se}^b = 0.2667$  $\theta_{se}^b = -20.63°$ |
|  | $V_{se}^c = 0.2667$  $\theta_{se}^c = -140.55°$ | $V_{se}^c = 0.2667$  $\theta_{se}^c = -140.63°$ |
| Number of iterations | 6 | 6 |

## 5.4 UPFC Modeling in Three-Phase Newton Power Flow in Polar Coordinates

**Table 5.8.** Power flow solutions for the unbalanced 5 bus system with UPFC Model II

| Case No. | 5 | 6 |
|---|---|---|
| Shunt converter transformer type | 3 | 8 |
| Shunt bus | $V_3^a = 1.0000$ $\theta_3^a = 30.54°$ | $V_3^a = 1.0000$ $\theta_3^a = 32.21°$ |
| | $V_3^b = 1.0000$ $\theta_3^b = -90.03°$ | $V_3^b = 1.0000$ $\theta_3^b = -91.56°$ |
| | $V_3^c = 1.0000$ $\theta_3^c = 150.48°$ | $V_3^c = 1.0000$ $\theta_3^c = 150.54°$ |
| Shunt converter | $V_{sh}^a = 0.9871$ $\theta sh_{sh}^a = 0.19°$ | $V_{sh}^a = 1.0099$ $\theta_{sh}^a = 60.82°$ |
| | $V_{sh}^b = 0.9873$ $\theta_{sh}^b = -119.32°$ | $V_{sh}^b = 0.9850$ $\theta_{sh}^b = -60.68°$ |
| | $V_{sh}^c = 0.9945$ $\theta_{sh}^c = 120.41°$ | $V_{sh}^c = 0.9750$ $\theta_{sh}^c = 181.34°$ |
| Series converter | $V_{se}^a = 0.3029$ $\theta_{se}^a = 93.93°$ | $V_{se}^a = 0.3261$ $\theta_{se}^a = 96.57°$ |
| | $V_{se}^b = 0.3029$ $\theta_{se}^b = -8.23°$ | $V_{se}^b = 0.1885$ $\theta_{se}^b = -8.99°$ |
| | $V_{se}^c = 0.3029$ $\theta_{se}^c = -144.83°$ | $V_{se}^c = 0.3193$ $\theta_{se}^c = -145.09°$ |
| Number of iterations | 6 | 7 |

**Table 5.9.** Power flow solutions for the unbalanced 5 bus system with UPFC Model III

| Case No. | 7 | 8 |
|---|---|---|
| Shunt converter transformer type | 3 | 8 |
| Shunt bus | $V_3^a = 1.0038$ $\theta_3^a = 30.58°$ | $V_3^a = 0.9848$ $\theta_3^a = 31.47°$ |
| | $V_3^b = 0.9978$ $\theta_3^b = -90.18°$ | $V_3^b = 0.9921$ $\theta_3^b = -91.53°$ |
| | $V_3^c = 0.9984$ $\theta_3^c = 150.69°$ | $V_3^c = 1.0240$ $\theta_3^c = 151.26°$ |
| Shunt converter | $V_{sh}^a = 0.9896$ $\theta_{sh}^a = 0.46°$ | $V_{sh}^a = 0.9894$ $\theta_{sh}^a = 60.52°$ |
| | $V_{sh}^b = 0.9896$ $\theta_{sh}^b = -119.54°$ | $V_{sh}^b = 0.9894$ $\theta_{sh}^b = -59.48°$ |
| | $V_{sh}^c = 0.9896$ $\theta_{sh}^c = 120.46°$ | $V_{sh}^c = 0.9894$ $\theta_{sh}^c = 180.52°$ |
| Series converter | $V_{se}^a = 0.3048$ $\theta_{se}^a = 93.44°$ | $V_{se}^a = 0.3108$ $\theta_{se}^a = 97.83°$ |
| | $V_{se}^b = 0.2117$ $\theta_{se}^b = -7.87°$ | $V_{se}^b = 0.1885$ $\theta_{se}^b = -7.21°$ |
| | $V_{se}^c = 0.3188$ $\theta_{se}^c = -144.33°$ | $V_{se}^c = 0.3388$ $\theta_{se}^c = -147.39°$ |
| Number of iterations | 6 | 7 |

### 5.4.7.2 Results for the Modified IEEE 118-Bus System

Further tests are carried out on the modified IEEE 118-bus system, which are as follows:

*Case 9*: Well-transposed transmission lines and the system with unbalanced load at bus 45 with 0.73+$j$0.22 p.u., 0.53+$j$0.22 p.u., 0.23+$j$0.22 p.u. for phase $a$, $b$, $c$ loading, respectively, and unbalanced load at bus 78 with 0.51+$j$0.26 p.u., 0.71+$j$0.26 p.u., 0.91+$j$0.26 p.u. for phase $a$, $b$, $c$ loading, respectively.

*Case 10*: As for case 9, there are two UPFCs installed on the transmission lines 30-38 and 68-81. The control model I is used for the two UPFC.

*Case 11*: Similar to case 10, but the control model II is used for the two UPFC.

*Case 12*: Similar to case 10, but the control model III is used for the two UPFC.

*Case 13*: Similar to case 10, but the control model I is used for the UPFC on line 30-38 and the control model II is used for the UPFC on line 68-81.

*Case 14*: Similar to case 10, but the control model I is used for the UPFC on line 30-38 and the control model III is used for the UPFC on line 68-81.

*Case 15*: Similar to case 10, but the control model II is used for the UPFC on line 30-38 and the control model I is used for the UPFC on line 68-81.

*Case 16*: Similar to case 10, but the control model II is used for the UPFC on line 30-38 and the control model III is used for the UPFC on line 68-81.

*Case 17*: Similar to case 10, but the control model III is used for the UPFC on line 30-38 and the control model I is used for the UPFC on line 68-81.

*Case 18*: Similar to case 10, but the control model III is used for the UPFC on line 30-38 and the control model II is used for the UPFC on line 68-81.

In cases 9 - 18, the active power control references of the UPFC are 140% of the base case power flows, respectively. It is assumed that (a) three separate series transformer units are used for each UPFC; (b) the shunt transformer of the UPFC on line 30-38 is a Wye-G/Delta three-phase transformer while the shunt transformer of the UPFC on line 68-81 is a Delta/Delta three-phase transformer. The test results are shown in Table 5.10. For all the cases above for the modified IEEE 118-bus system, the power flow algorithm can converge within 8 iterations.

**Table 5.10.** Results for the modified IEEE 118-bus system

| Case No. | Number of iterations |
|---|---|
| 9 | 6 |
| 10-18 | 8 |

## 5.5 Three-Phase Newton OPF in Polar Coordinates

Mathematically, as an example the objective function of a three-phase OPF may be minimizing the total operating cost as follows:

$$\text{Minimize } f(x) = \sum_{i}^{Ng}(\alpha_i * Pg_i^2 + \beta_i * Pg_i + \gamma_i) \tag{5.141}$$

while subject to the following constraints:

*Nonlinear equality constraints:*

$$\Delta P_i^p = -Pd_i^p - V_i^p \sum_{j \in i} \sum_{m=a,b,c} V_j^m (G_{ij}^{pm} \cos \theta_{ij}^{pm} + B_{ij}^{pm} \sin \theta_{ij}^{pm}) \tag{5.142}$$

$(p = a, b, c, \text{ and } i=1,2, \ldots, N)$

$$\Delta Q_i^p = -Qd_i^p - V_i^p \sum_{j \in i} \sum_{m=a,b,c} V_j^m (G_{ij}^{pm} \sin \theta_{ij}^{pm} - B_{ij}^{pm} \cos \theta_{ij}^{pm}) \tag{5.143}$$

$(p = a, b, c, \text{ and } i=1,2, \ldots, N)$

$$\begin{aligned}\Delta Pg_i &= -Pg_i \\ &- \sum_{p=a,b,c} \sum_{m=a,b,c} [V_i^p V_i^m (Gg_i^{pm} \cos \theta_i^{pm} + Bg_i^{pm} \sin \theta_i^{pm}) \\ &+ \sum_{p=a,b,c} \sum_{m=a,b,c} [V_i^p E_i^p (Gg_i^{pm} \cos(\theta_i^p - \delta_i^m) + Bg_i^{pm} \sin(\theta_i^p - \delta_i^m))\end{aligned} \tag{5.144}$$

$(i=1,2, \ldots, Ng)$

$$\begin{aligned}\Delta Qg_i &= -Qg_i \\ &- \sum_{p=a,b,c} \sum_{m=a,b,c} [V_i^p V_i^m (Gg_i^{pm} \sin \theta_i^{pm} - Bg_i^{pm} \cos \theta_i^{pm}) \\ &+ \sum_{p=a,b,c} \sum_{m=a,b,c} [V_i^p E_i^p (Gg_i^{pm} \sin(\theta_i^p - \delta_i^m) - Bg_i^{pm} \cos(\theta_i^p - \delta_i^m))\end{aligned} \tag{5.145}$$

$(i=1,2, \ldots, Ng)$

*Inequality constraints:*

$$(P_{ij}^p)^2 + (Q_{ij}^p)^2 \le (S_{ij}^{\max})^2 \tag{5.146}$$

$$P_i^{\min} \le Pg_i \le P_i^{\max} \quad (i = 1, 2, Ng) \tag{5.147}$$

$$Q_i^{\min} \le Qg_i \le Q_i^{\max} \quad (i = 1, 2, Ng) \tag{5.148}$$

$$t_i^{\min} \le t_i \le t_i^{\max} \quad (i = 1, 2, Nt) \tag{5.149}$$

$$V_i^{\min} \le V_i \le V_i^{\max} \quad (i = 1, 2, N) \tag{5.150}$$

where

$\alpha_i, \beta_i, \gamma_i$      coefficients of production cost functions of generator

| | |
|---|---|
| $\Delta P_i^p$ | bus active power mismatch equations |
| $\Delta Q_i^p$ | bus reactive power mismatch equations |
| $P_{ij}^p$ | active line power flow |
| $Q_{ij}^p$ | Reactive line power flow |
| $Pg$ | the vector of active power generation |
| $Qg$ | the vector of reactive power generation |
| $\theta g$ | the vector of generator internal bus voltage angle |
| $Vg$ | the vector of generator internal bus voltage magnitude |
| $\theta$ | the vector of bus voltage angle |
| $V$ | the vector of bus voltage magnitude |
| $t$ | the vector of transformer tap ratios |

$x = [Pg, Qg, \theta g, Vg, t, \theta, V]^T$ is the vector of variables

| | |
|---|---|
| $N$ | the number of system buses excluding the generator internal buses |
| $Ng$ | the number of generators |
| $Nt$ | the number of transformers |

The power flows $P_{ij}^p$ and $Q_{ij}^p$ are given by:

$$P_{ij}^p = V_i^p \sum_{m=a,b,c} V_j^m (G_{ij}^{pm} \cos\theta_{ij}^{pm} + B_{ij}^{pm} \sin\theta_{ij}^{pm}) \quad (p = a, b, c) \quad (5.151)$$

$$Q_{ij}^p = V_i^p \sum_{m=a,b,c} V_j^m (G_{ij}^{pm} \sin\theta_{ij}^{pm} - B_{ij}^{pm} \cos\theta_{ij}^{pm}) \quad (p = a, b, c) \quad (5.152)$$

In the three-phase OPF problem of (5.141)-(5.150), the SSSC and UPFC models with the extra equalities and inequalities, which have been presented in previous sections, can be included. The three-phase OPF problem may be solved by the nonlinear interior point methods that have been applied to the conventional OPF problems. With the integration of distributed generation into power networks, a three-phase OPF tool will be required in the operation, control and planning of power networks to ensure the security and reliability.

## 5.6 Appendix A - Definition of $Yg_i$

$Zg_i$ is the impedance matrix of a synchronous machine, which is given by:

$$Zg_i = T_{120}^{abc} \begin{bmatrix} z_1 & 0 & 0 \\ 0 & z_2 & 0 \\ 0 & 0 & z_0 \end{bmatrix} T_{abc}^{120}$$

$$= \frac{1}{3} \begin{bmatrix} z_0 + z_1 + z_2 & z_0 + az_1 + a^2 z_2 & z_0 + a^2 z_1 + az_2 \\ z_0 + a^2 z_1 + az_2 & z_0 + z_1 + z_2 & z_0 + az_1 + a^2 z_2 \\ z_0 + az_1 + a^2 z_2 & z_0 + a^2 z_1 + az_2 & z_0 + z_1 + z_2 \end{bmatrix} \quad (5.153)$$

where $T_{abc}^{120}$ and $T_{120}^{abc}$ are the transformation matrix of symmetrical components and its inverse matrix, respectively. $z_1$, $z_2$ and $z_0$ are the positive-, negative-, and zero-sequence impedances of a synchronous machine. $a = e^{j2\pi/3}$.

$Yg_i$ is the admittance matrix of a synchronous machine, which is given by:

$$Yg_i = (Zg_i)^{-1} \quad (5.154)$$

## 5.7 Appendix B - 5-Bus Test System

The 5 bus three-phase system is shown in Fig. 5.8. The system parameters are listed in Table 5.11 to Table 5.14.

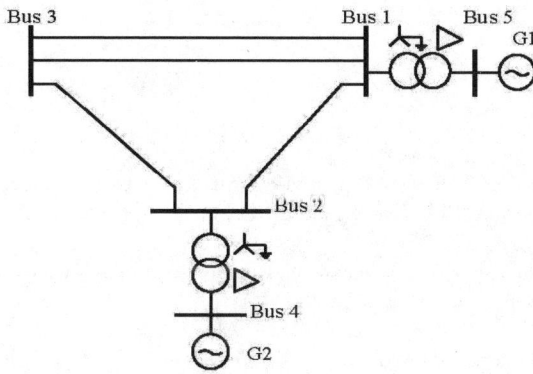

**Fig. 5.8.** 5-bus test system

**Table 5.11.** Generator data in p.u.

| Generator name | Bus No. | Sequence reactance $X_0$ | $X_1$ | $X_2$ | Power P | Voltage V |
|---|---|---|---|---|---|---|
| G1 | 5 | 0.02 | 0.20 | 0.04 | 21.0 | 1.061 |
| G2 | 4 | 0.02 | 0.20 | 0.04 | slack | 1.045 |

**Table 5.12.** Transformer data in p.u.

| Transformer | T1 & T2 |
|---|---|
| Connection | Wye-G/Delta |
| Leakage impedance | 0.0016+j0.015 |
| Primary tap | 1.0 |
| Secondary tap | 1.0 |

**Table 5.13.** Unbalanced line data for line 1-2, line 1-3 and line 2-3

| Series impedance matrix (p.u.) | | |
|---|---|---|
| Phase a | Phase b | Phase c |
| 0.0066 + j 0.0560 | 0.0017 + j 0.0270 | 0.0012 + j 0.0210 |
| | 0.0045 + j 0.0470 | 0.0014 + j 0.0220 |
| | | 0.0062 + j 0.0610 |
| Shunt admittance matrix (p.u.) | | |
| Phase a | Phase b | Phase c |
| j 0.150 | - j 0.030 | - j 0.010 |
| | j 0.250 | - j 0.020 |
| | | j 0.125 |

**Table 5.14.** Load data of the 5-bus system

| | Phase a | Phase b | Phase c |
|---|---|---|---|
| Bus 1 | 0.6 + j 0.3 | 0.6 + j 0.3 | 0.6 + j 0.3 |
| Bus 2 | 2.0 + j 0.8 | 2.0 + j 0.8 | 2.0 + j 0.8 |
| Bus 3 | 6.0 + j 3.0 | 6.3 + j 2.7 | 5.7 + j 3.3 |

# References

[1] El-Abiad AH, Tarsi DC (1967) Load flow study of untransposed EHV networks. In Proceedings of the IEEE Power Industry Computer Application (PICA) Conference, Pittsburgh, USA, pp 337-384
[2] Wasley RG, Shlash MA (1974) Newton-Raphson algorithm for three phase load flow. IEE Proceedings, vol 121, pp 631-638
[3] Birt KA, Graf JJ, McDonald JD, El-Abiad AH (1976) Three phase power flow program. IEEE Transactions on PAS, vol 95, pp 59-65
[4] Arrillaga J, Harker BJ (1978) Fast-decoupled three phase load flow. IEE Proceedings, vol 125, pp 734-740

[5] Laughton MA, Saleh AOM (1980) Unified phase coordinate load flow and fault analysis of polyphase networks. International Journal of Electrical Power and Energy Systems, vol 2, pp 181-192
[6] Arrillaga J, Arnold CP (1983) Computer Modelling of Electrical Power Systems, John Wiley & Sons
[7] Chen BK, Chen MS, Shoults RR, Liang CC (1990) Hybrid three phase load flow. Proc. IEE, pt C, vol 137, 177-185
[8] Allemong, Bennon RJ, Selent PW (1993) Multiphase power flow solutions using EMTP and Newton's method. IEEE Transactions on Power Systems, vol 8, no 4, pp 1455-1462
[9] Zhang XP, Chen H (1994) Asymmetrical three phase load flow study based on symmetrical component theory. IEE Proceedings– Generation, Transmission & Distribution, vol 143, no 3, pp 248-252
[10] Zhang XP (1996) Fast three phase load flow methods. IEEE Transactions on Power Systems, vol 11, no 3, pp 1547-1554
[11] Zhang XP, Chu W, Chen H (1996) Decoupled Asymmetrical Three Phase Load Flow Study by Parallel Processing. IEE Proceedings– Generation, Transmission & Distribution, vol 143, no 1, pp 61-65
[12] Garcia PAN, Pereira JLR, Carneiro S, da Costa VM, Martins N (2000) Three phase power flow calculations using current injection method. IEEE Transactions on Power Systems, vol 15, no 2, pp 508-514
[13] Sun DI, Abe S, Shoults RR, Chen MS, Eichenberger P, Farris D (1980) Calculation of energy losses in distribution system. IEEE Transactions on PAS, vol 90, no 4, pp 1347-1356
[14] Luo GX, Semlyen A (1990) A compensation based power flow method for weakly meshed distribution and transmission networks. IEEE Transactions on Power Systems, vol 5, no 4, pp 1309-1316
[15] Cheng CS, Shirmohammadi D (1995) A three phase power flow method for real-time distribution system analysis. IEEE Transactions on Power Systems, vol 10, no 2, pp 671-679
[16] Zhang F, Cheng CS (1997) A modified Newton method for radial distribution system power flow analysis. IEEE Transactions on Power Systems, vol 12, no 1, pp 389-397
[17] Chen TH, Chen MS, Hwang K-J, Kotas P, Chebli EA (1991) Distribution system power flow analysis – a rigid approach. IEEE Transactions on Power Delivery, vol 6, no 3, pp 1547-1554
[18] Zimmerman RD, Chiang HD (1995) Fast decoupled power flow for unbalanced radial distribution systems. IEEE Transactions on Power Systems, vol 10, no 4, pp 2045-2052
[19] Exposito GA, Ramos ER (1999) Reliable load flow technique for radial distribution networks. IEEE Transactions on Power Systems, vol 14, no 3, pp 1063-1069
[20] Lin W-M, Su Y-S, Chin H-C, Teng JH (1999) Three phase unbalanced distribution with minimal data preparation. IEEE Transactions on Power Systems, vol 14, no 3, pp 1178-1183
[21] Teng JH (2002) A modified Gauss-Seidel algorithm of three phase power flow analysis in distribution networks. International Journal of Electrical Power and Energy Systems, vol 24, pp 97-102
[22] Dillon WE, Chen MS (1972) Transformer modeling in unbalanced three-phase networks. Proceedings of IEEE Summer Meeting, Vancouver, Canada, July 1972

[23] Chen MS, Dillon WE (1974) Power system modeling. Proceedings of the IEEE, vol 62, no 7, pp 901-915
[24] Zhang XP, Xue CF, Godfrey KR (2004) Modelling of the static synchronous series compensator (SSSC) in three phase Newton power flow. IEE Proceedings– Generation, Transmission & Distribution, vol 151, no 4, pp 486-494
[25] Zhang XP (2005) The unified power flow controller models for three-phase power flow analysis. Electrical Engineering, doi: 10.1007/s00202-004-0283-x, online version available, to appear in 2005
[26] Schauder C, Gernhardt M, Stacey E, Lemak T, Gyugyi L, Cease TW, Edris A (1995) Development of a ±100MVar Static Condenser for voltage control of transmission systems. IEEE Transactions on Power Delivery, vol 10, no 3, pp 1486-1493
[27] Gyugyi L, Shauder CD, Sen KK (1997) Static synchronous series compensator: a solid-state approach to the series compensation of transmission lines. IEEE Transactions on Power Delivery, vol 12, no 1, pp 406-413
[28] Gyugyi L, Shauder CD, Williams SL, Rietman TR, Torgerson DR, Edris A (1995) The unified power flow controller: a new approach to power transmission control. IEEE Transactions on Power Delivery, vol 10, no 2, pp 1085-1093

# 6 Steady State Power System Voltage Stability Analysis and Control with FACTS

Voltage stability analysis and control become increasingly important as the systems are being operated closer to their stability limits including voltage stability limits. This is due to the fact that there is lack of network investments and there are large amounts of power transactions across regions for economical reasons in electricity market environments. It has been recognized that a number of the system blackouts including the recent blackouts that happened in North America and Europe are related to voltage instabilities of the systems.

For voltage stability analysis, a number of special techniques such as power flow based methods and dynamic simulations methods have been proposed and have been used in electric utilities [1]-[4]. Power flow based methods, which are considered as steady state analysis methods, include the standard power flow methods [5], continuation power flow methods [6]-[11], optimization methods [18]-[22], modal methods [2], singular decomposition methods [1], etc.

This chapter focuses on the methods for steady state power system voltage stability analysis and control with FACTS. The objectives of this chapter are summarized as follows:

1. to discuss steady state power system voltage stability analysis using continuation power flow techniques,
2. to formulate steady state power system voltage stability problem as an OPF problem,
3. to investigate FACTS control in steady state power system voltage stability analysis,
4. to discuss the transfer capability calculations using continuation power flow and optimal power flow methods,
5. to discuss security constrained OPF for transfer capability limit determination.

## 6.1 Continuation Power Flow Methods for Steady State Voltage Stability Analysis

### 6.1.1 Formulation of Continuation Power Flow

**Predictor Step.** To simulate load change, $Pd_i$ and $Qd_i$, may be represented by:

$$Pd_i = Pd_i^0(1+\lambda * KPd_i) \tag{6.1}$$

$$Qd_i^p = Qd_i^0(1+\lambda * KQd_i) \tag{6.2}$$

where $Pd_i^0$ and $Qd_i^0$ are the base case active and reactive load powers of phase $p$ at bus $i$. $\lambda$ is the loading factor, which characterize the change of the load. The ratio of $KPd_i^p / KQd_i^p$ is constant to maintain a constant power factor.

Similarly, to simulate generation change, $Pg_i$ and $Qg_i$, are represented as functions of $\lambda$ and given by:

$$Pg_i = Pg_i^0(1+\lambda * KPg_i) \tag{6.3}$$

$$Qg_i = Qg_i^0(1+\lambda * KQg_i) \tag{6.4}$$

where $Pg_i^0$ and $Qg_i^0$ are the total active and reactive powers of the generator of the base case. The ratio of $KPg_i / KQg_i$ is constant to maintain constant power factor for a PQ machine. For a PV machine, equation (6.4) is not required. For a PQ machine, when the reactive limit is violated, $Qg_i$ should be kept at the limit and equation (6.4) is also not required.

The nonlinear power flow equations are augmented by an extra variable $\lambda$ as follows:

$$f(x,\lambda) = 0 \tag{6.5}$$

where $f(x,\lambda)$ represents the whole set of power flow mismatch equations.

The predictor step is used to provide an approximate point of the next solution. A prediction of the next solution is made by taking an appropriately sized step in the direction tangent to the solution path.

To solve (6.5), the continuation algorithm with predictor and corrector steps can be used. Linearizing (6.5), we have:

$$df(x,\lambda) = f_x dx + f_\lambda d\lambda = 0 \tag{6.6}$$

In order to solve (6.6), one more equation is needed. If we choose a non-zero magnitude for one of the tangent vector and keep its change as $\pm 1$, one extra equation can be obtained:

$$t_k = \pm 1 \tag{6.7}$$

where $t_k$ is a non-zero element of the tangent vector $dx$.

Combining (6.6) and (6.7), we can get a set of equations where the tangent vector $dx$ and $d\lambda$ are unknown variables:

## 6.1 Continuation Power Flow Methods for Steady State Voltage Stability Analysis

$$\begin{bmatrix} f_x & f_\lambda \\ \hline & e_k & \end{bmatrix} \begin{bmatrix} dx \\ d\lambda \end{bmatrix} = \begin{bmatrix} 0 \\ \pm 1 \end{bmatrix} \tag{6.8}$$

where $e_k$ is a row vector with all elements zero except for $K^{th}$, which equals one. In (6.8), whether +1 or −1 is used depends on how the $K^{th}$ state variable is changing as the solution is being traced. After solving (6.8), the prediction of the next solution may be given by:

$$\begin{bmatrix} x^* \\ \lambda^* \end{bmatrix} = \begin{bmatrix} x \\ \lambda \end{bmatrix} + \sigma \begin{bmatrix} dx \\ d\lambda \end{bmatrix} \tag{6.9}$$

where * denotes the estimated solution of the next step while $\sigma$ is a scalar, which represents the step size.

**Corrector Step.** The corrector step is to solve the augmented Newton power flow equation with the predicted solution in (6.9) as the initial point. In the augmented Newton power flow algorithm an extra equation is included and $\lambda$ is taken as a variable. The augmented Newton power flow equation may be given by:

$$\begin{bmatrix} f(x,\lambda) \\ \hline x_k - \eta \end{bmatrix} = \begin{bmatrix} 0 \\ 0 \end{bmatrix} \tag{6.10}$$

where $\eta$, which is determined by (6.10), is the predicted value of the continuation parameter $x_k$. The determination of the continuation parameter is shown in the following solution procedure.

The corrector equation (6.10), which consists of a set of augmented nonlinear equations, can be solved iteratively by Newton's approach as follows:

$$\begin{bmatrix} f_x & f_\lambda \\ \hline & e_k & \end{bmatrix} \begin{bmatrix} \Delta x \\ \Delta \lambda \end{bmatrix} = -\begin{bmatrix} f(x,\lambda) \\ \hline x_k - \eta \end{bmatrix} \tag{6.11}$$

### 6.1.2 Modeling of Operating Limits of Synchronous Machines

Normally a generator terminal bus is considered as a PV bus, at which the voltage magnitude is specified while the rotor, stator currents and reactive power limits are being monitored according to the capability curve of the generator. The operating limits of a generator that should be satisfied are as follows:

$$I_a \leq I_a^{max} \tag{6.12}$$

$$I_f^{min} \leq I_f \leq I_f^{max} \quad \text{or} \quad E_f^{min} \leq E_f \leq E_f^{max} \tag{6.13}$$

$$Pg^{\min} \leq Pg \leq Pg^{\max} \tag{6.14}$$

$$Qg^{\min}(Pg) \leq Qg \leq Qg^{\max}(Pg) \tag{6.15}$$

where $I_a^{\max}$ is the current limit of the generator stator winding. $I_f^{\max}$ and $I_f^{\min}$ are the maximum and minimum current limits of the generator rotor winding, respectively, while $E_f^{\max}$ and $E_f^{\min}$ are the corresponding excitation voltage limits. $Pg^{\max}$ and $Pg^{\min}$ are the maximum and minimum reactive power limits determined by the capability curve, which are used in continuation power flow analysis. $Qg^{\max}$ and $Qg^{\min}$ are the maximum and minimum reactive power limits determined by the capability curve, which are usually the functions of active power generation.

When one of the inequalities above is violated, the variable is kept at the limit while the voltage control constraint is released. However, when more than one inequality is violated, the technique proposed in [12] can be applied to identify the dominant constraint, and then the dominant constraint is enforced while the other constraints are monitored.

### 6.1.3 Solution Procedure of Continuation Power Flow

The general solution procedure for the Continuation Three-Phase Power Flow is given as follows:

*Step 0:* Run three-phase power flow when $Pd_i$, $Qd_i$, $Pg_i$ and $Qg_i$ are set to $Pd_i^0$, $Qd_i^0$, $Pg_i^0$ and $Qg_i^0$, respectively. The initial point for tracing the PV curves is found.

*Step 1:* Predictor Step
    (a) Solve (6.8) and get the tangent vector $[d\mathbf{x}, d\lambda]^t$;
    (b) Use (6.9) to find the predicted solution of the next step.
    (c) Choose the continuation parameter by evaluating $x_k : t_k = \max(|dx_i|)$.
    (d) Check whether the critical point (maximum loading point) has been passed by evaluating the sign of $d\lambda$. If $d\lambda$ changes its sign from positive to negative, then the critical point has just passed.
    (e) Check whether $\lambda^* < 0$ (Note $0 \leq \lambda \leq \lambda_{\max}$). If this is true, go to Step 3.

*Step 2:* Corrector Step
    (a) According to the chosen continuation parameter to form the aug-

mented equation (6.10);
(b) Form and solve the Newton equation (6.11);
(c) Update the Newton solution and continue the iterations until the corrector step converges to a solution with a given tolerance;
(d) Go to Step 1.

*Step 3:* Output solutions of the PV curves.

For tracing the upper portion of PV curves, $\lambda$ may be taken as the continuation parameter. If, at the predictor step, $d\lambda$ is changed from positive to negative, then the critical point has just passed, and the continuation parameter may be changed from loading factor $\lambda$ to bus voltage magnitude. The bus voltage magnitude with the largest decrease may be chosen as the continuation parameter.

The negative voltage sensitivities, at or near the critical point, with respect to the loading factor $\lambda$ are very useful information in identifying the vulnerable system buses. The bigger the voltage sensitivities, the more vulnerable the system buses are.

The continuation power flow described above can be applied to two situations. The first situation is in the determination of system loadability limit while the second is in the determination of system transfer capability limit. If, in the analysis, voltage limits of load buses and thermal limits of transmission lines are not considered, the system loadability limit or the transfer capability limit is, in principle, corresponding to the system voltage stability limit. However, if voltage limits of load buses and thermal limits of transmission lines are considered, in principle the system loadability limit or the transfer capability limit may be lower than the corresponding system voltage stability limit.

## 6.1.4 Modeling of FACTS-Control in Continuation Power Flow

In principle, similar to the power flow analysis, the models for FACTS-devices such as SVC, TCSC, STATCOM, SSSC, UPFC, IPFC, GUPFC and VSC HVDC are applicable to the continuation power flow for the steady state voltage stability analysis. In addition to the FACTS-devices, other control devices such as explicit model of excitation systems, tap changer control may be considered.

## 6.1.5 Numerical Results

In the following, numerical results are carried out on the IEEE 30-bus system and the IEEE 118-bus system. The single-line diagram of the IEEE 30-bus system is shown in Fig. 2.2, while the single-line diagram of the IEEE 118-bus system is presented in Fig. 6.1.

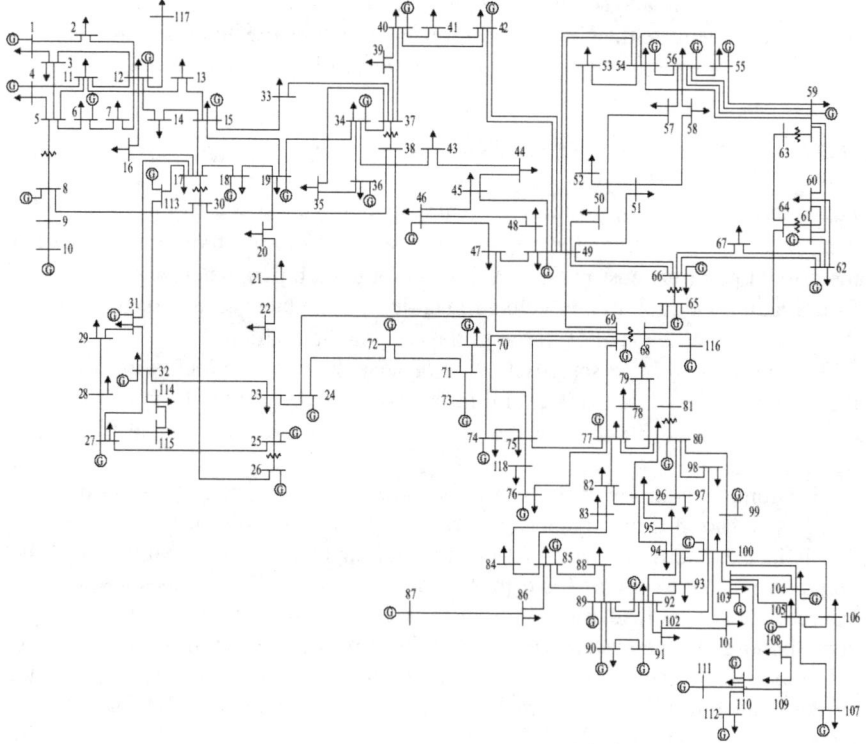

**Fig. 6.1.** IEEE 118-bus system

### 6.1.5.1 System Loadability with FACTS-Devices

Two cases for the IEEE 30-bus system and the IEEE 118-bus system have been studied. The maximum loading factors of these two cases are shown in Table 6.1.

For the IEEE 30-bus system, the candidate buses, at which STATCOMs are installed, are the buses with larger voltage sensitivities with respect to system loading factor $\lambda$ at the voltage collapse point or the nose point. It is found that three largest voltage sensitivities are at buses 28, 29 and 30. A STATCOM is installed at bus 29 of the IEEE 30-bus system. The maximum loading factor is given by Table 6.2, which shows an increase of the maximum voltage stability limit by 33%.

**Table 6.1.** Maximum loading factors

| Case No. | System | Maximum loading factor $\lambda_{max}$ |
|---|---|---|
| Case 1 | IEEE 30-bus system | 2.08 |
| Case 2 | IEEE 118-bus system | 2.25 |

## 6.1 Continuation Power Flow Methods for Steady State Voltage Stability Analysis

**Table 6.2.** Maximum loading factors

| Case No. | System | Maximum loading factor $\lambda_{max}$ | Increase of the maximum loading in percentage using FACTS control |
|---|---|---|---|
| Case 3 | IEEE 30-bus system | 2.77 | 33% |
| Case 4 | IEEE 118-bus system | 2.66 | 18% |

For the IEEE 118-bus system without STATCOM, it is found that the largest voltage sensitivities at the voltage collapse point or the nose point are at buses 38, 43, 44, 45. Four STATCOM are installed at these buses, respectively. The maximum loading factor for the IEEE 118-bus system with 4 STATCOMs is presented in Table 6.1, which shows an increase of the voltage stability limit by 18%.

It has been found that for case 3 and case 4, shunt reactive power control using STATCOM (or SVC) is very effective while series reactive power compensation control using SSSC and series-shunt reactive power compensation using UPFC are not effective.

### 6.1.5.2 Effect of Load Models

Without considering the frequency effect, a general static load model may be given by:

$$Pd_i^0 = Pd_i^{norm}(a_{i0} + a_{i1}V_i + a_{i2}V_i^2) \tag{6.16}$$

$$Qd_i^0 = Qd_i^{norm}(b_{i0} + b_{i1}V_i + b_{i2}V_i^2) \tag{6.17}$$

where subscript $i$ denotes the bus number. $Pd_i^{norm}$ and $Qd_i^{norm}$ are the active and reactive powers at nominal voltage. $a_{i0}$ and $b_{i0}$ represent the constant power components; $a_{i1}$ and $b_{i1}$ represent the constant current components; $a_{i2}$ and $b_{i2}$ represent the constant impedance components. The model in (6.16) and (6.17) is also known as ZIP model where Z represents impedance, I represents current, and P represents power. The parameters in (6.16) and (6.17) should satisfy the following equations:

$$a_{i0} + a_{i1} + a_{i2} = 1 \tag{6.18}$$

$$b_{i0} + b_{i1} + b_{i2} = 1 \tag{6.19}$$

In order to investigate the effects of different load models on voltage stability limits, cases 5-10 for the IEEE 30-bus system, which are presented in Table 6.3. The PV curves of bus 27 are shown in Fig. 6.2-Fig. 6.7, respectively.

**Table 6.3.** Case studies with different load models for the IEEE 30-bus system

| Case No. | Load model parameters |
|---|---|
| Case 5 | $a_{i0} = b_{i0} = 1.0$, $a_{i1} = b_{i1} = 0.0$, $a_{i2} = b_{i2} = 0.0$, PQ load |
| Case 6 | $a_{i0} = b_{i0} = 0.6$, $a_{i1} = b_{i1} = 0.4$, $a_{i2} = b_{i2} = 0.0$ |
| Case 7 | $a_{i0} = b_{i0} = 0.6$, $a_{i1} = b_{i1} = 0.0$, $a_{i2} = b_{i2} = 0.4$ |
| Case 8 | $a_{i0} = b_{i0} = 0.6$, $a_{i1} = b_{i1} = 0.2$, $a_{i2} = b_{i2} = 0.2$ |
| Case 9 | $a_{i0} = b_{i0} = 0.0$, $a_{i1} = b_{i1} = 1.0$, $a_{i2} = b_{i2} = 0.0$, Current load |
| Case 10 | $a_{i0} = b_{i0} = 0.0$, $a_{i1} = b_{i1} = 0.0$, $a_{i2} = b_{i2} = 1.0$, Impedance load |

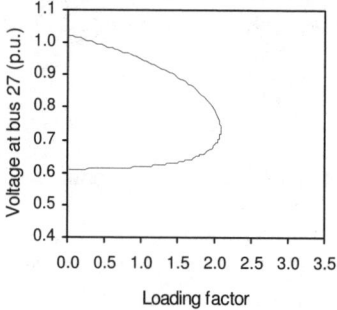

**Fig. 6.2.** PV curve at bus 27 for case 5

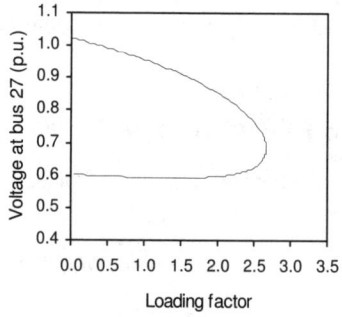

**Fig. 6.3.** PV curve at bus 27 for case 6

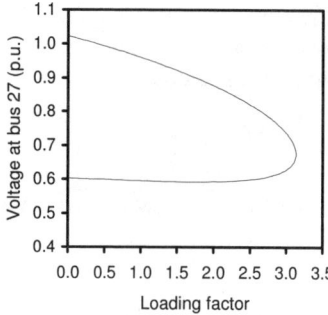

**Fig. 6.4.** PV curve at bus 27 for case 7

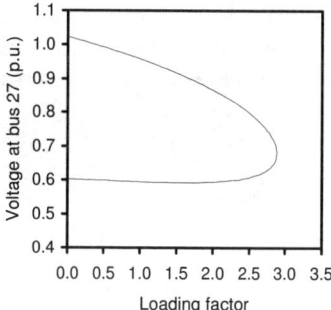

**Fig. 6.5.** PV curve at bus 27 for case 8

## 6.1 Continuation Power Flow Methods for Steady State Voltage Stability Analysis

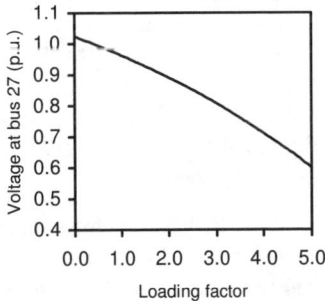

Fig. 6.6. PV curve at bus 27 for case 9

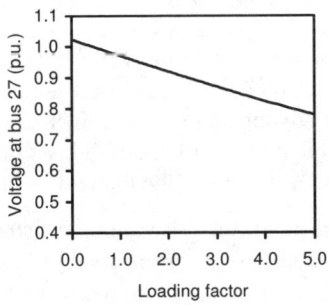

Fig. 6.7. PV curve at bus 27 for case 10

From Figures 6.2 to 6.5 and Table 6.3, it can be seen that for the constant power model, the maximum loading factor is the minimal. It should be pointed out, that for case 9 and case 10, there is no nose point available. From these examples, it is clear that load models play a very important role in voltage stability analysis.

### 6.1.5.3 System Transfer Capability with FACTS-Devices

Two cases for the IEEE 30-bus system without and with FACTS-devices have been studied. In the study, the IEEE 30 bus system was divided into two areas. The two areas are interconnected by intertie lines: 4-12, 6-9, 6-10, and 28-27 while buses 4, 6, 28 belong to the area 1. The power transfer from the area 1 to the area 2 has been investigated.

The maximum loading factors of these two cases are shown in Table 6.4. The system transfer capability limit here is limited by voltage stability limit while the load bus voltage limits and thermal limits of the transmission lines are not considered. As it has been discussed, the candidate buses, at which STATCOMs are installed, are the buses with larger voltage sensitivities with respect to system loading factor $\lambda$ at the voltage collapse point or the nose point. It is found that three largest voltage sensitivities are at buses 27, 29 and 30. A STATCOM is installed at bus 29 of the IEEE 30-bus system. The maximum loading factor is given by Table 6.4, which shows an increase of the maximum voltage stability limit by 35%.

Table 6.4. Maximum loading factors

| Case No. | System | Maixmum loading factor $\lambda_{max}$ |
|---|---|---|
| Case 11 | IEEE 30-bus system without FACTS | 2.15 |
| Case 12 | IEEE 30-bus system with FACTS | 2.90 |

It has been found that for case 11 and case 12, shunt reactive power control using STATCOM (or SVC) is very effective while series reactive power compensation control using SSSC and series-shunt reactive power compensation using UPFC are not effective. The reason is that the effectiveness of series FACTS control relies on its global optimal setting.

The advantage of the continuation method is that operating limits such as thermal, voltage and voltage stability limits can be fully taken into account. However, the disadvantages of the method include:

- Adjustment of generation, transformer tap positions, FACTS-controls, etc. in loadability or transfer capability calculations would be very difficult.
- It is heuristic in nature when voltage and thermal limits are considered in loadability or transfer capability calculations.
- The difficulty of implementation of global coordination. For instance, it is difficult to find optimal settings for series FACTS-devices and coordinate their controls.

In nature, continuation power flow belongs to power flow analysis. In principle, techniques that have been successfully applied to solve power flow problems should be applicable to continuation power flow calculations.

## 6.2 Optimization Methods for Steady State Voltage Stability Analysis

It has been well recognized that optimization methods can be applied to determine the system loadability and transfer capability. However, the definition and formulation of these in literature are not consistent since there are a few possible different formulations of these problems considering different combination of equipment, voltage and thermal constraints. In the following, different formulations for system loadability and transfer capability problems are discussed at first, then numerical examples are given.

### 6.2.1 Optimization Method for Voltage Stability Limit Determination

The maximum voltage stability limit can be formulated as a nonlinear optimization problem. The objective of the problem is to determine the maximum voltage stability limit for a power system considering either the increase of total system load for the case of loadability determination, or the increase of load at a specified region or buses for the case of transfer capability determination while satisfying generator bus voltage constraints and equipment constraints. The optimization problem may be formulated as follows:

$$\text{Maximize: } \lambda \qquad (6.20)$$

subject to:

$$g(x,u,\lambda) = 0 \qquad (6.21)$$

$$h_{min} \le h(x,u) \le h_{max} \qquad (6.22)$$

where
- **u** -  the set of control variables
- **x** -  the set of dependent variables
- **g(x,u)** -  the power flow equations, and control equality constraints for FACTS, transformers, generators, etc
- **h(x,u)** -  the limits of the control variables, operating limits of power system components such as generators, transformers and FACTS-devices, and voltage constraints at load buses

The problem in (6.20) (6.22) can be solved by nonlinear interior point methods [19]-[21]. In the problem in (6.20)-(6.22), the bus load may be represented by:

$$P_d = \lambda P_d^0 \qquad (6.23)$$

$$Q_d = \lambda Q_d^0 \qquad (6.24)$$

where $P_d^0$ and $Q_d^0$ are the base case bus active and reactive load powers, and it is assumed that a constant power factor is maintained. It should be pointed out that $\lambda$ defined here is different from $\lambda$ used in the continuation power flow analysis as shown in (6.1)-(6.4). In other words, $\lambda$ defined in (6.23) and (6.24) is corresponding to $\lambda+1$ in the continuation power flow analysis.

In (6.22), thermal limits of transmission line and voltage constraints at load buses are not included. The maximum voltage stability limit problem in (6.20)-(6.22) is very similar to the continuation power flow problem in section 6.1. The significant difference between the two methods is that the former can be only used to determine the voltage stability limit, while the latter is able to trace the bus PV curves, simulate control sequences and actions, and obtain sensitivity information along the PV curves. The advantages of the former are:

- Coordinated adjustment of control settings of generators, transformers and FACTS-devices, etc.
- Direct consideration of equipment limits and operating limits in the formulation.

## 6.2.2 Optimization Method for Voltage Security Limit Determination

The maximum loadability or transfer capability limit determination can be formulated as a nonlinear optimization problem. The objective of the problem is to determine the maximum system load increase for a power system considering either the increase of total system load for the case of loadability determination, or the increase of load at a specified region or buses for the case of transfer capability determination while satisfying bus voltage constraints and equipment constraints. The optimization problem may be formulated, which is very similar to problem in (6.20)-(6.22) except that now voltage constraints at load buses are also considered. The optimization problem can be solved by nonlinear interior point methods [19]-[21].

### 6.2.3 Optimization Method for Operating Security Limit Determination

The maximum loadability or transfer capability limit determination considering operating security constraints can be formulated as a nonlinear optimization problem. The objective of the problem is to determine the maximum system load increase for a power system considering either the increase of total system load for the case of loadability determination, or the increase of load at a specified region or buses for the case of transfer capability determination while satisfying all bus voltage constraints, thermal constraints of transmission lines, and equipment constraints. The optimization problem may be formulated as follows:

$$\text{Maximize:} \quad \lambda \quad (6.25)$$

subject to:

$$\mathbf{g(x,u,\lambda)} = 0 \quad (6.26)$$

$$\mathbf{h}_{min} \leq \mathbf{h(x,u)} \leq \mathbf{h}_{max} \quad (6.27)$$

where
- $\mathbf{u}$ - the set of control variables
- $\mathbf{x}$ - the set of dependent variables
- $\mathbf{g(x,u)}$ - the power flow equations, and control equality constraints for FACTS, transformers, generators, etc
- $\mathbf{h(x,u)}$ - the limits of the control variables, operating limits of power system components such as generators, transformers and FACTS-devices, voltage constraints at all buses, and thermal limits of transmission lines

The optimization problem in (6.25)-(6.27) can be solved by nonlinear interior point methods [19]-[21]. In (6.27), thermal limits of transmission lines and voltage constraints at all buses are included. In transfer capability calculations, when contingencies should be considered, a security-constrained transfer capability problem can be formulated, which will be discussed in section 6.3.

### 6.2.4 Optimization Method for Power Flow Unsolvability

As the requirements for satisfactory system operation, the region of feasible solutions, satisfying all constraints simultaneously, may not be able to converge. In other words, the power flow or optimal power flow problem is unsolvable. In this situation, the critical question is how to take control actions to restore the solvability of the power flow or optimal power problem. In the following, a robust nonlinear OPF formulation which introduces reactive slack variables and load shedding variables in the unsolvable problem is proposed to handle the infeasibility of a solution. It is formulated as:

$$\text{Minimize: } \begin{aligned} &\sum_{i}^{N} CQr0_i + CQr1_i * Qr_i + CQr2_i * Qr_i^2 \\ &+ \sum_{i}^{N} CQc0_i + CQc1_i * Qc_i + CQc2_i * Qc_i^2 \\ &+ \sum_{i}^{N} CPd0_i + CPd1_i * \Delta Pd_i + CPd2_i * \Delta Pd_i^2) \end{aligned} \quad (6.28)$$

subject to the following constraints:

$$Pg_i - Pd_i + \Delta Pd_i - P_i(V,\theta,T) = DP_i(x) = 0 \quad (6.29)$$
$$(i=1,2,3,N)$$

$$Qg_i - Qd_i + a_i * \Delta Pd_i + Qc_i - Qr_i - Q_i(V,\theta,T) = DQ_i(x) = 0 \quad (6.30)$$
$$(i=1,2,3,N)$$

$$h_j^{min} \le h_j(x) \le h_j^{max} \quad (6.31)$$
$$(j = 1, 2, ..., Nh)$$

where

$x = [V, \theta, T, Pg, Qg, Qr, Qc, \Delta Pd]^T$

$\Delta Pd_i, \Delta Qd_i$ - bus active and reactive load shedding, respectively

$DP_i, DQ_i$ - bus active and reactive power mismatch, respectively

$CPd0_i, CPd1_i, CPd2_i$ - bus load shedding cost coefficient

$a_i$ - constant ratio

$Qr_i, Qc_i$ - bus fictitious inductive and capacitive VAR injections, respectively

$CQr0_i, CQr1_i, CQr2_i$ - bus cost coefficients for fictitious inductive VAR injections

$CQc0_i, CQc1_i, CQc2_i$ - bus cost coefficients for fictitious capacitive VAR injections

$h_j^{min}, h_j^{max}$ - lower and upper limits of inequality

The main idea of the optimization problem for restoring unsolvability is to minimize the cost of control actions in (6.29) while satisfying voltage and thermal constraints and determining the optimal values of reactive power and load shedding controls. If the resulting fictitious inductive and capacitive VAr injections cost coefficients are set to very high values, the solvability of the power flow problem is restored by load shedding only. However, if the load shedding cost coefficients are set to very high values, the solvability is restored by reactive power compensation. For some unsolvable situations, the power flow solution may be restored by combination of reactive power and load shedding controls. The optimal solution of the problem indicates the minimum cost of the control actions should be taken to make the power flow problem solvable.

### 6.2.5 Numerical Examples

Test cases are carried out on the IEEE 30-bus system and IEEE 118-bus system. For all cases tested, the convergence criteria are:

1. Complementary gap $Cgap \leq 5.0e^{-4}$
2. Barrier parameter $\mu \leq 1.0e^{-4}$
3. Maximum mismatch of the Newton equation $\| \mathbf{b} \|_{\infty} \leq 1.0e^{-4}$ p.u.

#### 6.2.5.1 IEEE 30-Bus System Results

In the study, the IEEE 30-bus system was divided into two areas. The two areas are interconnected by tie-lines: 4-12, 6-9, 6-10, and 28-27 while buses 4, 6, 28 belong to the area 1. The transfer from the area 1 to the area 2 has been carried out. The two cases are presented as follows:

*Case 1*: This is a case for transfer capability computation without FACTS-devices.

*Case 2*: This is similar to the case 1 except that there is a SSSC installed on line 2-1.

For case 1, the transfer capabilities considering the voltage stability limit, voltage security limit and operating security limit, respectively, as discussed in previous sections, are shown in Fig.6.8. The corresponding models are referred to model 1 for voltage stability limit, model 2 for voltage security limit and model 3 for operating security limit. In Fig. 6.8, vertical axis shows the transfer capability (TC), which is described by $\lambda$. Comparing the transfer capabilities shown in Fig. 6.8, it can be seen that the voltage stability limit is bigger than the voltage security limit and operating security limit, and the voltage security limit is bigger than the operating security limit.

The transfer capabilities considering the operating security limits for case 1 and 2 are shown in Fig. 6.9. For case 2, the transfer capability has been increased by around 5%.

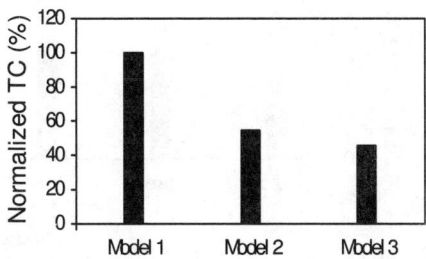

**Fig. 6.8.** The system transfer capabilities of case 1 using different models

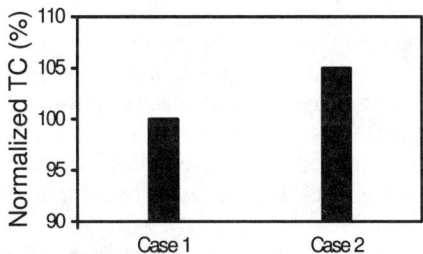

**Fig. 6.9.** The system transfer capabilities for case 1 and case 2

### 6.2.5.2 IEEE 118-Bus System Results

In the study on the IEEE 118-bus system, it is assumed that the whole system includes two areas, which are interconnected by tie-lines: 15-33, 19-34, 30-38, and 23-24 while buses 15, 19, 30, 23 belong to the area 1. The transfer capability from the area 1 to the area 2 has been carried out. Three cases are presented as follows:

*Case* 3: This is a case of the IEEE 118-bus system without FACTS-devices.

*Case* 4: This is similar to the Case 3 except that there is a GUPFC installed at bus 30 and on lines 30-38, 30-8.

*Case* 5: This is similar to the Case 4 except that there is a second UPFC further installed at bus 25 and on line 25-27.

The normalized transfer capabilities for the cases 3-5 are shown in Fig. 6.10. It can be seen from this figure that the transfer capabilities can be increased using FACTS-devices.

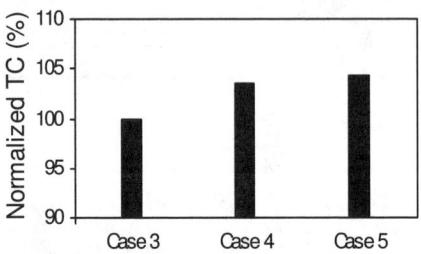

**Fig. 6.10.** The system transfer capabilities for cases 3-5

## 6.3 Security Constrained Optimal Power Flow for Transfer Capability Calculations

It has been well recognized that in the operation of electric power markets, determination of the transfer capabilities of the transmission system is a very important analysis function. Transfer capability of electric power systems is limited by a number of different mechanisms, including thermal, voltage and stability constraints, which is characterized by the so-called Available Transfer Capability (ATC) [13][14]. The comprehensive definition of the transfer capability is referred to [14].

The Transfer Capability (TC) computation methods in literature can be classified into, (a) DC power flow calculation method (or linear method) (b) repeated power flow calculation method (c) continuation power flow method (d) OPF and security-constrained OPF methods. The DC power flow calculation method has been implemented in a commercial software product called MUST [15]. The advantage of such method is its simplicity in terms of formulation and computation. In the method, voltage and voltage stability limits are not considered. It has been recognized that neglecting the reactive power and voltage influence in TC may generate errors that in certain conditions could drive the computation to be wrong or at least give inaccurate results. The repeated power flow calculation method has been proposed [16]. The repeated power flow calculation method is heuristic in nature. The computational effort of the method is significantly higher compared with other methods. The continuation method for TC has been reported and implemented in commercial software products [17]. The advantage of the continuation method is that various operating limits such as thermal, voltage and voltage stability limits can be fully taken into account. The disadvantage of this method is that adjustment of generation, transformer tap positions, FACTS controls, etc in TC calculations would be very difficult if not impossible. Similar to the DC power flow calculation method and repeated power flow calculation method, continuation power flow method could not be used in an integrated contingency-constrained analysis framework. In such a situation, sequential heuristic TC calculations are used instead. Solution of TC by a successive linear programming based OPF has been proposed [18]. Non-linear interior point OPF algorithms for TC calculations have also been proposed [19]-[21]. However, most OPF based TC methods are based on single state calculations. It is recognized that the single state optimization based approaches have difficulty to deal with control actions such as preventive and/or corrective controls. The deficiencies of the current TC computational methods are:

- lack of couplings between base case and contingencies,
- lack of adequate consideration of reactive power/voltage effects and/or voltage stability effects,
- lack of modeling of FACTS-devices in transfer capability determination.

In order to handle the deficiencies of the current TC computation methods, in [22] the TC computation problem has been formulated as a general contingency-constrained optimization problem and has been solved by the nonlinear interior point optimization algorithms. The TC computational method proposed has the following features:

- Considering various operating limits and contingency constraints.
- Incorporating corrective or/and preventive control actions in the united framework.
- Modeling of FACTS-devices.
- Solving simultaneously the base case and contingencies in a united optimization framework.

### 6.3.1 Unified Transfer Capability Computation Method with Security Constraints

A unified transfer capability computation problem with security constraints may be formulated as:

*Objective function:*

$$\max f(\mathbf{y}) = \lambda \text{ or } \min -f(\mathbf{y}) \tag{6.32}$$

*subject to the following constraints:*
*Base case constraints:*

$$\mathbf{g}_0(\mathbf{y}_0) = \mathbf{0} \tag{6.33}$$

$$\mathbf{h}_0^{\min} \leq \mathbf{h}_0(\mathbf{y}_0) \leq \mathbf{h}_0^{\max} \tag{6.34}$$

*Contingency constraints:*

$$\mathbf{g}_i(\mathbf{y}_i) = \mathbf{0} \quad (i = 1, 2, ..., Nc) \tag{6.35}$$

$$\mathbf{h}_i^{\min} \leq \mathbf{h}_i(\mathbf{y}_0) \leq \mathbf{h}_i^{\max} \quad (i = 1, 2, ..., Nc) \tag{6.36}$$

where subscripts 0 and $i$ indicate base case and contingencies, respectively. $Nc$ is the total number of contingencies. $\mathbf{g}_0(\mathbf{y}_0)$ and $\mathbf{h}_0(\mathbf{y}_0)$ are base case equalities and inequalities, respectively. While $\mathbf{g}_i(\mathbf{y}_i)$ and $\mathbf{h}_i(\mathbf{y}_i)$ are equalities and inequalities respectively for contingency $i$. $\mathbf{y} = [\mathbf{x}, \mathbf{u}, \lambda]^T$ is the system variable vector. $\mathbf{u}$ is the control variable vector with preventive control actions. $\lambda$ is a scalar parameter, which represents the loading factor. Modeling of FACTS-devices in power system network analysis can be found in [23]-[25].

Without loss of generality, preventive control actions may be formulated as:

*Preventive control actions:*

$$\mathbf{u}_0 = \mathbf{u}_i \quad (i=1,2,\ldots Nc) \tag{6.37}$$

where $\mathbf{u}_0$, $\mathbf{u}_i$ are base case and contingency control vectors respectively.

The problem in (6.32)-(6.37) is a unified security constrained transfer capability computation problem. In this problem, bus load may be represented by:

$$P_d = \lambda P_d^0 \tag{6.38}$$

$$Q_d = \lambda Q_d^0 \tag{6.39}$$

where $P_d^0$ and $Q_d^0$ are base case bus active and reactive load powers, and it is assumed that a constant power factor is maintained.

## 6.3.2 Solution of Unified Security Constrained Transfer Capability Problem by Nonlinear Interior Point Method

Mathematically, the unified transfer capability computation problem is an optimization problem, which may be solved by nonlinear interior point methods. The nonlinear OPF problem given in (6.32)-(6.37) can be solved by the nonlinear interior point methods [26][27], which include three important achievements in optimization. Those achievements are Fiacco & McCormick's barrier method for optimization with inequalities, Lagrange's method for optimization with equalities and Newton's method for solving nonlinear equations [26]-[28].

By applying Fiacco & McCormick's barrier method, the unified OPF problem (6.32)-(6.37) can be transformed into the following equivalent OPF problem:

*Objective:*

$$\min \left\{ \begin{array}{l} -f(\mathbf{y}) - \mu \sum_{j=1}^{Nh} \ln(sl_{0j}) - \mu \sum_{j=1}^{Nh} \ln(su_{0j}) \\ -\mu \sum_{i=1}^{Nc} \sum_{j=1}^{Nh} \ln(sl_{ij}) - \mu \sum_{i=1}^{Nc} \sum_{j=1}^{Nh} \ln(su_{ij}) \end{array} \right. \tag{6.40}$$

*subject to the following equality constraints:*

## 6.3 Security Constrained Optimal Power Flow for Transfer Capability Calculations

$$g_0(y_0) = 0 \tag{6.41}$$

$$h_0(y_0) - sl_0 - h_0^{min} = 0 \tag{6.42}$$

$$h_0(y_0) + su_0 - h_0^{max} = 0 \tag{6.43}$$

$$g_i(y_i) = 0 \quad (i = 1, 2, \ldots, Nc) \tag{6.44}$$

$$h_i(y_i) - sl_i - h_i^{min} = 0 \quad (i = 1, 2, \ldots, Nc) \tag{6.45}$$

$$h_i(y_i) + su_i - h_i^{max} = 0 \quad (i = 1, 2, \ldots, Nc) \tag{6.46}$$

$$u_0 - u_i = 0 \quad (i = 1, 2, \ldots, Nc) \tag{6.47}$$

where $\mu > 0$, $sl_0 > 0$, $su_0 > 0$, $sl_i > 0$ and $su_i > 0$. $Nh$ is the number of double sided inequalities.

The Lagrangian function for equalities optimization of problem (6.40)-(6.47) is:

$$\begin{aligned}
L = &-f(y) - \mu \sum_{j=1}^{Nh} \ln(sl_{0j}) - \mu \sum_{j=1}^{Nh} \ln(su_{0j}) \\
&- \mu \sum_{i=1}^{Nc}\sum_{j=1}^{Nh} \ln(sl_{ij}) - \mu \sum_{i=1}^{Nc}\sum_{j=1}^{Nh} \ln(su_{ij}) \\
&- \lambda g_0^T g_0(y_0) - \pi l_0^T (h_0(y_0) - sl_0 - h_0^{min}) \\
&- \pi u_0^T (h_0(y_0) + su_0 - h_0^{max}) \\
&- \sum_i^{Nc} \lambda g_i^T g_i(y_i) - \sum_i^{Nc} \pi l_i^T (h_i(y_i) - sl_i - h_i^{min}) \\
&- \sum_i^{Nc} \pi u_i^T (h_i(y_i) + su_i - h_i^{max}) \\
&- \sum_i^{Nc} \lambda u_i^T (u_0 - u_i)
\end{aligned} \tag{6.48}$$

where $\mu > 0$, $\mathbf{sl}_0 > \mathbf{0}$, $\mathbf{su}_0 > \mathbf{0}$, $\mathbf{sl}_i > \mathbf{0}$ and $\mathbf{su}_i > \mathbf{0}$. $\lambda\mathbf{g}_0$, $\pi\mathbf{l}_0$ and $\pi\mathbf{u}_0$ are dual variable vectors to equalities (6.41), (6.42) and (6.43), respectively. $\lambda\mathbf{g}_i$, $\pi\mathbf{l}_i$ and $\pi\mathbf{u}_i$ are dual variable vectors to equalities (6.44), (6.45) and (6.46), respectively. $\lambda\mathbf{u}_i$ is dual variable vector to equalities (6.47), which represent the constraints of the preventive controls. In (6.48), transformer tap ratios are treated as continuous variables.

The Karush-Kuhn-Tucker (KKT) first order conditions for the Lagrangian function of (6.48) are:

$$\nabla_{\mathbf{y}_0} L_\mu = -\nabla_{\mathbf{y}_0} f(\lambda) - \nabla_{\mathbf{y}_0} \mathbf{g}_0(\mathbf{y}_0)^T \lambda\mathbf{g}_0 \\ - \nabla_{\mathbf{y}_0} \mathbf{h}_0(\mathbf{y}_0)^T \pi\mathbf{l}_0 - \nabla_{\mathbf{y}_0} \mathbf{h}_0(\mathbf{y}_0)^T \pi\mathbf{u}_0 \\ - \sum_i^{Nc} \nabla_{\mathbf{y}_0} \mathbf{u}_0^T \lambda\mathbf{u}_i \quad (6.49)$$

$$\nabla_{\lambda\mathbf{g}_0} L_\mu = -\mathbf{g}_0(\mathbf{y}_0) = \mathbf{0} \quad (6.50)$$

$$\nabla_{\pi\mathbf{l}_0} L_\mu = -(\mathbf{h}_0(\mathbf{y}_0) - \mathbf{sl}_0 - \mathbf{h}_0^{min}) \quad (6.51)$$

$$\nabla_{\pi\mathbf{u}_0} L_\mu = -(\mathbf{h}_0(\mathbf{y}_0) + \mathbf{su}_0 - \mathbf{h}_0^{max}) \quad (6.52)$$

$$\nabla_{\mathbf{sl}_0} L_\mu = \mu\mathbf{e} - \mathbf{SL}_0 \Pi\mathbf{L}_0 \quad (6.53)$$

$$\nabla_{\mathbf{su}_0} L_\mu = \mu\mathbf{e} + \mathbf{SU}_0 \Pi\mathbf{U}_0 \quad (6.54)$$

$$\nabla_{\mathbf{y}_i} L_\mu = -\nabla_{\mathbf{y}_i} f(\lambda) - \nabla_{\mathbf{y}_i} \mathbf{g}_i(\mathbf{y}_i)^T \lambda\mathbf{g}_i \\ - \nabla_{\mathbf{y}_i} \mathbf{h}_i(\mathbf{y}_i)^T \pi\mathbf{l}_i - \nabla_{\mathbf{y}_i} \mathbf{h}_i(\mathbf{y}_i)^T \pi\mathbf{u}_i \\ + \sum_i^{Nc} \nabla \mathbf{u}_i^T \lambda\mathbf{u}_i \quad (6.55)$$

$$\nabla_{\lambda\mathbf{g}_i} L_\mu = -\mathbf{g}_i(\mathbf{y}_i) = \mathbf{0} \quad (6.56)$$

$$\nabla_{\pi\mathbf{l}_i} L_\mu = -(\mathbf{h}_i(\mathbf{y}_i) - \mathbf{sl}_i - \mathbf{h}_i^{min}) \quad (6.57)$$

$$\nabla_{\pi\mathbf{u}_i} L_\mu = -(\mathbf{h}_i(\mathbf{y}_i) + \mathbf{su}_i - \mathbf{h}_i^{max}) \quad (6.58)$$

$$\nabla_{\mathbf{sl}_i} L_\mu = \mu\mathbf{e} - \mathbf{SL}_i \Pi\mathbf{L}_i \quad (6.59)$$

$$\nabla_{\mathbf{su}_i} L_\mu = \mu\mathbf{e} + \mathbf{SU}_i \Pi\mathbf{U}_i \quad (6.60)$$

## 6.3 Security Constrained Optimal Power Flow for Transfer Capability Calculations

$$\nabla_{\lambda u_i} L_\mu = -(\mathbf{u}_0 - \mathbf{u}_i) \tag{6.61}$$

$$\nabla_\lambda I'_\mu = -\frac{\partial f(\lambda)}{\partial \lambda} - \nabla_\lambda \mathbf{g}_0(\mathbf{y}_0)^T \mathbf{e} - \nabla_\lambda \mathbf{g}_i(\mathbf{y}_i)^T \mathbf{e} \tag{6.62}$$

where $i = 1, 2, ..., Nc$. $\mathbf{SL}_0 = diag(sl_{0j})$, $\mathbf{SU}_0 = diag(su_{0j})$, $\mathbf{\Pi L}_0 = diag(\pi l_{0j})$, $\mathbf{\Pi U}_0 = diag(\pi u_{0j})$, $\mathbf{SL}_i = diag(sl_{ij})$, $\mathbf{SU}_i = diag(su_{ij})$, $\mathbf{\Pi L}_i = diag(\pi l_{ij})$, $\mathbf{\Pi U}_i = diag(\pi u_{ij})$.

The nonlinear equations (6.49)-(6.62) in polar coordinates can be solved simultaneously. The simultaneous equations can be linearized and expressed in a compact Newton form:

$$\mathbf{A} \Delta \mathbf{x} = -\mathbf{b} \tag{6.63}$$

where $\mathbf{A} = \dfrac{\partial \mathbf{b}}{\partial \mathbf{x}}$ . $\mathbf{x} = [\mathbf{x}_0, \mathbf{x}_i, \lambda \mathbf{u}_i, \lambda]^T$ . $\mathbf{b} = [\mathbf{b}_0, \mathbf{b}_i, \mathbf{bu}_i, b_\lambda]^T$ . $\mathbf{X}_0$ and $\mathbf{X}_i$ are given by:

$$\mathbf{x}_0 = [\mathbf{sl}_0, \mathbf{su}_0, \pi \mathbf{l}_0, \pi \mathbf{u}_0, \mathbf{y}_0, \lambda \mathbf{g}_0]^T \tag{6.64}$$

$$\mathbf{x}_i = [\mathbf{sl}_i, \mathbf{su}_i, \pi \mathbf{l}_i, \pi \mathbf{u}_i, \mathbf{y}_i, \lambda \mathbf{g}_i]^T \tag{6.65}$$

and $\mathbf{b}_0$ $\mathbf{b}_i$, $\mathbf{bu}_i$ and $b_\lambda$ are given by:

$$\mathbf{b}_0 = [\nabla_{\mathbf{sl}_0} L_\mu, \nabla_{\mathbf{su}_0} L_\mu, \nabla_{\pi \mathbf{l}_0} L_\mu, \nabla_{\pi \mathbf{u}_0} L_\mu, \nabla_{\mathbf{y}_0} L_\mu, \nabla_{\lambda \mathbf{g}_0} L_\mu]^T \tag{6.66}$$

$$\mathbf{b}_i = [\nabla_{\mathbf{sl}_i} L_\mu, \nabla_{\mathbf{su}_i} L_\mu, \nabla_{\pi \mathbf{l}_i} L_\mu, \nabla_{\pi \mathbf{u}_i} L_\mu, \nabla_{\mathbf{y}_i} L_\mu, \nabla_{\lambda \mathbf{g}_i} L_\mu]^T \tag{6.67}$$

$$\mathbf{bu}_i = \nabla_{\lambda \mathbf{u}_i} L_\mu \tag{6.68}$$

$$b_\lambda = \nabla_\lambda L_\mu \tag{6.69}$$

The security constrained TC problem can be solved iteratively via the Newton equation in (6.63), and at each iteration the solution can be updated as follows:

$$\mathbf{sl}_0[k+1] = \mathbf{sl}_0[k] + \sigma \alpha_p \Delta \mathbf{sl}_0[k] \tag{6.70}$$

$$\mathbf{su}_0[k+1] = \mathbf{su}_0[k] + \sigma \alpha_p \Delta \mathbf{su}_0[k] \tag{6.71}$$

$$\mathbf{y}_0[k+1] = \mathbf{y}_0[k] + \sigma \alpha_p \Delta \mathbf{y}_0[k] \tag{6.72}$$

$$\pi \mathbf{l}_0[k+1] = \pi \mathbf{l}_0[k] + \sigma \alpha_d \Delta \pi \mathbf{l}_0[k] \tag{6.73}$$

$$\pi \mathbf{u}_0[k+1] = \pi \mathbf{u}_0[k] + \sigma \alpha_d \Delta \pi \mathbf{u}_0[k] \tag{6.74}$$

$$\mathbf{sl}_i[k+1] = \mathbf{sl}_i[k] + \sigma \alpha_p \Delta \mathbf{sl}_i[k] \tag{6.75}$$

$$\mathbf{su}_i[k+1] = \mathbf{su}_i[k] + \sigma\alpha_p \Delta\mathbf{su}_i[k] \tag{6.76}$$

$$\mathbf{y}_i[k+1] = \mathbf{y}_i[k] + \sigma\alpha_p \Delta\mathbf{y}_i[k] \tag{6.77}$$

$$\boldsymbol{\pi}\mathbf{l}_i[k+1] = \boldsymbol{\pi}\mathbf{l}_i[k] + \sigma\alpha_d \Delta\boldsymbol{\pi}\mathbf{l}_i[k] \tag{6.78}$$

$$\boldsymbol{\pi}\mathbf{u}_i[k+1] = \boldsymbol{\pi}\mathbf{u}_i[k] + \sigma\alpha_d \Delta\boldsymbol{\pi}\mathbf{u}_i[k] \tag{6.79}$$

$$\boldsymbol{\lambda}\mathbf{u}_i[k+1] = \boldsymbol{\lambda}\mathbf{u}_i[k] + \sigma\alpha_d \Delta\boldsymbol{\lambda}\mathbf{u}_i[k] \tag{6.80}$$

$$\lambda[k+1] = \lambda[k] + \sigma\alpha_p \Delta\lambda[k] \tag{6.81}$$

where $i = 1, 2, \ldots, Nc$. $k$ is the iteration count. Parameter $\sigma \in [0.995\text{-}0.99995]$. $\alpha_p$ and $\alpha_d$ are the primal and dual step-length parameters, respectively. The step-lengths are determined as follows:

$$\alpha p_0 = \min\left[\min\left(\frac{sl_{0j}}{-\Delta sl_{0j}}\right), \min\left(\frac{su_{0j}}{-\Delta su_{0j}}\right), 1.00\right] \tag{6.82}$$

$$\alpha d_0 = \min\left[\min\left(\frac{\pi l_{0j}}{-\Delta \pi l_{0j}}\right), \min\left(\frac{\pi u_{0j}}{-\Delta \pi u_{0j}}\right), 1.00\right] \tag{6.83}$$

$$\alpha p_i = \min\left[\min\left(\frac{sl_{ij}}{-\Delta sl_{ij}}\right), \min\left(\frac{su_{ij}}{-\Delta su_{ij}}\right), 1.00\right] \tag{6.84}$$

$$\alpha d_i = \min\left[\min\left(\frac{\pi l_{ij}}{-\Delta \pi l_{ij}}\right), \min\left(\frac{\pi u_{ij}}{-\Delta \pi u_{ij}}\right), 1.00\right] \tag{6.85}$$

$$i = 1, 2, \ldots, Nc$$

for those $\Delta sl<0$, $\Delta su<0$, $\Delta \pi l<0$ and $\Delta \pi u>0$. $\alpha_p$ and $\alpha_d$ are determined by:

$$\alpha_p = \min[\alpha p_0, \alpha p_i] \quad (i = 1, 2, \ldots, Nc) \tag{6.86}$$

$$\alpha_d = \min[\alpha d_0, \alpha d_i] \quad (i = 1, 2, \ldots, Nc) \tag{6.87}$$

The Barrier parameter $\mu$ can be evaluated by:

$$\mu = \frac{\beta \times Cgap}{2 \times Nh \times (Nc+1)} \tag{6.88}$$

where $\beta \in [0.01\text{-}0.2]$ and $Cgap$ is the complementary gap for the transfer capability calculation problem with security constraints. It can be determined by:

$$Cgap = (\mathbf{sl}_0)^T \pi \mathbf{l}_0 - (\mathbf{su}_0)^T \pi \mathbf{u}_0$$
$$+ \sum_{i=1}^{Nc} [(\mathbf{sl}_i)^T \pi \mathbf{l}_i - (\mathbf{su}_i)^T \pi \mathbf{u}_i] \qquad (6.89)$$

### 6.3.3 Solution Procedure of the Security Constrained Transfer Capability Problem

The solution procedure of the nonlinear interior point optimization algorithm for the unified security constrained transfer capability problem is summarized as follows:

*Step 0*: Set iteration count $k = 0$, $\mu = \mu 0$, and initialize the optimization solution

*Step 1*: If KKT conditions (6.49)–(6.62) are satisfied and the complementary gap is less than a tolerance, output results. Otherwise go to step 2

*Step 2*: Form and solve Newton equation in (6.63)

*Step 3*: Update Newton solution (6.70)–(6.81)

*Step 4*: Compute complementary gap (6.89)

*Step 5*: Determine barrier parameter (6.88)

*Step 6*: Set $k=k+1$, and go to step 1

### 6.3.4 Numerical Results

Test cases are carried out on the IEEE 30-bus system. The IEEE 30-bus system has 6 generators, 4 OLTC transformers and 37 transmission lines. The single-line diagram of the IEEE 30-bus system is shown in Fig. 2.2. For all cases tested, the convergence criteria are:

1. Complementary gap $Cgap \leq 5.0e^{-4}$

2. Barrier parameter $\mu \leq 1.0e^{-4}$

3. Maximum mismatch of the Newton equation $\| \mathbf{b} \|_\infty \leq 1.0e^{-4}$ p.u.

#### 6.3.4.1 IEEE 30-Bus System Results

In the simulations, it is assumed that all generators except the generator at bus 1 are using preventive control of active power generation while other control resources are using corrective controls. Bus 1 is the slack bus.

In the study, the IEEE 30-bus system was divided into two areas. The two areas are interconnected by tie-lines 4-12, 6-9, 6-10, and 28-27 while buses 4, 6, 28 belong to the area 1. The power transfer from the area 1 to the area 2 has been investigated. The single state cases are presented as follows:

*Case 1*: This is a base case for the transfer capability computation.

*Case 2*: This is similar to Case 1 except that there is an outage of line 5-7.

*Case 3*: This is similar to Case 1 except that there is an outage of line 24-25.

The transfer capabilities of Case 1-3 on the IEEE 30 bus system are shown in Table 6.5.

**Table 6.5.** Transfer Capability Results of Single State Cases

| Case No. | Case 1 | Case 2 | Case 3 |
|---|---|---|---|
| Transmission line outage description | None | Line 5-7 | Line 24-25 |
| $\lambda$ | 1.64 | 1.49 | 1.63 |
| Number of iterations | 14 | 15 | 16 |

The transfer capability results of cases with security constraints on the IEEE 30-bus system are presented as follows:

*Case 4*: This is a case for the transfer capability computation including one $N$-1 contingency with line 5-7 outage.

*Case 5*: This is a case for the transfer capability computation including one $N$-1 contingency with line 24-25 outage.

*Case 6*: This is a case for the transfer capability computation with two contingencies. The first contingency is the outage of line 5-7 while the second one is the outage of line 24-25.

The transfer capabilities of Case 4-6 are shown in Table 6.6. From this Table, it can be seen the CPU time for the transfer capability calculations with security constraints is proportional to the total number of base case and contingencies.

Cases 7 and 8 are presented to show the security constrained transfer capability computation with FACTS. Cases 7 and 8 are corresponding to Cases 4 and 6, respectively except that an UPFC is installed between buses 3 and 4. The test results of Cases 7 and 8 are shown in Table 6.7. The UPFC solutions of Cases 7 and 8 are given by Table 6.8. In the calculations, the UPFC is using corrective controls. The results indicate that UPFC taking corrective control actions can improve the transfer capability effectively.

A further case, case 9, is carried out with 8 contingencies included in the transfer capability computation. The algorithm converges in 15 iterations. The $N$-1 contingencies are outages of lines 2-6, 4-6, 5-7, 6-7, 10-21, 12-15, 12-16 and 24-25, respectively. It is found for this case $\lambda = 1.05$.

It can also be seen that in the above cases, the more contingencies, the less system transfer capability is available.

### 6.3.4.2 Discussion of the Results

From these results on the IEEE 30-bus system, it can be seen:

1. Numerical results demonstrate the feasibility of the proposed unified optimization framework for transfer capability computation with security constraints.
2. The computation framework is general, which can simultaneously take voltage, thermal and voltage stability limits as well as any electricity transaction constraints into consideration.
3. The optimization framework of transfer capability computation with security constraints can be solved by nonlinear interior point methods.
4. FACTS devices can be modeled as corrective control devices in the calculations.
5. In addition, electricity transaction constraints may be taken into consideration.

**Table 6.6.** Transfer Capability Results of Cases with Security Constraints

| Case No. | Case 4 | Case 5 | Case 6 |
|---|---|---|---|
| Transmission line outage description | Base case and one N-1 contingency with line 5-7 outage | Base case and one N-1 contingency with line 24-25 outage | Base case and two N-1 contingencies: with line 24-25 outage and line 5-7 outage |
| $\lambda$ | 1.49 | 1.63 | 1.49 |
| Number of iterations | 14 | 14 | 15 |
| Normalised CPU time | 100% | 100% | 150% |

**Table 6.7.** Transfer Capability Results of Cases with FACTS

| Case No. | Case 7 | Case 8 |
|---|---|---|
| Transmission line outage description | Base case and one N-1 contingency with line 5-7 outage | Base case and two N-1 contingencies with line 24-25 outage and line 5-7 outage, respectively |
| $\lambda$ | 1.75 | 1.64 |
| Number of iterations | 28 | 32 |

**Table 6.8.** UPFC solutions of Cases 7 and 8

| Case 7 | Case 8 |
|---|---|
| Base case:<br>Shunt converter:<br>$Vsh = 0.9786$ p.u., $\theta sh = -10.61°$ | Base case:<br>Shunt converter:<br>$Vsh = 0.9596$ p.u., $\theta sh = -7.54°$ |
| Series converters:<br>$Vse = 0.1829$ p.u., $\theta se = 118.05°$ | Series converters:<br>$Vse = 0.1033$ p.u., $\theta se = 179.54°$ |
| Contingency with line 5-7 outage:<br>Shunt converter:<br>$Vsh = 0.9707$ p.u., $\theta sh = -12.86°$ | Contingency with line 5-7 outage:<br>Shunt converter:<br>$Vsh = 0.9615$ p.u., $\theta sh = -7.51°$ |
| Series converters:<br>$Vse = 0.2503$ p.u., $\theta se = -114.30°$ | Series converters:<br>$Vse = 0.1006$ p.u., $\theta se = -179.67°$ |
| | Contingency with line 24-25 outage:<br>Shunt converter:<br>$Vsh = 0.9685$ p.u., $\theta sh = -9.85°$ |
| | Series converters:<br>$Vse = 0.1280$ p.u., $\theta se = -145.05°$ |

# References

[1] Mansour Y, editor (1993) Suggested Techniques for Voltage Stability Analysis. Publication No 93 TH0620-5WR, IEEE Power Engineering Society
[2] Kundur P (1994) Power System Stability and Control. EPRI Power Engineering Series, McGraw-Hill
[3] Taylor CW (1994) Power System Voltage Stability. EPRI Power Engineering Series, McGraw-Hill
[4] Van Cutsem T, Vournas C (1998) Voltage Stability of Electric Power Systems. Kluwer Academic Publishers
[5] Tinney WF, C.E. Hart (1967) Power flow solution by Newton's method. *IEEE Trans. on Power App. Syst.*, vol 86, no 11, pp1449-1456
[6] Huneault M, Fahmideh-Vodani A, Juman M, Galiana FG (1985) The continuation method in power system optimization: applications to economy security functions. IEEE Transactions on PAS, vol 104, no 1, pp 114-124
[7] Huneault M, Galiana FG (1990) An investigation of the solution to the optimal power flow problem incorporating continuation methods. IEEE Transactions on Power Systems, vol 5, no 1, pp 103-110

[8] Iba K, Suzuki H, Egawa M, Watanabe T (1990) Calculation of critical loading condition with nose curve using homotopy continuation method. IEEE Transactions on Power Systems, vol 5, no 1, pp 103-110

[9] Ajjarapu V, Christy C (1992) The continuation power flow: a tool for steady state voltage stability analysis. IEEE Transactions on Power Systems, vol 7, no 1, pp 416 – 423

[10] Canizares CA, Alvarado FL (1993) Point of collapse and continuation methods for large ac/dc systems. IEEE Transactions on Power Systems, vol 8, no 1, pp 1-8

[11] Chiang HD, Shah KS, Balu N (1995) CPFLOW: a practical tool for tracing power system steady-state stationary behavior due to load and generation variations. IEEE Transactions on Power Systems, vol 10, no 2, pp 623-634

[12] Zhang XP, Handschin E, Yao M (2004) Multi-control functional static synchronous compensator (STATCOM) in power system steady state operations. Journal of Electric Power Systems Research, vol 72, no 3, pp 269-278

[13] Sauer P W (1997) Technical challenges of computing available transfer capability (ATC) in electric power systems. 30th Hawaii International Conference on System Science, Maui, Hawaii

[14] Transmission Transfer Capability Task Force (1996) Available transfer capability definitions and determination. North America Reliability Council, Princeton, New Jersey

[15] Gisin BS, Obessis MV, Mitsche JV (2000) Practical methods for transfer limit analysis in the power industry deregulated environment. IEEE Trans. on Power Systems, vol 15, no 3, pp 955-961

[16] Gravener MH, Nwankpa C (1999) Available transfer capability and first order sensitivity. IEEE Trans. on Power Systems, vol 14, no 2, pp 512-518

[17] Ejebe GC, Tong J, Waight J G, et al (1998) Available transfer capability calculations. IEEE Trans. on Power Systems, vol 13, no 4, pp 1521-1527

[18] Xia F, Meliopoulos APS (1996) A Methodology for probabilistic simultaneous transfer capability analysis. IEEE Trans. on Power Systems, vol 11, no 3, pp 1269-1278

[19] Irisarri GD, Wang X, et al (1997) Maximum loadability for power systems using interior point nonlinear optimization method. IEEE Transactions on Power Systems, vol 12, no 1, pp 162-172

[20] Mello JCO, Melo ACG, Granville S (1997) Simultaneous transfer capability assessment by combining interior point methods and monte carlo simulation. IEEE Trans. on Power Systems, vol 12, no 2, pp 736-742

[21] Zhang XP, Handschin E (2002) Transfer capability computation of power systems with comprehensive modelling of facts controllers. 14th Power System Computation Conference (PSCC), Sevilla, Spain

[22] Zhang XP (2005), (Paper for the Invited Session: Operation of Mega Grids), Transfer capability computation with security constraints. 15th Power System Computation Conference (PSCC), Liege, Belgium

[23] Zhang XP, Handschin E (2001) Advanced implementation of UPFC in a nonlinear interior point OPF. IEE Proceedings - Generation, Transmission and Distribution, vol 148, no 5, pp 489-496

[24] Zhang XP, Handschin E, Yao M (2001) Modeling of the generalized unified power flow controller in a nonlinear interior point OPF. IEEE Trans. on Power Systems, vol 16, no 3, pp 367-373

[25] Zhang XP, Handschin E (2001) Optimal power flow control by converter based FACTS controllers. 7th International Conference on AC-DC Power Transmission, IEE, Savoy Place, London, UK

[26] Granville S (1994) Optimal reactive power dispatch through interior point methods. IEEE Transactions on Power Systems, vol 9, no 1, pp 136-146

[27] Wu YC, Debs A, Marsten RE (1994) A direct nonlinear predictor-corrector primal-dual interior point algorithm for optimal power flows. IEEE Transactions on Power Systems, vol 9, no 2, pp 876-883

[28] El-Bakry AS, Tapia RA, Tsuchiya T, Zhang Y (1996) On the formulation and theory of the newton interior-point method for nonlinear programming. Journal of Optimization Theory and Applications, vol 89, no 3, pp 507-541

# 7 Steady State Voltage Stability of Unbalanced Three-Phase Power Systems

This chapter discusses the recent developments in steady state unbalanced three phase voltage stability analysis and control with FACTS. The objectives of this chapter are:

1. to review steady state voltage stability analysis methods in unbalanced three-phase power systems;
2. to introduce the continuation three-phase power flow technique that can be used for steady state unbalanced three-phase voltage stability analysis;
3. to examine the PV curves of unbalanced three-phase power systems;
4. to reveal the interesting phenomena of voltage stability of unbalanced three-phase power systems;
5. to investigate the impact of FACTS controls on voltage stability limit of unbalanced three-phase power systems.

## 7.1 Steady State Unbalanced Three-Phase Power System Voltage Stability

Voltage stability has been recognized as a very important issue for operating power systems when the continuous load increase along with economic and environmental constraints has led to systems to operate close to their limits including voltage stability limit. In the past, various methodologies have been proposed for voltage stability analysis [1]-[4]. Among the voltage stability analysis methods, the continuation power flow methods have been considered as one of the useful tools [5]-[11]. However, in the literature only the application of the continuation power flow methods in voltage stability analysis of positive-sequence power systems has been described.

Due to the following reasons, a continuation three-phase power flow may be required: (a) there are unbalances of three-phase transmission lines in high voltage transmission networks; (b) there are unbalanced three-phase loads; (c) in addition, there are single-phase or two-phase lines in distribution networks; (d) there are single-phase or two-phase loads; (e) there may also be possible unbalanced three-phase structures and control of transformers and FACTS-devices. In addition to the reasons above, with the recent integration of large amount of distributed generation into power networks, new voltage stability analysis tools, which should have the modeling capability of unbalanced networks, become increasingly important. Furthermore, it is recognized that voltage stability analysis should be able to deal with asymmetrical contingencies such as single-phase and two-phase transmission line outages, etc. It is known that the single-phase continuation power flow is not able to deal with unbalanced networks and loads and can not deal with single-phase and two-phase outages of unbalanced transmission lines.

In the light of the above considerations, in this chapter, a continuation three-phase power flow approach for voltage stability analysis of unbalanced three-phase power systems [12] is presented. In addition, voltage stability control by FACTS is also discussed.

## 7.2 Continuation Three-Phase Power Flow Approach

### 7.2.1 Modeling of Synchronous Machines with Operating Limits

The modeling of synchronous machines in three-phase power flow analysis has been discussed in chapter 5. The operating limits of synchronous machines, which play very important role in voltage stability analysis, should be considered. In the following, the operating constraints of synchronous machines are presented and incorporation of the limits in three-phase power flow and continuation three-phase power flow analysis is discussed.

In Fig. 5.1, $V_i^a = V_i^a \angle \theta_i^a$, $V_i^b = V_i^b \angle \theta_i^b$, $V_i^c = V_i^c \angle \theta_i^c$, which are the three-phase voltages at the generator terminal bus, are expressed in phasors in polar coordinates. Similarly, the voltages at the generator internal bus may be given by $E_i^a = E_i^a \angle \delta_i^a$, $E_i^b = E_i^b \angle \delta_i^b$, $E_i^c = E_i^c \angle \delta_i^c$. In fact the voltages at the generator internal bus are balanced, we have $E_i^a = E_i^b = E_i^c$ and $\delta_i^a = \delta_i^b + 120° = \delta_i^c - 120°$. Therefore, in the following derivation of the power flow equations of the generator, $\delta_i^a$ and $E_i^a$ can be considered as independent state variables of the internal generator bus while $\delta_i^b$ and $E_i^b$, $\delta_i^c$ and $E_i^c$ are dependent state variables and can be represented by $\delta_i^a$ and $E_i^a$.

For a PV machine, the total reactive power $Qg_i$ at its terminal bus should be within its operating limits:

$$Qg_i^{min} \leq Qg_i \leq Qg_i^{max} \qquad (7.1)$$

where $Qg_i^{min}$ and $Qg_i^{max}$ are the lower and upper reactive limits, respectively. In addition, due to the limitation of the field current, the following constraint should hold

$$E_i^a \leq E_i^{max} \qquad (7.2)$$

where $E_i^{max}$ is the maximum limit of the internal voltage of the machine, which corresponds to the maximum filed current. $E_i^a$ is the actual voltage magnitude at the internal bus.

For a PQ machine, the positive-sequence voltage $V_i^1$ at its terminal bus should be within its operating limits:

$$V_i^{min} \leq V_i^1 \leq V_i^{max} \qquad (7.3)$$

where $V_i^{min}$ and $V_i^{max}$ are the upper and lower voltage limits, respectively. In addition, the field current constraint as given by (7.2) is also applicable.

The basic constraint enforcement principle of a synchronous machine is that, when an inequality constraint, such as a current or voltage or reactive power inequality constraint, is violated, the constraint is enforced by being kept at its limit, while the voltage or reactive power control constraint of the synchronous machine is released. In other words, enforcing an inequality constraint and releasing an equality constraint must form a pair. In case there are two or more inequality constraints of a synchronous machine being violated in the same time, the strategy proposed in [16] can be used. The reactive power constraint in (7.1) and current constraint in (7.2) of a machine are considered as internal constraints while the voltage constraint in (7.3) is considered as external constraint. Generally, an internal constraint has priority to be enforced if both the internal and external constraints are violated simultaneously. In case the internal and external constraints cannot be enforced within the limits simultaneously, the external constraint should be released.

### 7.2.2 Three-Phase Power Flow in Polar Coordinates

The power mismatch equations at buses except generator internal buses, which are given by (5.51) and (5.52), are presented as follows:

$$\Delta P_i^p = -Pd_i^p - V_i^p \sum_{j \in i} \sum_{m=a,b,c} V_j^m (G_{ij}^{pm} \cos \theta_{ij}^{pm} + B_{ij}^{pm} \sin \theta_{ij}^{pm}) = 0 \qquad (7.4)$$

$$\Delta Q_i^p = -Qd_i^p - V_i^p \sum_{j \in i} \sum_{m=a,b,c} V_j^m (G_{ij}^{pm} \sin \theta_{ij}^{pm} - B_{ij}^{pm} \cos \theta_{ij}^{pm}) = 0 \qquad (7.5)$$

where $i = 1, 2, \ldots, N$. $Pd_i^p$ and $Qd_i^p$ are the active and reactive load powers of phase $p$ at bus $i$, respectively.

The power mismatch equations at generator internal buses (for the case of PQ machine), which are given by (5.53) and (5.54), are presented as follows:

$$\Delta Pg_i = -Pg_i$$
$$- \sum_{p=a,b,c} \sum_{m=a,b,c} [V_i^p V_i^m (Gg_i^{pm} \cos\theta_i^{pm} + Bg_i^{pm} \sin\theta_i^{pm})$$
$$+ \sum_{p=a,b,c} \sum_{m=a,b,c} [V_i^p E_i^p (Gg_i^{pm} \cos(\theta_i^p - \delta_i^m) + Bg_i^{pm} \sin(\theta_i^p - \delta_i^m))] \quad (7.6)$$

$$\Delta Qg_i = -Qg_i$$
$$- \sum_{p=a,b,c} \sum_{m=a,b,c} [V_i^p V_i^m (Gg_i^{pm} \sin\theta_i^{pm} - Bg_i^{pm} \cos\theta_i^{pm})$$
$$+ \sum_{p=a,b,c} \sum_{m=a,b,c} [V_i^p E_i^p (Gg_i^{pm} \sin(\theta_i^p - \delta_i^m) - Bg_i^{pm} \cos(\theta_i^p - \delta_i^m))] \quad (7.7)$$

where $i = 1, 2, ..., Ng$. $Ng$ is the number of generators. In three-phase power flow calculations, $Pg_i$ and $Qg_i$, which are specified, are the active and reactive generation powers of the generator at bus $i$, respectively. For the case of PV and slack machine, two constraint equations can also be obtained. Modeling of other power system components is referred to [14][15].

A number of three-phase power flow methods [17]-[26], etc. have been proposed since 1960s. In the following, the three-phase Newton power flow algorithm in polar coordinates, which is similar to that proposed in [19], will be used. The nonlinear equations (7.4)-(7.7) can be combined and expressed in compact form:

$$\mathbf{F}(\mathbf{x}) = \mathbf{0} \quad (7.8)$$

where $\mathbf{F}(\mathbf{x})$ represents the whole set of power flow mismatch and machine terminal constraint equations. $\mathbf{x}$ is the state variable vector and given by $\mathbf{x} = [\mathbf{\theta}^a, \mathbf{V}^a, \mathbf{\theta}^b, \mathbf{V}^b, \mathbf{\theta}^c, \mathbf{V}^c, \mathbf{\delta}^a, \mathbf{E}^a]^t$. The Newton equation is given by:

$$\mathbf{J}(\mathbf{x})\Delta\mathbf{x} = -\mathbf{F}(\mathbf{x}) \quad (7.9)$$

where $\mathbf{F}(\mathbf{x}) = [\Delta\mathbf{P}^a, \Delta\mathbf{Q}^a, \Delta\mathbf{P}^b, \Delta\mathbf{Q}^b, \Delta\mathbf{P}^c, \Delta\mathbf{Q}^c, \Delta\mathbf{P}g^a, \Delta\mathbf{Q}g^a]^t$, $\mathbf{J}(\mathbf{x}) = \dfrac{\partial \mathbf{F}(\mathbf{x})}{\partial \mathbf{x}}$ is the system Jacobian matrix.

### 7.2.3 Formulation of Continuation Three-Phase Power Flow

**Predictor Step.** To simulate three-phase load change, $Pd_i^p$ and $Qd_i^p$, which are shown in (7.4) and (7.5), may be represented by:

$$Pd_i^p = Pd0_i^p (1 + \lambda * KPd_i^p) \quad (7.10)$$

$$Qd_i^p = Qd0_i^p (1 + \lambda * KQd_i^p) \quad (7.11)$$

where $Pd0_i^p$ and $Qd0_i^p$ are the base case active and reactive load powers of phase $p$ at bus $i$. $\lambda$ is the loading factor, which characterize the change of load. The ratio of $KPd_i^p / KQd_i^p$ is constant to maintain constant power factor.

Similarly, to simulate generation change, $Pg_i$ and $Qg_i$, which are shown in (7.5) and (7.6), are represented as functions of $\lambda$ and given by:

$$Pg_i = Pg0_i(1 + \lambda * KPg_i) \qquad (7.12)$$

$$Qg_i = Qg0_i(1 + \lambda * KQg_i) \qquad (7.13)$$

where $Pg0_i$ and $Qg0_i$ are the total active and reactive powers of the generator of the base case. The ratio of $KPg_i / KQg_i$ is constant to maintain constant power factor for a PQ machine. For a PV machine, equation (7.13) is not required. For a machine, when the reactive limit is violated, $Qg_i$ should be kept at the limit and equation (7.13) is also not required.

The nonlinear equations (7.9) are augmented by an extra variable $\lambda$ as follows:

$$\mathbf{F}(\mathbf{x}, \lambda) = \mathbf{0} \qquad (7.14)$$

where $\mathbf{F}(\mathbf{x}, \lambda)$ represents the whole set of power flow mismatch equations.

The predictor step is used to provide an approximate point of the next solution. A prediction of the next solution is made by taking an appropriately sized step in the direction tangent to the solution path.

To solve (7.14), the continuation algorithm with predictor and corrector steps can be used. Linearizing (7.14), we have:

$$d\mathbf{F}(\mathbf{x}, \lambda) = \mathbf{F}_x d\mathbf{x} + \mathbf{F}_\lambda d\lambda = \mathbf{0} \qquad (7.15)$$

In order to solve (7.15), one more equation is needed. If we choose a non-zero magnitude for one of the tangent vector and keep its change as $\pm 1$, one extra equation can be obtained:

$$t_k = \pm 1 \qquad (7.16)$$

where $t_k$ is a non-zero element of the tangent vector $d\mathbf{x}$.

Combining (7.15) and (7.16), we can get a set of equations where the tangent vector $d\mathbf{x}$ and $d\lambda$ are unknown variables:

$$\begin{bmatrix} \mathbf{F}_x & \mathbf{F}_\lambda \\ \hline e_k \end{bmatrix} \begin{bmatrix} d\mathbf{x} \\ d\lambda \end{bmatrix} = \begin{bmatrix} \mathbf{0} \\ \pm 1 \end{bmatrix} \qquad (7.17)$$

where $e_k$ is a row vector with all elements zero except for $K^{th}$, which equals one. In (7.17), whether +1 or –1 is used depends on how the $K^{th}$ state variable is changing as the solution is being traced. After solving (7.17), the prediction of the next solution may be given by:

$$\begin{bmatrix} \mathbf{x}^* \\ \lambda^* \end{bmatrix} = \begin{bmatrix} \mathbf{x} \\ \lambda \end{bmatrix} + \sigma \begin{bmatrix} d\mathbf{x} \\ d\lambda \end{bmatrix} \tag{7.18}$$

where * denotes the estimated solution of the next step while $\sigma$ is a scalar, which represents the step size.

**Corrector Step.** The corrector step is to solve the augmented Newton power flow equation with the predicted solution in (7.18) as the initial point. In the augmented Newton power flow algorithm an extra equation is included and $\lambda$ is taken as a variable. The augmented Newton power flow equation may be given by:

$$\begin{bmatrix} \mathbf{F}(\mathbf{x}, \lambda) \\ x_k - \eta \end{bmatrix} = \begin{bmatrix} 0 \\ 0 \end{bmatrix} \tag{7.19}$$

where $\eta$, which is determined by (7.18), is the predicted value of the continuation parameter $x_k$. The determination of the continuation parameter is shown in the following solution procedure.

The corrector equation (7.19), which consists a set of augmented nonlinear equations, can be solved iteratively by Newton's approach as follows:

$$\begin{bmatrix} \mathbf{F_x} & \mathbf{F}_\lambda \\ \mathbf{e}_k & \end{bmatrix} \begin{bmatrix} \Delta \mathbf{x} \\ \Delta \lambda \end{bmatrix} = -\begin{bmatrix} \mathbf{F}(\mathbf{x}, \lambda) \\ x_k - \eta \end{bmatrix} \tag{7.20}$$

### 7.2.4 Solution of the Continuation Three-Phase Power Flow

The general solution procedure for the Continuation Three-Phase Power Flow is given as follows:

*Step 0:*  Run three-phase power flow when $Pd_i^p$, $Qd_i^p$, $Pg_i$ and $Qg_i$ are set to $Pd0_i^p$, $Qd0_i^p$, $Pg0_i$ and $Qg0_i$, respectively. The initial point for tracing the PV curves is found.

*Step 1 - Predictor Step:*
  (a) Solve (7.17) and get the tangent vector $[d\mathbf{x}, d\lambda]^t$;
  (b) Use (7.18) to find the predicted solution of the next step.
  (c) Choose the continuation parameter by evaluating $x_k : t_k = \max(|dx_i|)$.
  (d) Check whether the critical point (maximum loading point) has been passed by evaluating the sign of $d\lambda$. If $d\lambda$ changes its sign from positive to negative, then the critical point has just passed.

(e) Check whether $\lambda^* < 0$ (Note $0 \le \lambda \le \lambda_{max}$). If this is true, go to Step 3.

*Step 2 - Corrector Step:*
   (a) According to the chosen continuation parameter to form the augmented equation (7.19);
   (b) Form and solve the Newton equation (7.20);
   (c) Update the Newton solution and continue the iterations until the corrector step converges to a solution with a given tolerance;
   (d) Go to Step 1.

*Step 3:* Output solutions of the PV curves.

## 7.2.5 Implementation Issues of Continuation Three-Phase Power Flow

### 7.2.5.1 The Structure of Jacobian Matrix

The structures of the Jacobian matrix (7.17) and the Jacobian matrix (7.20) are very similar. In comparison to the 4 by 4 Jacobian blocks in single-phase power flow analysis, the Jacobian matrix blocks of $\mathbf{F_x}$ in three-phase power flow analysis become 12 by 12 matrix blocks for all buses except internal buses of generators while the Jacobian blocks of the internal buses of generators are 4 by 4, 4 by 12, 12 by 4 matrix blocks. Similar to that of single-phase power flow analysis, the equations (7.17) and (7.20) of three-phase power systems can be solved by sparse matrix techniques.

### 7.2.5.2 Improvement of Computational Speed

In order to improve the computational speed for tracing the PV curves, in the implementation, the three-phase power flow calculations may be used with gradually increasing system load until the three-phase power flow cannot converge. Then the above continuation three-phase power flow approach can be used to trace the remaining parts of the PV curves. Using the three-phase continuation power flow, a small predictor step may be used at the vicinity of the point of voltage collapse while a large step may be used otherwise.

In the present implementation of the continuation three-phase power flow algorithm, the tangent method is used at the predictor step. It has been recognized that the tangent method may be more reliable than the secant method. In addition, the tangent method can produce an approximate left eigenvector at the saddle node bifurcation point. However, as far as computational time is concerned, the secant method may be more attractive [10] since using the method, solution of (7.17) is not needed.

On the other hand, the solution of (7.17) may be significantly improved by using the sparse vector method [27] in the implementation. Since the only one non-zero element of the right-hand vector in (7.17) is at the bottom, the forward substitution is not needed at all.

### 7.2.5.3 Comparison of Balanced Three-Phase Systems and Single-Phase Systems

It should be pointed out that for a balanced system, a three-phase power flow solution is, in principle, exactly identical to that of the equivalent positive-sequence power flow while positive-and zero-sequence voltage components of the former are zero and at any bus, phase $a$, $b$ and $c$ voltages are balanced and interdependent except 120° shifting between them. In other words, for a balanced three-phase power system, any phase voltage at a bus can completely characterize the positive-sequence voltage at that bus. Mathematically, the balanced three-phase system can be decoupled into the equivalent positive-, negative-, and zero-sequence networks, and the singularity resulting from the equivalent positive-sequence network can be solved by choosing any phase voltage at the bus as the continuation parameter. This means, in nature, the reason and solution of the singularity of the balanced three-phase power flow are exactly the same to that of the equivalent positive-sequence power flow. The difference between them is whether the system is represented in three-phase or single-phase coordinates.

### 7.2.6 Numerical Results

In this paper, numerical results are carried out on a 5-bus system and a modified IEEE 118-bus system. The single-line diagram of the 5-bus system I and the system data are presented in the Appendix of Chapter 5. For the modified IEEE 118-bus system, 54 three-phase Wye-Grounded/Delta transformers are inserted between the original network and 54 generators, and negative- and zero-sequence parameters of transmission lines are amended. The modified IEEE 118-bus system consists of 172 three-phase buses (or 516 single-phase buses). In the studies, loads are represented by P and Q powers.

#### 7.2.6.1 Results for the 5-Bus System without Line Outages

The following cases on the 5-bus network have been studied:

*Case 1*: Balanced network and the whole system with balanced load.

*Case 2*: Balanced network with unbalanced load at Bus 3 with 6.0+$j$3.0 p.u., 6.3+$j$2.7 p.u., 5.7+$j$3.3 p.u. for phase $a$, $b$, $c$ loading, respectively.

*Case 3*: Unbalanced network and the whole system with balanced load.

*Case 4*: Unbalanced network with unbalanced load at Bus 3 with 6.0+$j$3.0 p.u., 6.3+$j$2.7 p.u., 5.7+$j$3.3 p.u. for phase $a$, $b$, $c$ loading, respectively.

The PV curves of cases 1 - 4 are shown in Fig. 7.1 to Fig. 7.4. It is known that the tracing direction of the PV curves of a single-phase or positive-sequence system is clockwise. From Fig. 7.1, it can be seen that for the balanced three-phase power

## 7.2 Continuation Three-Phase Power Flow Approach

system, the three PV curves at any bus are exactly the same and the tracing direction of these PV curves is clockwise.

As expected, the three PV curves for phase $a$, phase $b$ and Phase $c$ at Bus 3 of case 1 are exactly the same. Furthermore, these PV curves have very similar pattern to that of a single-phase or positive-sequence power system. That is each PV curve consists of a high voltage portion and a low voltage portion. As the loading factor $\lambda$ is increasing between 0 and $\lambda_{max}$, the operating point of the system is moving from the initial point to the maximum loading point or the point of voltage collapse, which is corresponding to the higher voltage portion of the PV curve. After the point of voltage collapse, the loading factor $\lambda$ is decreasing from $\lambda_{max}$ to 0, which is corresponding to the lower voltage portion of the PV curve. It is known that any points on the lower voltage portion are unstable.

Having discussed the PV curves for single-phase and examined also the PV curves of balanced three-phase systems, the PV curves of unbalanced three-phase systems are to be discussed here. It has been found that the three PV curves for phase $a$, phase $b$ and Phase $c$ at Bus 3 for any of case 2 - 4, are not the same. Examining the PV curves at Bus 3 of case 2 shown in Fig. 7.2, it was interestingly found:

- In the PV curves of Phases $a$ and $c$, the voltages are decreasing when $\lambda$ is increasing between 0 and $\lambda_{max}$. The tracing direction of these two PV curves is clockwise and the patterns of these two PV curves are very similar to that of single-phase or balanced three-phase power systems.
- However, in the PV curve of phase $b$, the voltage is decreasing till at a point close to the point of $\lambda_{max}$, then the voltages become increasing. The tracing direction of the PV curve is anti-clockwise. In this PV curve, the 'higher voltage' portion is corresponding to the unstable power flow solutions while the 'lower voltage' portion is corresponding to the stable power flow solutions.

Further examining the PV curves of case 3 and 4 shown in Fig. 7.3 and Fig. 7.4, respectively, it can be found:

- The patterns of the PV curves of the unbalanced three-phase systems are quite different from that of the balanced three-phase systems. At least one of the PV curves at a bus has the clockwise tracing direction and the voltage of the phase is much lower than that of the other phases while the PV curves of the other phases may have the anti-clockwise tracing direction.
- However, when the network and load are balanced, the three PV curves at any bus merge into one as seen in Fig. 7.1. Then a positive-network analysis is sufficient.
- Voltage stability analysis of unbalanced three-phase power systems are much more complex than that of single-phase positive sequence power systems or balanced three-phase power systems.

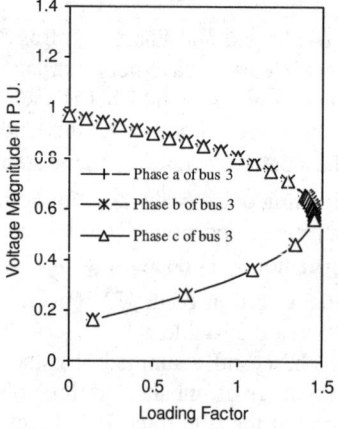

**Fig. 7.1.** PV curves of bus 3 for case 1     **Fig. 7.2.** PV curves of bus 3 for case 2

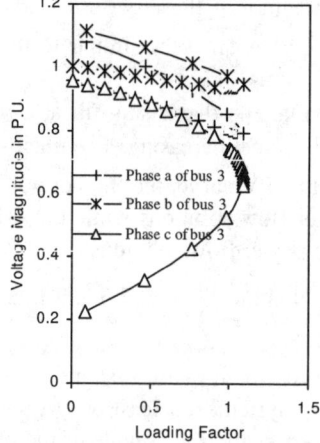 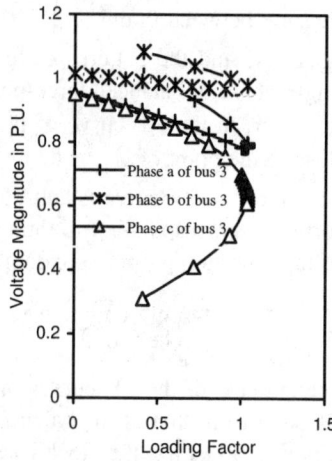

**Fig. 7.3.** PV curves of bus 3 for case 3     **Fig. 7.4.** PV curves of bus 3 for case 4

The maximum loading factors for cases 1-4 are shown in Table 7.1. From Table 7.1, it can be clearly seen that unbalanced network and load can significantly affect the system loading capability.

**Table 7.1.** Maximum loading factors without line outages

| Case No. | Maximum loading factor $\lambda_{max}$ | Lowest voltage magnitude (in p.u.) at maximum loading point | Bus no of the lowest voltage | Phase of the lowest voltage |
|---|---|---|---|---|
| 1 | 1.4530 | 0.5818 | 3 | a, b, c |
| 2 | 1.3127 | 0.6095 | 3 | a |
| 3 | 1.0864 | 0.6465 | 3 | c |
| 4 | 1.0374 | 0.6299 | 3 | c |

### 7.2.6.2 Results for the 5-Bus System with Line Outages

In order to investigate the voltage stability of unbalanced three-phase systems where there are transmission line outages, the following cases were carried out:

*Case 5*: This is similar to case 4 but Phase *a* of one of the double lines between Bus 1 and Bus 3 is open-circuited.

*Case 6*: This is similar to case 4 but Phase *b* of one of the double lines between Bus 1 and Bus 3 are open-circuited.

*Case 7*: This is similar to case 4 but the Phase *c* of one of the double lines between Bus 1 and Bus 3 are open-circuited.

*Case 8*: This is similar to case 4 but Phase *a* and Phase *b* of one of the double lines between Bus 1 and Bus 3 is open-circuited.

*Case 9*: This is similar to case 4 but Phase *b* and Phase *c* of one of the double lines between Bus 1 and Bus 3 are open-circuited.

*Case 10*: This is similar to case 4 but Phase *c* and Phase *a* of one of the double lines between Bus 1 and Bus 3 are open-circuited.

*Case 11*: This is similar to case 4 but Phase *a*, Phase *b* and Phase *c* of one of the double lines between Bus 1 and Bus 3 are open-circuited.

The maximum loading factors for cases 5-11 are shown in Table 7.2. From this table, it can be seen:

- Surprisingly the maximum loading factor of case 6 (with line 1-3 outage on phase *b*) is greater than that of case 4 without any line outage. This means that for the unbalanced three-phase system studied, case 6 with single–phase line outage is less serious than case 4 without line outages in the point of view of voltage stability.

**Table 7.2.** Maximum loading factors with line outages

| Case No. | Type of line outage | Maximum loading factor $\lambda_{max}$ | Lowest voltage magnitude (in p.u.) at maximum loading point | Bus no of the lowest voltage | Phase of the lowest voltage |
|---|---|---|---|---|---|
| 5 | Phase a | 0.6878 | 0.6183 | 3 | a |
| 6 | Phase b | 1.0627 | 0.6329 | 3 | c |
| 7 | Phase c | 0.5317 | 0.6182 | 3 | c |
| 8 | Phases a and b | 0.5681 | 0.6067 | 3 | a |
| 9 | Phases b and c | 0.5987 | 0.6329 | 3 | c |
| 10 | Phases a and c | 0.5176 | 0.6147 | 3 | c |
| 11 | Phases a, b and c | 0.5665 | 0.6407 | 3 | c |

- Among the single-phase line outages of cases 5-7, case 6 is less serious than the other two cases. The maximum loading factor of case 6 is about two times that of the other two cases, respectively.
- Cases 8 and 9 with two-phase line outages have larger loading factors than case 7 with one-phase line outage. This means that for the unbalanced three-phase system studied, the two-phase line outages of cases 8 and 9 are less serious than the one-phase line outage of case 7 in the point of view of voltage stability.
- Case 11 with three-phase line outage has larger loading factor than case 7 with one-phase line outage and case 10 with two-phase line outages. This means that the three-phase line outage of case 11 is less serious than the one-phase line outage of case 7 and the two-phase line outage of case 10 in the point of view of voltage stability.

The above observations are very interesting phenomena from the unbalanced three-phase systems, which are quite different from that of single-phase systems or balanced three-phase systems. Due to the combinations of the complexity of unbalanced load and network, it is not easy to explain qualitatively the above observations. Instead, we try to show numerical results to reveal the possible reasons for the above phenomena. The three-phase power flow results of cases 4 - 11 at particular loading levels are shown in Table 7.3 and Table 7.4 respectively. In the Tables, $V_3^1$, $V_3^2$ and $V_3^0$ are the positive-, negative-, zero-sequence voltage magnitudes at bus 3, respectively while voltage sensitivities, which are the largest in magnitude for the corresponding cases, are shown in the last column of these two Tables.

**Table 7.3.** Power flow results at $\lambda = 0.5$ (voltage and power in P.U.)

| Case No. | $V_3^1$ | $V_3^2$ | $V_3^0$ | $P_{loss}$ | $Q_{loss}$ | Normalized $Q_{loss}$ | $\frac{\partial V}{\partial \lambda}$ |
|---|---|---|---|---|---|---|---|
| 4 | 0.9117 | 0.0378 | 0.0403 | 1.33 | 8.89 | 100% | -0.21 |
| 5 | 0.8837 | 0.0567 | 0.0857 | 1.62 | 11.09 | 125% | -0.41 |
| 6 | 0.8983 | 0.0276 | 0.0154 | 1.45 | 10.00 | 113% | -0.20 |
| 7 | 0.8602 | 0.1037 | 0.1133 | 1.76 | 12.77 | 144% | -1.14 |
| 8 | 0.8672 | 0.0636 | 0.1193 | 1.85 | 13.04 | 147% | -0.74 |
| 9 | 0.8621 | 0.0727 | 0.0599 | 1.74 | 12.51 | 141% | -0.59 |
| 10 | 0.8373 | 0.0910 | 0.1312 | 2.02 | 14.74 | 166% | -1.51 |
| 11 | 0.8452 | 0.0788 | 0.0946 | 1.99 | 14.22 | 160% | -0.70 |

**Table 7.4.** Power flow results at $\lambda = 1.03$ (voltage and power in P.U.)

| Case No. | $V_3^1$ | $V_3^2$ | $V_3^0$ | Lowest phase voltage | $P_{loss}$ | $Q_{loss}$ | $\frac{\partial V}{\partial \lambda}$ |
|---|---|---|---|---|---|---|---|
| 4 | 0.8042 | 0.0887 | 0.1087 | 0.6629 | 3.18 | 24.70 | -1.94 |
| 6 | 0.7966 | 0.0540 | 0.0333 | 0.6971 | 3.38 | 26.39 | -0.88 |

The voltage sensitivities can be considered as an indicator of voltage instability. In principle, the larger the voltage sensitivity is, the lower the maximum loading factor will be. The voltage sensitivities in Table 7.3 and Table 7.4 correlate well with the maximum loading factors of case 4 – 11. In addition, it has been found that for most of the cases, high power losses and negative- and zero-sequence voltage components are associated with large voltage sensitivities.

Comparing the results of case 4 and case 6 in Table 7.3 and Table 7.4, it can be seen:

- At $\lambda = 0.5$, the largest voltage sensitivity of case 4 is larger than that of case 6.
- As $\lambda$ is increased to 1.03, the voltage sensitivity of case 4 becomes much larger than that of case 6. The voltage sensitivities indicate that case 4 is more vulnerable to voltage instability and hence a lower maximum loading factor is expected for this case.

### 7.2.6.3 Results for the Modified IEEE 118-Bus System

The following four cases were carried out on the modified IEEE 118-bus system with balanced network and loads:

*Case 12*: This is the base case system.

*Case 13*: This is similar to case 12 but one phase of the line between Bus 68 and Bus 81 is open-circuited.

*Case 14*: This is similar to case 12 but two phases of the line between Bus 68 and Bus 81 are open-circuited.

*Case 15*: This is similar to case 12 but three-phases of the line between Bus 68 and Bus 81 are open-circuited.

The maximum loading factors for cases 12-15 on the modified IEEE 118-bus system are shown in Table 7.5.

**Table 7.5.** Maximum loading factors on the modified IEEE 118-bus system

| Case No. | Maximum loading factor $\lambda_{max}$ | Type of line outage |
|---|---|---|
| 12 | 0.6910 | None |
| 13 | 0.6599 | One phase outage of line 68-81 |
| 14 | 0.5998 | Two phase outage of line 68-81 |
| 15 | 0.5010 | Three-phase outage of line 68-81 |

### 7.2.6.4 Reactive Power Limits

For case 4, with generator reactive power limits applied, the PV curves of bus 3 are shown in Fig. 7.5. From Fig. 7.5 it can be found:

- In comparison of Fig. 7.5 to Fig. 7.4, as expected, the maximum loading factor is decreased when the generator reactive power limits are applied.
- The significant reduction in the voltage magnitudes of phases *a* and *b* can be seen when the generator reactive power limits are encountered. Phase *c* voltage magnitudes are actually reduced as well.

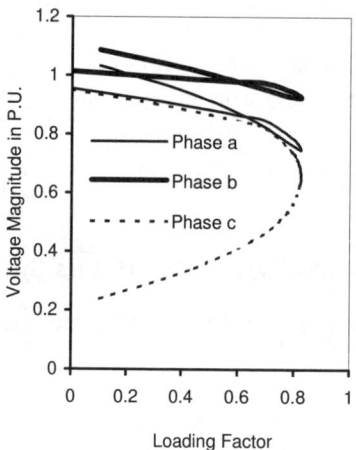

**Fig. 7.5.** PV curves of bus 3 for case 4 with generator reactive power limits

- The effect of the reactive power limits is the reduction in bus voltage magnitudes. In other words, the three PV curves move down since the generator terminal bus voltage cannot hold up to the setting point any longer. Noting the tracing direction of the PV curves of phases $a$ and $b$ is anti-clockwise, the effect of the reactive power limits on the voltage magnitudes of these two phases is more significant and delays the voltage rise of phases $a$ and $b$. Hence, the effect causes the "higher portion" of the PV curves fall and cross the "lower portion" of these.

A continuation three-phase power flow approach for voltage stability analysis of unbalanced three-phase power systems has been proposed. The approach can take into account the unbalances of both network and load. In addition, it can also deal with various transmission line outages.

Numerical examples have demonstrated the approach proposed is effective. Some very interesting results using PQ loads have been obtained:

- When network and load are balanced, the PV curves of phase $a$, phase $b$ and phase $c$ at a bus are identical as expected and the pattern of these is very similar to that of the PV curves of single-phase power systems. In this situation, a single-phase positive network analysis is sufficient.
- However, when network and load are unbalanced, the patterns of the PV curves are very interesting, which have not been observed and discussed in the past. It has been found that with unbalanced network and load, at least one of the PV curves at a bus is very similar to that of single-phase or balanced three-phase systems and the tracing direction of the PV curve is clockwise while the rest of the PV curves (or curve) at the bus have the anti-clockwise tracing direction. For those PV curves with anti-clockwise tracing direction, the higher voltage portion of the PV curves is corresponding to the unstable power flow solutions while the lower voltage portion of the PV curves is corresponding to the stable power flow solutions. The characteristic is unique to unbalanced three-phase power systems.
- It has been found for unbalanced power systems that (a) the maximum loading factor of the system with a single-phase line outage may be greater than that of the system without any line outages; (b) the maximum loading factor of the system with a two-phase line outage may not be necessarily less than that of the system with a single-phase line outage; (c) similarly the maximum loading factor of the system with a three-phase line outage may not be necessarily less than that of the system with a single-phase line outage or a two-phase line outage. The phenomena have been explained based on numerical analyses. Basically, the maximum loading factor is dependent on the degree of unbalance, which is characterized by the magnitudes of negative- and zero-sequence voltages. The degree of unbalance itself is determined by the combination of unbalanced network and loading conditions.

The phenomena observed above reveal that the voltage stability mechanisms of three-phase power systems are much more complex than that of single-phase power systems. This clearly indicates that a continuation three-phase power flow is needed when there are unbalanced network and load existing in a power system. Otherwise, the results may be unrealistic and could not be able to characterize accurately the voltage stability problem of unbalanced power systems.

Similar to that in conventional continuation power flow analysis, reactive power limits of generators play a very important role in the determination of PV curves of three-phase power systems. Basically consideration of the reactive power limits will decrease the maximum loading factor and affect the shape of the PV curves. When the reactive power limits are taken into account, there is the reduction in voltage magnitudes, and subsequently this will affect the shape of the PV curves. It has been found that the effect of the reactive power limits on the PV curves whose tracing direction is anti-clockwise is more significant than on those whose tracing direction is clockwise.

The present results are based on the PQ load model. Further research is needed to investigate the effect of voltage dependent load models on the voltage stability of unbalanced three-phase power systems.

The continuation three-phase power flow approach will be a very useful tool for voltage stability of unbalanced three-phase power systems. The approach can also be used to investigate multiple power flow solutions of unbalanced three-phase power systems.

As distributed generators are increasingly connected to power systems, the continuation three-phase power flow approach may become an important tool to evaluate the unbalanced system operation conditions including contingencies.

## 7.3 Steady State Unbalanced Three-Phase Voltage Stability with FACTS

In this section, the effects of FACTS controls on the steady state voltage stability limit of unbalanced three-phase power systems are investigated. FACTS-devices considered here are STATCOM, SSSC and UPFC [28][29]. The modeling of these FACTS in three-phase power flow analysis is referred to chapter 5.

### 7.3.1 STATCOM

Cases 1-4 with a STATCOM installed at the middle of transmission line 1-3, which are corresponding to case 1-4 in section 7.2.6.2, respectively, have been studied:

*Case 1*: Balanced network and the whole system with balanced load.

*Case 2*: Balanced network with unbalanced load at Bus 3 with 6.0+$j$3.0 p.u., 6.3+$j$2.7 p.u., 5.7+$j$3.3 p.u. for phase $a$, $b$, $c$ loading, respectively.

*Case 3*: Unbalanced network and the whole system with balanced load.

## 7.3 Steady State Unbalanced Three-Phase Voltage Stability with FACTS

*Case 4*: Unbalanced network with unbalanced load at Bus 3 with 6.0+*j*3.0 p.u., 6.3+*j*2.7 p.u., 5.7+*j*3.3 p.u. for phase *a*, *b*, *c* loading, respectively.

The STATCOM models used in the studies are the three-phase model with symmetrical components control, and three single-phase units with independent phase control. The former is referred to model 1 while the latter is referred to model 2. Assuming that the voltage control reference of the STATCOM is 1.05 p.u. for both models.

The maximum loading factors for cases 1-4 are shown in columns 3 and 4 in Table 7.6. For the sake of comparison, in Table 7.6 the maximum load factors for cases 1-4 without STATCOM as discussed in section 7.2.6.1 are listed in column 2. From the table, for both model 1 and model 2, the maximum loading factors have been improved significantly. When both the network and load are balanced, the maximum loading factors are the same for both model 1 and model 2. However, when either the network or load is unbalanced, the maximum loading factors for model 1 are bigger than that for model 2. The reason is that model 1 can balance the bus voltages well in comparison to model 2 since the former can simultaneously control three single-phase voltages while the latter can only control positive sequence voltage.

For case 4, the relationship between voltage control reference of the STATCOM and the maximum loading factor is investigated, which is given by Fig. 7.6. From this figure, it can be found that the higher the voltage control reference, the bigger the system maximum loading factor.

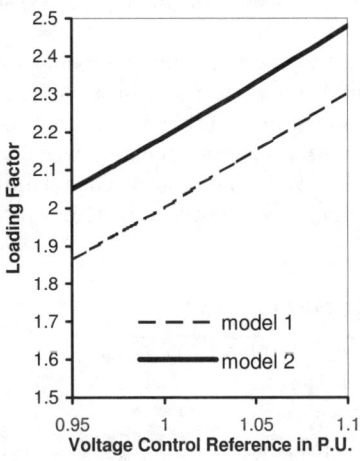

**Fig. 7.6.** The relationship between the voltage control reference and the maximum loading factor

**Table 7.6.** Maximum loading factors with STATCOM

| Case No. | Maximum loading factor $\lambda_{max}$ without STATCOM | Maximum loading factor $\lambda_{max}$ with STATCOM (Model 1) | Maximum loading factor change in percentage with STATCOM (Model 1) | Maximum loading factor $\lambda_{max}$ with STATCOM (Model 2) | Maximum loading factor change in percentage with STATCOM (Model 2) |
|---|---|---|---|---|---|
| 1 | 1.4530 | 2.6962 | 86% | 2.6962 | 86% |
| 2 | 1.3127 | 2.3695 | 81% | 2.5351 | 93% |
| 3 | 1.0864 | 2.1913 | 102% | 2.3630 | 117% |
| 4 | 1.0374 | 2.1517 | 107% | 2.3313 | 125% |

### 7.3.2 SSSC

Cases 5-8 with a SSSC installed at the middle of transmission line 1-3, which are corresponding to case 1-4 in section 7.3.1, respectively, have been studied. The SSSC models used in the studies are the three-phase model with symmetrical components control, and the three single-phase units with independent phase power flow control. The models have been discussed in section 5.3 of chapter 5. The three-phase model with symmetrical components control is referred to model 1 while the three single-phase units with independent phase power flow control is referred to model 2. Assuming that the total three-phase power flow control reference for model 1 is 6.5*3 p.u. while the single phase power flow control reference is 6.5 p.u.

The maximum loading factors with SSSC control are shown in Table 7.7. From this table, it can be found that proper power flow control using SSSC can increase the maximum loading factors. From Table 7.7, it can be seen that as the unbalance of network and load increases, the maximum loading factor in percentage change increases using the SSSC power flow control. The relationship between the power flow control reference for SSSC model 1 and the maximum loading factor is shown in Fig. 7.7.

**Table 7.7.** Maximum loading factors with SSSC

| Case No. | Maximum loading factor $\lambda_{max}$ without SSSC | Maximum loading factor $\lambda_{max}$ with three-phase SSSC (Model 1) | Maximum loading factor change in percentage with SSSC (Model 1) | Maximum loading factor $\lambda_{max}$ with single phase SSSC (Model 2) | Maximum loading factor change in percentage with SSSC (Model 2) |
|---|---|---|---|---|---|
| 5 | 1.4530 | 1.8884 | 30% | 1.8884 | 30% |
| 6 | 1.3127 | 1.8155 | 38% | 1.8000 | 37% |
| 7 | 1.0864 | 1.6918 | 56% | 1.6969 | 56% |
| 8 | 1.0374 | 1.6746 | 61% | 1.7277 | 67% |

## 7.3 Steady State Unbalanced Three-Phase Voltage Stability with FACTS

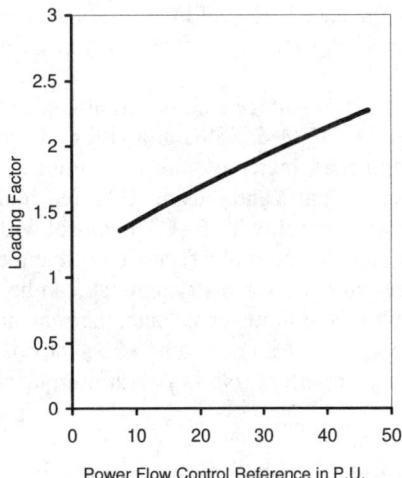

**Fig. 7.7.** The relationship between the power flow control reference and the maximum loading factor

### 7.3.3 UPFC

Cases 9-12 with a UPFC installed at the middle of transmission line 1-3, which are corresponding to case 1-4 in section 7.3.1, respectively, have been studied. The UPFC model used in the studies is the three-phase model with symmetrical components control. The model has been discussed in section 5.3 of chapter 5. Assuming that the total three-phase power flow control reference for model 1 is 7.5*3 p.u. while the single phase power flow control reference is 7.5 p.u. The voltage control reference for both models is 1.05 p.u. The maximum loading factors with the UPFC are shown in Table 7.8.

**Table 7.8.** Maximum loading factors with UPFC

| Case No. | Maximum loading factor $\lambda_{max}$ without UPFC | Maximum loading factor $\lambda_{max}$ with three-phase UPFC | Increase of maximum loading factor in percentage |
|---|---|---|---|
| 9 | 1.4530 | 1.8654 | 28% |
| 10 | 1.3127 | 1.8468 | 41% |
| 11 | 1.0864 | 1.8676 | 72% |
| 12 | 1.0374 | 1.8478 | 78% |

From Table 7.8, it can be found that proper power flow and voltage control using UPFC can increase the maximum loading factors. In particular, when the network and load are unbalanced, the UPFC control can effectively increase the maximum loading factor in percentage change with respect to that without UPFC control.

It has been found that the voltage stability limit can be improved significantly using FACTS such as STATCOM, SSSC and UPFC. In particular, when the unbalance of network and load increases, the maximum loading factor change in percentage can increase significantly using FACTS control. The continuation three-phase power flow approach with FACTS control will be a useful tool to investigate the voltage stability control of unbalanced three-phase power system. The continuation three-phase power flow approach can be also used to determine the security operating limits in terms of voltage, thermal and voltage stability limits. With the integration of distribution generation into power grids, such a tool will pay an increasingly important role in operation, planning and control of distributed power grids.

## References

[1] Mansour Y, editor (1993) Suggested Techniques for Voltage Stability Analysis. Publication No 93 TH0620-5WR, IEEE Power Engineering Society
[2] Kundur P (1994) Power System Stability and Control. EPRI Power Engineering Series, McGraw-Hill
[3] Taylor CW (1994) Power System Voltage Stability. EPRI Power Engineering Series, McGraw-Hill
[4] Van Cutsem T, Vournas C (1998) Voltage Stability of Electric Power Systems. Kluwer Academic Publishers
[5] Huneault M, Fahmideh-Vodani A, Juman M, Galiana FG (1985) The continuation method in power system optimization: applications to economy security functions. IEEE Transactions on PAS, vol 104, no 1, pp 114-124
[6] Huneault M, Galiana FG (1990) An investigation of the solution to the optimal power flow problem incorporating continuation methods. IEEE Transactions on Power Systems, vol 5, no 1, pp 103-110
[7] Iba K, Suzuki H, Egawa M, Watanabe T (1990) Calculation of critical loading condition with nose curve using homotopy continuation method. IEEE Transactions on Power Systems, vol 5, no 1, pp 103-110
[8] Ajjarapu V, Christy C (1992) The continuation power flow: a tool for steady state voltage stability analysis. IEEE Transactions on Power Systems, vol 7, no 1, pp 416-423
[9] Canizares CA, Alvarado FL (1993) Point of collapse and continuation methods for large ac/dc systems. IEEE Transactions on Power Systems, vol 8, no 1, pp 1-8
[10] Chiang HD, Shah KS, Balu N (1995) CPFLOW: a practical tool for tracing power system steady-state stationary behavior due to load and generation variations. IEEE Transactions on Power Systems, vol 10, no 2, pp 623-634

[11] Ejebe GC, Tong J, Waight JG, Frame JG, Wang X, Tinny WF (1998) Available transfer capability calculations. IEEE Transactions on Power Systems, vol 13, no 4, pp 1521-1527
[12] Zhang XP, P Ju, Handschin E (2005) Continuation three-phase power flow: a tool for voltage stability analysis of unbalanced three-phase power systems. IEEE Transactions on Power Systems, vol 20, no 3, pp 1320-1329
[13] Zhang XP, Handschin E, Yao M (2004) Multi-control functional static synchronous compensator (STATCOM) in power system steady state operations. Journal of Electric Power Systems Research, vol 72, no 3, pp 269-278
[14] Chen MS, Dillon WE (1974) Power system modeling. Proceedings of the IEEE, vol 62, no 7, pp 901-915
[15] Dillon WE, Chen MS (1972) Transformer modeling in unbalanced three-phase networks. Proceedings of IEEE Summer Meeting, Vancouver, Canada, July 1972
[16] Zhang XP, Handschin E, Yao M (2004) Multi-control functional static synchronous compensator (STATCOM) in power system steady state operations. Journal of Electric Power Systems Research, vol 72, no 3, pp 269-278
[17] El-Abiad AH, Tarsi DC (1967) Load flow study of untransposed EHV networks. In Proceedings of the IEEE Power Industry Computer Application (PICA) Conference, Pittsburgh, USA, pp 337-384
[18] Wasley RG, Shlash MA (1974) Newton-Raphson algorithm for three phase load flow. IEE Proceedings, vol 121, no 7, pp 631-638
[19] Birt KA, Graf JJ, McDonald JD, El-Abiad AH (1976) Three phase power flow program. IEEE Transactions on PAS, vol 95, no 1, pp 59-65
[20] Arrillaga J, Harker BJ (1978) Fast-decoupled three phase load flow. IEE Proceedings, vol 125, no 8, pp 734-740
[21] Laughton MA, Saleh AOM (1980) Unified phase coordinate load flow and fault analysis of polyphase networks. International Journal of Electrical Power and Energy Systems, vol 2, no 4, pp 181-192
[22] Chen BK, Chen MS, Shoults RR, Liang CC (1990) Hybrid three phase load flow. Proc. IEE, pt C, vol 137,no 3, pp 177-185
[23] Allemong, Bennon RJ, Selent PW (1993) Multiphase power flow solutions using EMTP and Newton's method. IEEE Transactions on Power Systems, vol 8, no 4, pp 1455-1462
[24] Zhang XP, Chen H (1994) Asymmetrical three phase load flow study based on symmetrical component theory. IEE Proceedings– Generation, Transmission & Distribution, vol 143, no 3, pp 248-252
[25] Zhang XP (1996) Fast three phase load flow methods. IEEE Transactions on Power Systems, vol 11, no 3, pp 1547-1554
[26] Zhang XP, Chu W, Chen H (1996) Decoupled Asymmetrical Three Phase Load Flow Study by Parallel Processing. IEE Proceedings– Generation, Transmission & Distribution, vol 143, no 1, pp 61-65
[27] Tinney WF, Brandvajn V, Chan SM (1985), Sparse vector methods. IEEE Transactions on PAS, vol 104, no 2, pp 295-301
[28] Zhang XP, Xue CF, Godfrey KR (2004) Modelling of the static synchronous series compensator (SSSC) in three phase Newton power flow. IEE Proceedings– Generation, Transmission & Distribution, vol 151, no 4, pp 486-494
[29] Zhang XP (2005) Unified power flow controller models for three-phase power flow analysis. Electrical Engineering, DOI: 10.1007/s00202-004-0283-x

# 8 Congestion Management and Loss Optimization with FACTS

This chapter focuses on power flow controlling FACTS-devices and their benefits in market environments. These devices have a significant influence on congestion management and loss reduction. Especially the speed of FACTS-devices provides an additional benefit in comparison to conventional power flow control methods. However, to earn these benefits a special post-contingency operation strategy has to be applied which will be explained in this chapter.

The aim of this chapter is beside the analyses of the qualitative benefits as well to assess the quantitative economic benefits. In particular, we

- analyse under which conditions fast load flow controlling devices like DFC or UPFC allow for a reduction of total system cost,
- estimate the amount of this reduction exemplarily for a realistic scenario within the UCTE system.

In this chapter 'Load Flow Controller' (LFC) is used as a general term for power flow controlling devices like Dynamic Flow Controller (DFC), Unified Power Flow Controller (UPFC) and Phase Shifting Transformer (PST). The acronym DFC is used exemplarily for all kinds of fast and dynamic power flow controllers.

## 8.1 Fast Power Flow Control in Energy Markets

### 8.1.1 Operation Strategy

The liberalisation of electricity markets has led and continues to lead to an increase in volume and volatility of cross-border power exchanges. As a consequence, particularly the transmission networks are operated closer to their technical limits. At least indirectly, some of the numerous major blackouts of the recent years have been related to this development.

Beside strict regulations [1], there are several new technologies with the aim to enable transmission system operators (TSOs) to cope with these challenges by reaching optima in terms of maximum transmission capacity, minimum cost and ensuring of network security. Among the most promising of these innovations are FACTS-devices for power flow control such as DFC or UPFC.

## 8 Congestion Management and Loss Optimization with FACTS

Shifting power flows between areas of a power system means to deviate from the natural power flow. The target for doing this is to increase the power flow over a line or corridor with free capacity or to decrease the flow in an overloaded part of the system. The benefit is measured as increase of the total or available transfer capability (TTC or ATC), which considers the N-1-criteria. The drawback is normally increased losses in the system.

Traditionally the set values of power flow control devices, usually phase shifting transformers, are predetermined to be optimal for all expected contingencies. This means, that the maximum transfer for the expected most critical contingency is increased. The benefit is the difference $TTC_2$-$TTC_1$ in Fig. 8.1. The system is prepared for this contingency, but it is running almost all the time in a non-optimal way according to losses or other criteria.

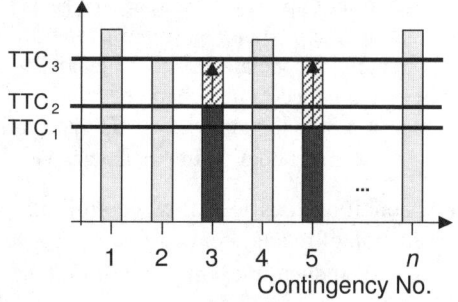

**Fig. 8.1.** Total Transfer Capability without ($TTC_1$), with PST ($TTC_2$) and DFC ($TTC_3$)

In comparison to this traditional approach, a fast controllable power flow control device opens up opportunities to change the set values within or even below a seconds time range to adapt to just occurring contingencies. The fast power flow control may, in principle, result in the following advantages for Transmission System Operators:

- Already during undisturbed network operation slow load flow controllers like PSTs have to be set such that after any contingency all technical quantities remain within their admissible limits. With a fast flow controller, the (N-1)-security criterion can also be fulfilled if after the contingency the DFC is shifted to relieve any overloaded transmission lines or transformers. Even for fast evolving instabilities the DFC is fast enough to reach a stable operation point. If the system has a certain overload capacity and a PST would in principle be fast enough, the DFC would provide more flexibility. The result is that the power system operates loss optimal most of the time. Only in emergency situations the DFC changes to a new set value according to the concrete contingency.
- With a preset target value - the usual operating practice with PSTs - one setting needs to satisfy all contingency situations. By using the DFCs' ability of fast, post-contingency switching, the amount and direction of load flow control can

be dynamically adapted to the actual location of the fault or the overloaded network element. Depending on the network and market conditions, this may enable TSOs to provide additional transmission capacity without compromising network security.Two contingencies, which would require contradictory control actions, can be handled with one device. In Fig. 8.1, contingencies 3 and 5 would require contradictory actions to increase the TTC value. The DFC adapts its action to the respective contingency just after its occurrence. This gives the additional benefit $TTC_3 - TTC_2$.
- Besides, power electronic devices allow for an improvement of network stability.

This chapter focuses on the first two benefits and shows how the reduction of losses as well as an increase of transmission capacity leads to a decrease of total system cost.

## 8.1.2 Control Scheme

Due to the wide-area influence that load flow controlling devices have on the transmission system, the practical realisation of the above advantages requires the provision and utilisation of distant wide area power, current and voltage measurements. In both cases of the previous section - loss reduction and transmission capacity increase - an automatic control scheme needs to be implemented. The time scales of changing the set values depend on which kind of stability boundary is limiting the transfer capability. In case of thermal limitations a certain overload over a couple of minutes can be accepted, but the speed of the action increases the flexibility to react on changing situations.

There are two principle options to automate the control scheme:

- The information on the most severe contingencies, for instance line outages, must be transmitted to the controller. The controller has a set of pre-defined post-contingency set-values, which are used according to the specific contingency. The calculation of these pre-defined values must be done frequently to be as accurate as possible to the actual situation.
- As an alternative, not the contingency itself, e.g. the outage of a line, is measured, but the effect on the parts of the system leading to the limitation. In this case, the flows on the parts or lines, which tend to be overloaded after the contingencies, need to be measured and transferred to the controller. The controller automatically controls the flow of the most critical line to its defined maximum.

In both cases the control scheme is based on rules, which guarantee well defined and unambiguous actions of the DFC (see as well chapters 10, 11 and 12). The second control scheme has the advantage of a higher accuracy, because the effect of the contingency is directly measured and no pre-calculated set-values are required except the maximum flows over the lines.

The required speed for the communication depends on the desired control speed. The fastest and most accurate control system would be a wide area control system based on time-synchronized phasor measurements [2]. With such a system, specific algorithms to identify actual limitations of corridors or lines can be applied as input variables for the Dynamic Flow Controller [3]. Wide area control schemes for these applications will be discussed separately in chapter 12.

## 8.2 Placement of Power Flow Controllers

At first we investigate which fundamental prerequisites need to be fulfilled such that the DFC provides more transmission capacity than the PST. This analysis is done on the basis of simple four node networks. In a second step, we perform an exemplary quantification of the DFC's annual benefit for a realistic network scenario in order to verify the fundamental findings.

According to the basic approach from section 8.1, using a slow LFC means to have one tap position that meets all network constraints in all (N-1)-cases. Using DFC for each topology of the (N-1)-criterion separate tap positions can be used, which are applied in the post-contingency cases. When considering a single topology, there is in general a range of admissible tap positions. Only when the transmission volume and hence the loading of the network exceeds a certain level, the admissible tap range becomes empty, meaning that the slow LFC is no longer able to maintain network security.

Using a PST at least one common tap position needs to exist in all topologies. In other words, there must be a non-empty intersection of the admissible tap ranges. A DFC yields a benefit if a compromise between tap positions for different topologies is necessary. This is the case when for a higher transmission volume admissible tap ranges still exist for each topology, but no tap position can be found that is admissible for all topologies.

From this we can conclude that a benefit of a DFC compared to a PST can only be achieved if two requirements are fulfilled:

- two different (N-1)-topologies are limiting the transmission volume, not including the DFC outage,
- the DFC needs to have sensitivities with opposite signs on two 'limiting' lines (i.e. lines that are fully loaded in the critical topologies).

Therefore, the admissible tap ranges in the relevant (N-1)-topologies are the key measure to assess whether the DFC yields a benefit compared to the PST.

A four node network in Figure 8.2 with three lines in the prevailing transmission direction has been developed for the purpose of illustrating the principle of this approach. Power is injected at the lower three nodes and has its sink at the fourth node on top. With this network we have created three scenarios with different installations of LFCs. In the first scenario an LFC is placed crossways to the transmission direction between a double line and a single circuit.

## 8.2 Placement of Power Flow Controllers

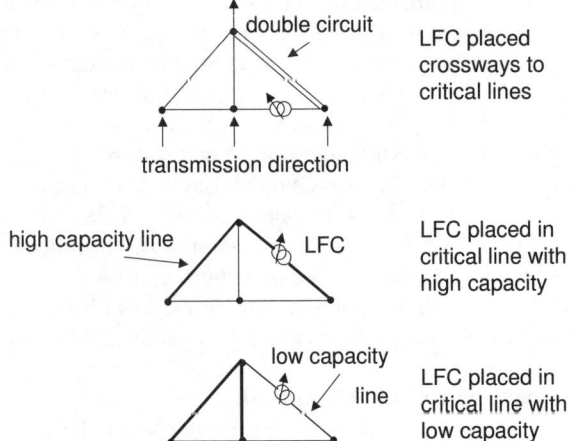

Fig. 8.2. Possible locations of LFCs - schematic illustration

In both remaining scenarios the LFC is installed in one of the lines in main transmission direction, with the difference that it is a strong line in the second scenario and a weak line in the third one.

In the first schematic network configuration shown in Figure 8.3 the LFC is placed crossways to critical lines, which satisfies the prerequisites of two limiting topologies not including the LFC outage and sensitivities with opposite sign on two limiting lines. In case of a line outage of the single line on the left the LFC would be used to relieve the central line, which is illustrated by an arrow. If one circuit of the double line trips, the LFC will aim at relieving the remaining circuit to prevent an overload.

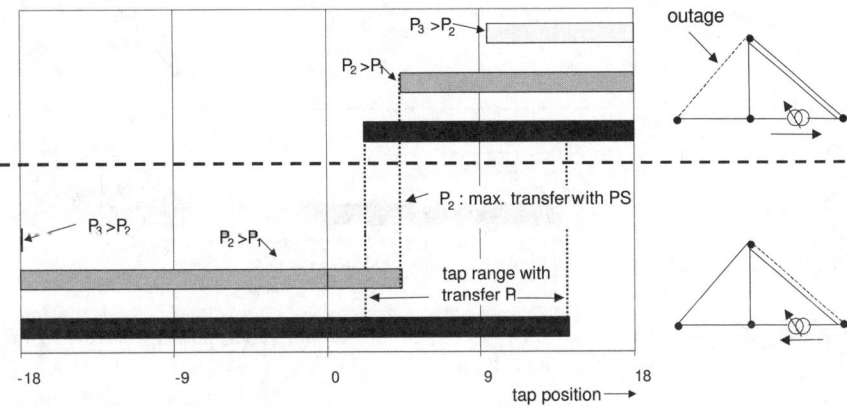

Fig. 8.3. Admissible tap ranges for two critical outages and varied power transfers – LFC placed crossways to critical lines

Hence, the circular flow injected by the LFC is in opposite directions for the two critical topologies. When using a PST a compromise between tap positions for these two topologies will be necessary which limits the amount of transfer. With low transfer $P_1$ the admissible tap ranges overlap and several common tap positions can be found that fulfil all network constraints (black bars in Fig. 8.3). When increasing the transfer the admissible tap ranges will shrink up to the point when they have just one single tap position in common (grey bars). The according transfer $P_2$ is the highest transfer that can be achieved using a PST. For any transfer higher than $P_2$ a DFC is required because the admissible tap ranges (white bars) have then become mutually exclusive. The maximum transfer $P_3$ is achieved when the DFC reaches its maximum or minimum tap setting for at least one of the critical topologies. The gain of transmission capacity by using a DFC instead of a PST is $P_3$-$P_2$.

In the second schematic network configuration in Figure 8.4 the LFC is placed in a line with high transmission capacity. This means that the LFC contingency is among the critical contingencies. Hence, the resulting tap range is only relevant for a single critical topology, meaning that a PST is equivalent to a DFC.

To avoid the LFC being among the critical contingencies, it is now placed in a low capacity line while the two other lines leading to the sink node are strong ones in this scenario of Figure 8.5. However, in case of an outage the relief of critical lines is achieved by tap changing in the same direction for both topologies as it is indicated by the arrows in the Figure. Consequently, the admissible tap ranges always overlap, which means that again a DFC is equivalent to a PST.

Even a scenario with two LFCs, one in each strong line, can be traced back to a superposition of the previous two scenarios making a PST equivalent to a DFC.

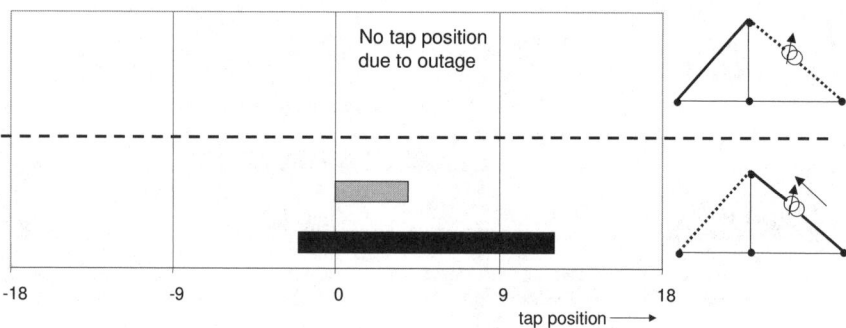

**Fig. 8.4.** Admissible tap ranges for two critical outages and varied power transfers - LFC placed in critical line with high capacity

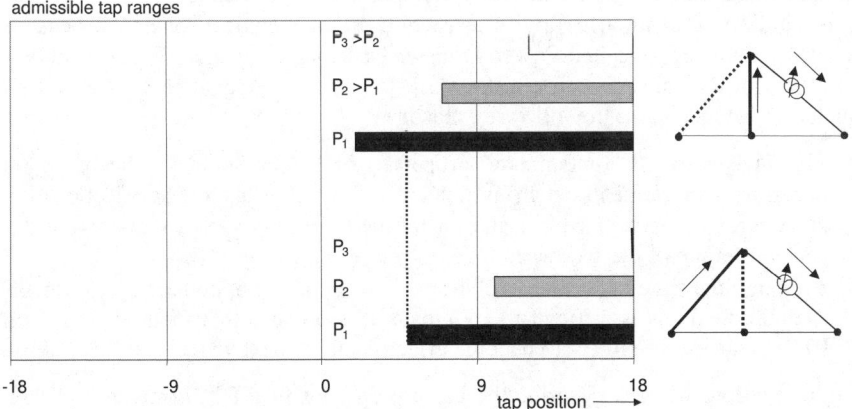

**Fig. 8.5.** Admissible tap ranges for two critical outages and varied power transfers – LFC placed in critical line with low capacity

The fundamental analysis on the basis of the schematic networks shows that an increase of transmission capacity can only be achieved under special circumstances. To achieve non overlapping admissible tap ranges the following conditions need to be fulfilled:

- DFC placed crossways to 2 critical lines, and
- two critical topologies exist excluding the DFC.

To confirm this conclusion under realistic UCTE network conditions an analysis of different DFC locations for an exemplary network situation has been carried out, which will be presented in section 8.4.

## 8.3 Economic Evaluation Method

Both advantages to be analysed - loss reduction and increase of transmission capacity - relate to the topic of cross-border congestion management, because this is the primary reason for the installation of load flow controlling devices. Therefore, we first discuss how to include load flow controllers in network models used for congestion management, in particular for the allocation of transmission rights.

### 8.3.1 Modelling of LFC for Cross-Border Congestion Management

Various different methods for the allocation of transmission rights have been implemented by the TSOs in recent years. In the EU this development has been accelerated by the coming-into-force of the related EC regulation 1228/2003 in mid-

2004 [1]. There is a clear tendency towards solutions that are based on intensified coordination among TSOs and between TSOs and other actors, with the aim of better utilisation of the network infrastructure. In the technical sphere, this can be achieved by using so-called Power Transfer Distribution Factor- (PTDF-) models to represent the transmission constraints [4]. The following method is based on such PTDF-models, for the following reasons:

- The analysis of the fundamental properties of LFCs should not be based on transitory arrangements (such as bilateral and/or non market-based allocation procedures for transmission rights in meshed grids) - also in view of the expected lifetime of these devices.
- Although the pace of further evolution of the actually applied congestion management methods is difficult to anticipate, it is obvious that methods based on PTDF-matrices are likely to become effective in the next years.

In this section, we first describe the basic properties of a PTDF-model and then discuss how 'slow' and 'fast' load flow controllers can be included therein.

### 8.3.1.1 Basic Network Model

The cross-border transmission capability of for instance the UCTE network is mostly restricted by the admissible line currents, which, given the relatively constant voltage level, can approximately be expressed in terms of active power flows.

The PTDF-model is based on a linearization of the steady-state load flow equations, which is valid with acceptable accuracy for the context of congestion management: The power flow on each transmission line (or transformer) has an approximately constant sensitivity with respect to the export/import balance of a given network zone (corresponding to a trade area, usually one TSO's control area). Therefore, the limited transmission capability of the network can be expressed through a set of inequalities that link the maximum admissible flow on the lines (or transformers) to the zonal balances:

$$\begin{pmatrix} \text{Topo. 1} & \begin{matrix} \text{line 1} \\ \vdots \\ \text{line m} \end{matrix} & S \\ \text{Topo. 2} & \begin{matrix} \text{line 1} \\ \text{line m} \end{matrix} & S \end{pmatrix} \cdot \begin{pmatrix} \text{zone 1} \\ \vdots \\ \text{zone n} \end{pmatrix} \Delta P \leq \begin{pmatrix} \text{line 1} \\ \text{line m} \\ \text{line 1} \\ \text{line m} \end{pmatrix} P^{max} \quad (8.1)$$

The (N-1)-network security criterion can be reflected in the PTDF-model by computing the sensitivities for each contingency topology and combining the results to a large set of inequalities. This makes sure that a set of zonal balances (i.e. power exchanges between the zones) must not lead to a violation of line flow limits in any of the considered topologies.

### 8.3.1.2 Inclusion of 'Slow' LFC

In principle, load flow controlling devices have a similar effect on the network as the zonal balances: They alter the flow on the lines and transformers. It can be shown that this influence is also approximately linear, i.e. the incremental power flows due to tap changes of load flow controllers (LFCs) can be superimposed on those induced by the zonal balances, and they are proportional to the tap setting. (Note that we are using the term 'tap change' here and in the following, although power electronic load flow controllers can be designed such that they allow for continuous shifting. In the PTDF-model, this can be easily reflected by allowing the 'tap position' to have continuous values instead of integers in the case of conventional PSTs.) Consequently, the PTDF matrix needs to be extended by one column per LFC:

$$\begin{pmatrix} \text{Topo. 1} & \begin{matrix} \text{line 1} \\ \vdots \\ \text{line m} \end{matrix} & S & S_{tap} \\ \text{Topo. 2} & \begin{matrix} \text{line 1} \\ \text{line m} \end{matrix} & S & S_{tap} \end{pmatrix} \begin{pmatrix} \text{zone 1} \\ \vdots \\ \text{zone n} \\ \text{LFC 1} \\ \text{LFC 2} \end{pmatrix} \begin{pmatrix} \Delta P \\ \text{-----} \\ \Delta tap_{LFC1} \\ \Delta tap_{LFC2} \end{pmatrix} \leq \begin{pmatrix} \text{line 1} \\ \text{line m} \\ \text{line 1} \\ \text{line m} \end{pmatrix} P^{max} \quad (8.2)$$

Like with the basic model, the integration of sensitivities for all contingency topologies ensures that one set of tap positions (and one set of zonal balances) satisfies all contingency conditions, thus reflecting properly the requirements of the 'slow' PSTs.

### 8.3.1.3 Inclusion of 'Fast' LFC

The difference between 'slow' and 'fast' LFCs is that the latter can be shifted to an individual tap position after each contingency. When modelling the network constraints, this means that each topology may have its individual set of tap settings. For example, tap settings applying to topology 1 have no effect in topology 2, 3 etc. This is reflected by blocks of zeros in the PTDF-matrix:

$$\begin{pmatrix} \text{Topo. 1} & \begin{matrix} \text{line 1} \\ \vdots \\ \text{line m} \end{matrix} & S & S_{tap} & 0 & S_{tap} & 0 \\ \text{Topo. 2} & \begin{matrix} \text{line 1} \\ \text{line m} \end{matrix} & S & 0 & S_{tap} & 0 & S_{tap} \end{pmatrix} \begin{pmatrix} \text{zone 1} \\ \vdots \\ \text{zone n} \\ \text{LFC 1} \\ \text{LFC 1} \\ \text{LFC 2} \\ \text{LFC 2} \end{pmatrix} \begin{pmatrix} \Delta P \\ --- \\ \Delta tap_{LFC1} \\ \Delta tap_{LFC1} \\ \Delta tap_{LFC2} \\ \Delta tap_{LFC2} \end{pmatrix} \leq \begin{pmatrix} \text{line 1} \\ \text{line m} \\ \text{line 1} \\ \text{line m} \end{pmatrix} P^{max} \quad (8.3)$$

## 8.3.2 Determination of Cross-Border Transmission Capacity

Algorithms calculating bilateral cross-border transmission capacity as well as coordinated mechanisms for multi-zone capacity allocation determine the maximum cross-border power exchange that is admissible within the limitations imposed by the transmission network. Mathematically, this can be expressed as an optimisation problem in which the PTDF-model constitutes the principal part of the constraints. A comparison between PSTs and DFCs can then be achieved by simply switching between the models described by equations (8.2) and (8.3), respectively.

The specification of the objective function reflects the context of exchange maximisation (e.g. bilateral capacity calculation, co-ordinated explicit auctioning or implicit auctioning). For this study, two methods are appropriate, depending on the focus of the investigations:

- For the increase of transmission capacity by DFC in comparison to PST (section 8.2), the amount of power exchange in a fixed direction (e.g. from country A to country B) forms the objective function. This means that the zonal balance in A contributes positively and in B negatively to the objective function, whereas all other balances are set to zero. Optimisation variables are the zonal balances and the LFC settings. Such a procedure is based on the assumption that (in a given trading interval) the regarded power transfer direction is economically beneficial. This allows to isolate the effect of having either fast or slow LFCs and avoids confusion by superposition with interdependent effects that are difficult to trace in detail.
- In a market with several trading zones, the most beneficial transfer direction is volatile. Moreover, there might be interdependency between the optimal transfer direction and the PTDF-model variant for PST or DFC. Therefore, the estimations of loss reduction as well as of the economic welfare gain through LFCs are carried out without prescribing such a direction. Rather, the zonal balances are a result of the variable unit commitment, and the LFCs' tap positions are used as degrees of freedom in an optimization with the objective function of minimal total generation cost. The methods used for these analyses besides the PTDF-model are described in the following section.

### 8.3.3 Estimation of Economic Welfare Gain through LFC

Severe transmission congestions have occurred since the liberalisation of the electricity supply sector as a consequence of increasing cross-border power transfers. The congestion hinders free energy trades and leads to different regional electricity prices at the national power markets. In an ideal market, the economic benefit of additional transmission capacity is determined by the reduction of generation costs due to an additional power transfer from the area with lower marginal costs into the area with higher marginal costs of generation. The associated costs of additional transmission capacity consist of investment and maintenance costs of network reinforcement, as well as costs of network losses. The maximum of social welfare can be reached by maximising the difference between the benefit and the associated costs.

In the following the reduction of generation costs due to an increased transmission capacity is estimated applying a market simulation based on a generation dispatch model. This model optimises generation plant dispatch and transmission capacity usage by minimising total generation costs in the system. A part of the UCTE system is considered for which the model comprises major Western European countries. On the generation side, the thermal and hydro generation as well as the wind power injection are considered. The transmission network including LFCs is modelled as linearized transmission constraints (see section 8.3.1).

Fig. 8.6 shows an overview of the developed methodology to evaluate the benefit of LFCs. The influences of LFCs on the transmission capacity and thus the generation dispatch are quantified by the market simulation. Hourly unit commitment, generation costs, cross-border energy exchange, as well as the setting of LFCs are the essential results of the market simulation. With this information a load flow calculation can be carried out to compute the corresponding hourly network losses.

This procedure shall be done both for conventional PSTs and for the fast DFCs. Finally we can compare the change of generation costs, as well as of network loss cost.

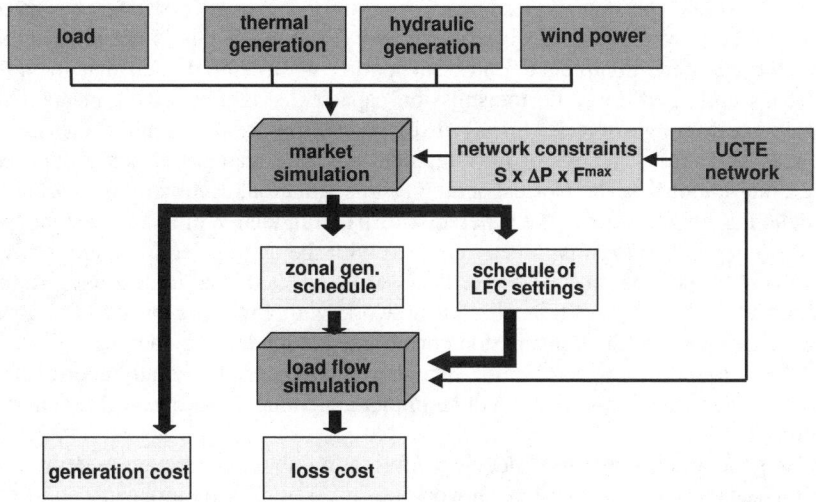

**Fig. 8.6.** Overview of the methodology applied to quantify the economic welfare gain through DFCs

### 8.3.3.1 Generation Cost - Market Simulation

Owing to the problem size the optimisation is solved by linear programming. The objective function represents minimisation of variable generation cost to cover the load in the entire system. Besides the hourly output of generation units, the setting of LFCs is a part of the optimisation variables, too. The optimisation constraints consist of technical properties of power plants, transmission network restrictions and many other technical conditions, such as load balance and system reserve requirements.

The generation dispatch model uses the system marginal cost as the price estimator. The electricity price in a perfectly competitive market shall be equal to the system marginal cost at all times. In a market with only a few players, electricity prices are subject to the strategic behaviour of dominant market participants. In this case, electricity prices can be raised above the system marginal cost. While this market power phenomenon is more or less evident in a number of European regional markets today, it can be expected to be mitigated in the medium term as a consequence of efforts by the European Commission, the national regulators and other stakeholders. For example, mitigating market power is one of the main goals of improving congestion management methods in the European transmission system [4]. Hence we can expect that market prices will tend to approach the marginal cost level in the future. Besides, a simulation based on marginal cost minimisation is ideal to reveal the socio-economic welfare gain that could be achieved by DFCs.

While the cross-border transmission capacity often limits the international power transfers, the power grids inside a country are usually strong enough to support free power transfers within the country. Although this is not always exactly the case, the continental European market is structured such that in most countries unlimited domestic transmission capacity is assumed. This means that any internal congestion is managed by corrective re-dispatching measures, whereas ex-ante restrictions of inter-regional power exchanges are only imposed on an international scale. Corresponding to this situation, each national market is modelled as a trade zone in the generation dispatch model. Within a trade zone the whole generation system is directly coupled with the aggregated load on a fictive lossless hub, thus the internal power flows are neglected. The trade zones are interconnected through the transmission network with limited capacity. The load flow behaviours and the transmission constraints are modelled as described in section 8.3.1, based on the assumption that step by step a multilaterally coordinated allocation of transmission rights will be implemented in the European Union in the future.

The generation model uses detailed information about generation systems and the interconnected transmission network of seven Western European countries: Austria, Belgium, Switzerland, the Netherlands, Germany, France and Italy (Figure 8.7). The Netherlands and Belgium are merged to one region in order to save computing time, as network congestion between these two countries seems to be quite rare, according to transmission capacity auctioning results.

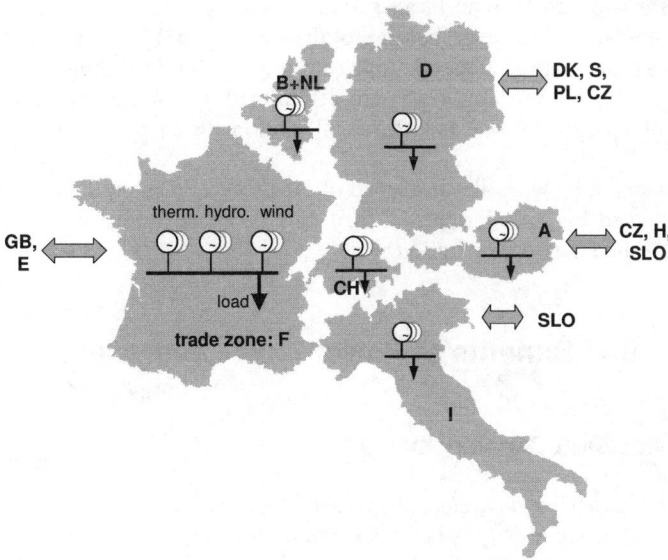

**Fig. 8.7.** Countries considered in the market simulation

The surrounding countries - Spain, Great Britain, the Scandinavian countries as well as the Eastern European countries - are not modelled in detail. Their interactions with those countries that are modelled in detail are considered as predefined power exchange schedules.

Exogenous inputs such as load curves, fuel prices, available generation capacities and the transmission network aim at mirroring the situation of year 2003. All input data have been taken from public sources. For simplification a single network topology is used for the entire year, which is sufficient to show under which circumstances DFCs can have advantages compared to conventional PSTs. As we are interested here in a comparison between new DFCs and conventional PSTs, not the absolute value, but only the differences of simulation results are considered, which reduces further the impact of inaccuracies of the input data.

### 8.3.3.2 Loss Cost – Load Flow simulation

As mentioned above, the investigation is based on a single load flow dataset drawn from public sources. In order to calculate the transmission losses it is necessary to update this load flow model according to the hourly market simulation results.

These results contain the aggregated generation and load within each trade zone for each simulated point of time. To update the load flow model the deviations of these zonal values from the base case are distributed to all the generation units and loads proportional to their power injection/take-off in the base case. This constitutes a decoupling of the market simulation model and the load flow model.

The transmission constraints in all relevant (N-1)-situations are considered in the market simulation. However, because of the short duration of post-contingency situations only the normal network topology is used for load flow calculation. Hence, the yearly transmission losses are determined by adding up the transmission losses obtained from one load flow calculation for each simulated point of time.

The comparison between the influence of DFCs and PSTs on the transmission losses is achieved by considering that they may be set to different tap positions during undisturbed network operation.

## 8.4 Quantified Benefits of Power Flow Controllers

### 8.4.1 Transmission Capacity Increase

As a scenario we have the selected the European region around Belgium, the Netherlands and Luxembourg. The reason is on the one hand the actual presence of congestion and on the other hand the fact that there are already LFCs in operation at two substations (Meeden and Gronau) at the Dutch-German border. Figure 8.8 gives an overview of the principal transmission lines in that area. A more detailed description of the congestion situation can be found in [5].

Before performing the one year market simulation to estimate the economic benefit an appropriate location for the DFC needs to be identified. Today the existing PSTs in the considered Benelux region are placed in tie lines, so we considered first DFCs as replacements for the existing PSTs at the Dutch-German border. As result of the fundamental analysis in section 8.2 the use of DFCs for transmission increase in a tie line is unlikely to provide a benefit towards PSTs.

For the part of UCTE-system a location for a DFC to increase the transmission capacity more than a PST could only be found under some specific network development assumptions. We had to reduce the thermal limits of the double circuit Vigy-Uchtelfangen by 40 % to make it a critical line. Also neglecting limitations of 220-kV-lines (e.g. tie line Moulaine-Aubange) was necessary. The DFC and in comparison the PST is placed in the branch Vigy-Moulaine.

For the analysis we first regard a single network situation and a fixed transmission direction from France to Belgium/Netherlands (BNL). This is equivalent to calculating the available transfer capacity (ATC) from France to BNL. In this modified scenario BNL imports 3900 MW with 200 MW remaining import capacity from France to BNL without any LFCs. By installing a PST a total import of 4300 MW can be achieved. Considering a DFC we have evaluated a maximum power import of 4700 MW which gives and additional benefit of 400 MW by the speed of the device. The result of the tap calculation is shown in Figure 8.9.

The next step is to determine how often this benefit is achieved during a year and what total profit will be gained. Therefore we perform a one year market simulation for this network situation.

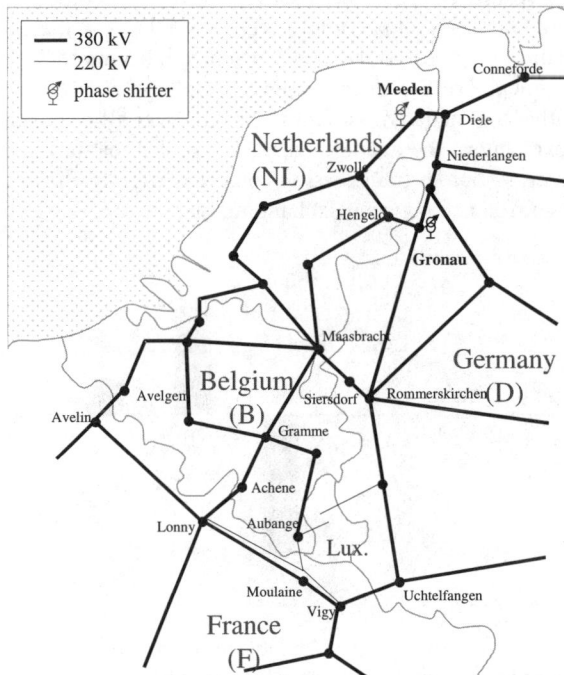

**Fig. 8.8.** Overview of the transmission network in the vicinity of Belgium and the Netherlands

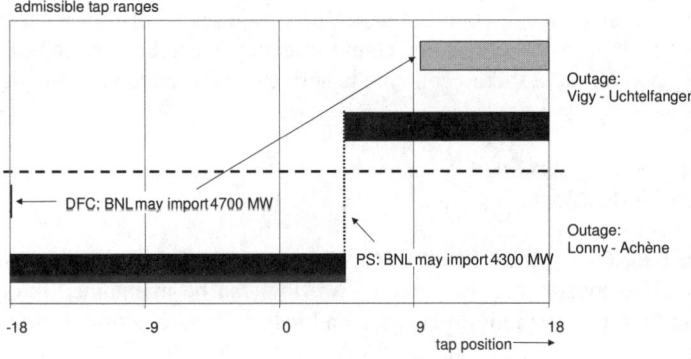

**Fig. 8.9.** Admissible tap ranges for two critical outages and varied power transfers - DFC placed crossways to critical lines

The simulation results of Figure 8.10 show that total generation cost can be reduced by 3.8 Mio. €/a. The comparison between the DFC and the PST simulation show that a variation of cross-border power exchange occurs in 53 % of the time slots. As a consequence of the inter-temporal couplings due to the hydro power plants, both positive and negative variations occur. Over all time slots with variations of power exchanges, the average increase (i.e. the net sum of increases and decreases) of the cross-border transmission volume is slightly above 200 MW, and the average price difference between exchanging market zones is 3.76 €/MWh.

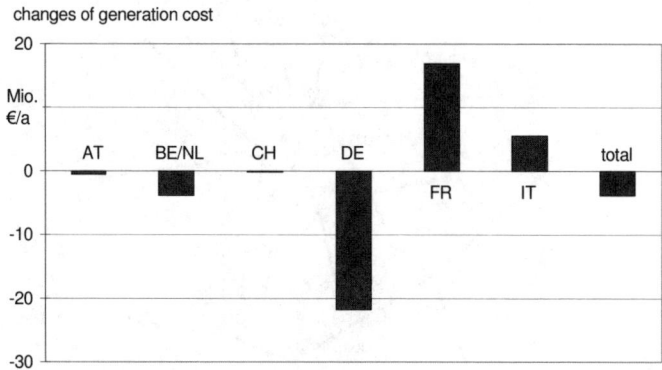

**Fig. 8.10.** Changes of generation cost through DFC in comparison to PST – Scenario: Reduction of thermal limit on Vigy-Uchtelfangen, neglect of 220-kV-limitations, DFC/PST placed in branch Vigy-Moulaine

The simulation also shows that the changes of generation cost are quite different in the affected countries. France for example bears an increase of generation cost caused by a higher amount of power export. On average this would also result in higher marginal cost in France, since it is likely that at least during some hours of the year the increased amount of generation requires to operate significantly more costly generators. So only some countries earn a 'local' benefit from the DFC installation, but in total a socio-economic benefit can be achieved for this fictitious scenario.

### 8.4.2 Loss Reduction

Because of their fast controllability DFCs allow to react to a contingency after its occurrence. This means that the natural load flow can be maintained throughout most of the time by operating in 'neutral' and loss optimal position during undisturbed network operation. In contrast to that the schedule of PST settings must be followed in all network situations according to the most critical contingencies which might come. From this difference a reduction of network losses can be assumed.

With the model that has been introduced in section 8.3.3 a simulation of an ideal market is carried out taking into account load and generation as well as the

UCTE-network with its constraints (Figure 8.11). Load flow calculations are performed determining the loss cost regarding the zonal generation schedule and the schedule of LFC-settings resulting from the market simulation. The comparison between the utilisation of DFCs or PSTs for load flow control is achieved by following the schedule of hourly post-contingency tap settings (PST) or maintaining the neutral setting (DFC), respectively.

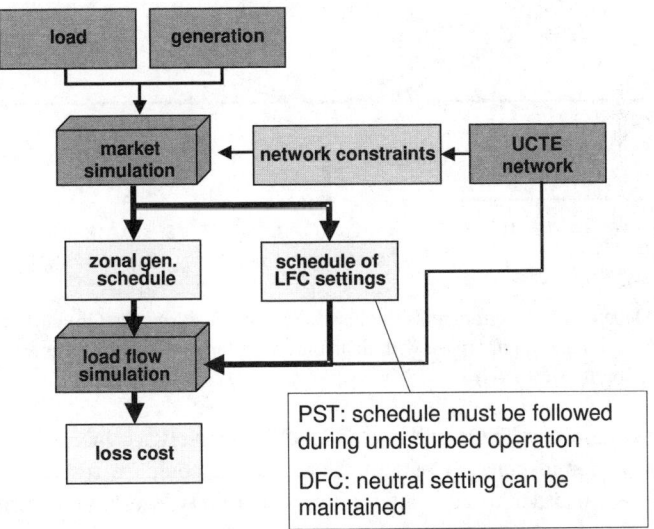

**Fig. 8.11.** Methodology for determination of loss reduction by using DFC instead of PST

Strictly speaking, the loss-minimal tap setting can, depending on the network situation, differ from the neutral setting. This would mean that, in theory, also the DFC would have to follow an hourly schedule. In order to assess the relevance of this aspect for the comparison between DFC and PST, we first perform a sensitivity analysis regarding the dependency of transmission losses on tap positions of PSTs in the two locations Gronau and Meeden.

We first regard the level of transmission losses for 2 exemplary load situations (Figure 8.12, curves, right axis). While the loss minimal tap position is zero or near zero in both peak load and off-peak situations, maximum losses occur at the lower and upper boundary of the tap range.

A subsequent one year market simulation identifies market optimal tap positions at the lower boundary of the tap range in most of the hours (Figure 8.12, columns, left axis). In comparison to the difference of the incremental losses between loss minimal and market optimal tap positions, the difference of incremental losses between the two loss minimal tap positions (peak and off-peak) is almost negligible. Therefore, neutral setting (i.e. a tap position of zero) is an adequate assumption for the default DFC-setting in undisturbed operation.

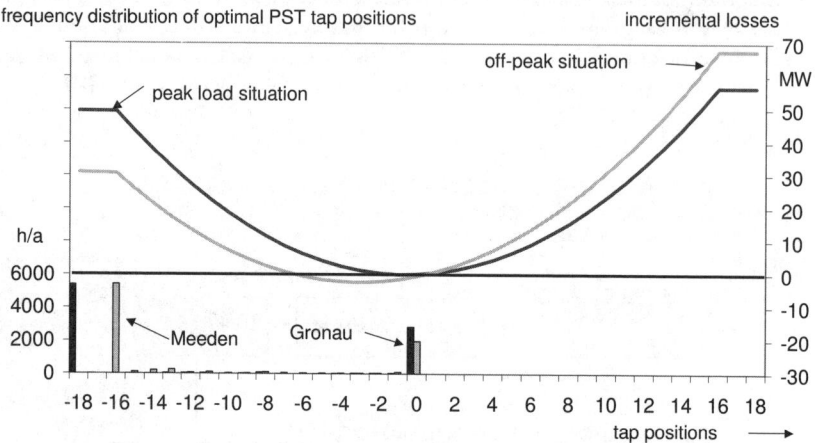

**Fig. 8.12.** Dependency of transmission losses on tap positions in Gronau and Meeden (curves, right axis) and comparison with distribution of tap positions resulting from market simulation (columns, left axis)

It is not certain that TSOs will use an LFC exclusively for increasing transmission capacity. For instance, a part of the tap range could be reserved for emergency actions. In order to account for this uncertainty, we have determined the yearly loss reduction for different restrictions on the share of the tap range to be used for capacity increase. For the monetary assessment we have assumed a loss price of 34 €/MWh, taken from European Energy Exchange (EEX) in November 2004 as the average price for base load for 2005 and 2006.

With full admissible tap range (i.e. the entire tap range is used for capacity maximisation) the loss reduction is 265 GWh/a, which constitutes a monetary benefit of 9.0 Mio. €/a with our assumption of 34 €/MWh. Yet, the simulation results also indicate a strong dependency of the loss reduction on tap range restrictions (Figure 8.13): For example, a bisection of the admissible tap range results in a decrease of the loss cost reduction by 80 %, meaning a drop to 1.8 Mio €/a in monetary terms. Obviously the uncertainty which share of the maximum tap range will be used by the TSOs has a significant influence on the profitability of DFCs. Figure 8.13 also shows a break-down of the loss changes per country. In the considered scenario the complete loss savings are gained in the zone Belgium/Netherlands (BNL), whereas all other zones will get a (smaller) increase of losses. We assume, however, that this finding cannot be generalised, but is depending on the specific scenario and might also be affected by the inevitable roughness of the simulation models.

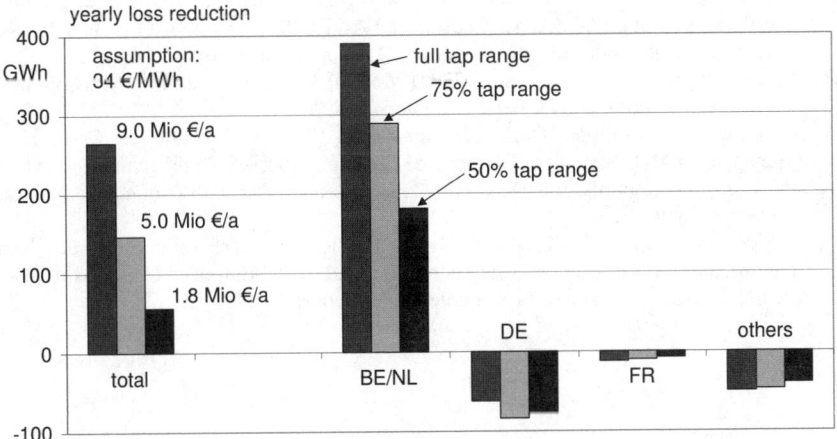

**Fig. 8.13.** Yearly loss reduction through DFCs (instead of PSTs) in Gronau and Meeden for different tap range restrictions

The energy market simulations of real situations in the European UCTE-systems quantify the benefits of the DFC in comparison to the PST. The control speed can primarily be used to reduce the system losses by operating in normal conditions according to optimal losses and adapt the operational point only in case of a contingency. The benefit in the case shown here was up to 9.0 Mio. €/a. This means that the dynamic capability would justify roughly a double price for the DFC in comparison to a 1000-MVA-PST with a payback in one year.

The first case for transmission capability increase has shown significant benefits only under special assumptions. The selection of a location where contradictory control actions are required might only occur in situations where the market actions change the power flow direction over the interconnection.

In conclusion, with specific operation strategy and placement significant benefits can be achieved by using fast controllable FACTS-devices in comparison to conventional power flow controllers. The capability for stability increase was neglected so far and will be discussed in the following two chapters. After that in chapters 10 to 12 a complete control scheme will be defined enabling to earn the benefits shown here.

## References

[1] European Parliament and Council of the European Union, "Regulation on conditions for access to the network for cross-border exchanges in electricity" Regulation (EC) No 1228/2003 of 26 June 2003, Official Journal of the European Union, L 176/1, 15 July 2003

[2] Larsson M, Rehtanz C, Westermann D (2004) Improvement of Cross-border Trading Capabilities through Wide-area Control of FACTS. IREP Symposium, Bulk Power System Dynamics and Control VI, Cortina D'Ampezzo, Italy
[3] Larsson M, Rehtanz C, Bertsch J (2003) Monitoring and Operation of Transmission Corridors. IEEE Bologna Power Tech, Italy
[4] Consentec and Frontier Economics (2004) Analysis of Cross-Border Congestion Management Methods for the EU Internal Electricity Market. Study commissioned by the European Commission, DG TREN, Final Report, Aachen/London (available at www.consentec.de)
[5] IAEW and Consentec (2001) Analysis of Electricity Network Capacities and Identification of Congestion. Study commissioned by the European Commission, DG TREN, Final Report, Aachen (available at www.consentec.de)

# 9 Non-Intrusive System Control of FACTS

For the implementation of FACTS-Devices, especially for controllable transmission paths in an AC-system, intensive planning studies and redesign of control and protection systems have to be executed. Adverse control interactions with other controllers and a lack of optimization potential due to predefined devices have to be considered. Applying a control architecture, which enables the operation of a new FACTS-device and especially a controlled transmission path without affecting the rest of the system, can eliminate these problems. This non-intrusiveness is the key issue of the so-called Non-Intrusive System Control (NISC) architecture. In this chapter the basic requirements and structure of this new control architecture are described first. A second focus is given to the problem of controller interactions in abnormal operation situations where the NISC architecture helps to avoid malfunctioning or adverse reactions.

## 9.1 Requirement Specification

Power system control analysis and design methodologies are mainly aiming at the assessment of single devices by means of their systemic behavior. In particular in the area of devices enhancing the flexibility of power systems (FACTS-devices) the corresponding design techniques are dedicated to either steady-state operation or power system dynamic improvement. In the spot of application studies normally FACTS-devices are considered as stand-alone solutions. These approaches are limited to a given device functionality rather than considering and designing the entire system on functional basis. The design of a solution for a transmission problem by starting from a functional specification offers more degrees of freedom. Herein, impedance control, voltage and current injection are considered as single functions. However, this design process demands a corresponding portfolio of modularized components comprising switched elements as well as power electronic subsystems. The device requirements as a result of the design process needs to be mapped to select the specific FACTS-devices out of the available portfolio.

Beyond these hardware related issues the design of a proper control and system integration methodology is needed. Most of the known approaches demand to consider the entire system, i.e. detailed knowledge of the structure and parameters of all other network components is mandatory for the design process. This is not only related to a huge effort during the design phase but also more and more limited due to the deregulation. Since transmission as such becomes a competitive instrument the availability of planning data cannot basically be assured. Especially for congested transmission paths between utility or country borders it is hard to get complete system planning data for the entire system.

Furthermore, the design methods may yield a complete set of new parameters for all controllers of the entire system. Both, new controlled and uncontrolled AC-transmission paths will always affect the dynamics and behavior of the rest of system. In conclusion, it is mandatory to provide a system behavior that is not inadvertently affecting the entire system. Exceptions are related to the provision of certain control functions as ancillary services.

The proposed control architecture, called Non-Intrusive System Control (NISC) avoids complete system redesign. It enables a most effective system expansion and more effective network utilization by considering the needed transmission functions first. In a second step the hardware modules are assembled accordingly. The goal of the NISC-architecture is to simplify the design process so that the new controlled transmission paths can be designed without extensive system studies. For the operation of a new transmission path the NISC-architecture avoids adverse control interactions within the entire system without causing a redesign of already implemented controllers. Those are automatic voltage regulators, power system stabilizers etc. Additionally, the proposed architecture allows for a proper reaction on critical events and avoids insufficient and hence wrong operation after the power system state changes. Both, normal and abnormal operation situations are considered at the same time. In contrast, if the entire control systems would have been designed according to global parameterization for a fixed topology maloperations and adverse control interactions may occur [1], [2].

After describing the general approach of NISC, the different aspects of the NISC design methodology are discussed to more detailed extend in the following.

### 9.1.1 Modularized Network Controllers

The expansion of an electric power network means adding a new part to the system or upgrading an existing part for the transmission of electric power. Mostly this is limited to a connection of two points of a given network or between two networks (included are also 3 point connections or the interconnection of a new independent power producer). If this connection is supposed to be controllable or the controllability of a given transmission system is suggested to certain extensions, transformer based, especially phase shifter, or power electronic based subsystems are installed. In particular the latter ones are integrated into the system to enable power flow control, reactive power compensation or ancillary services like damping of oscillations. Ideally, a controllable transmission line can be modeled as a system comprising sending end, receiving end and an intermediate coupling. In the ideal case both ends show a decoupled behavior. Figure 9.1 shows the principle structure of such a transmission interconnection.

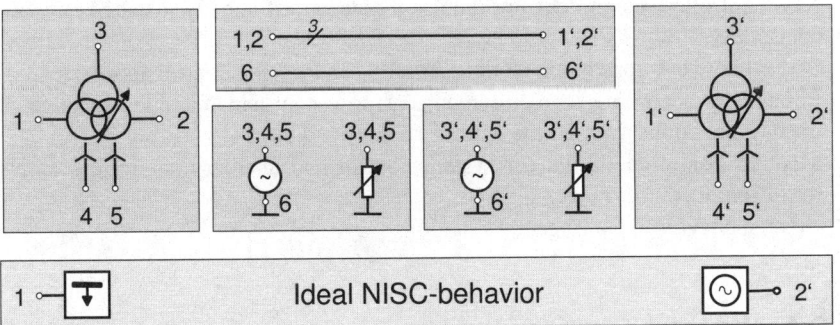

**Fig. 9.1.** Model of a controllable transmission line with the NISC-approach and underlying building block philosophy

Against this background the NISC-architecture as control philosophy demands a certain amount of controllability. This can be achieved by the FACTS-devices introduced in chapter 1 based on controlled impedances or voltage sources and transformers. In addition special designs could be considered like a four conductor transmission line with symmetry compensation [3] or transmission lines with a certain surge impedance in order to avoid bulk series compensation equipment [4]. Furthermore, controlled series resonance circuits can be added for decoupling the sending and receiving end in terms for short circuit current contributions [5]. As a result the transmission path can be designed according to a building block concept and hence a huge variety of controllers can be created based on the basic FACTS-elements.

### 9.1.2 Controller Specification

Conventional controller designs for controllable transmission paths demand to incorporate the entire system. In most of the cases this results in a redesign of other network controllers. The controllers should follow the desired functionality independent of hardware configuration of new transmission elements. Easy scalability to different control ranges and flexibility to add ancillary services is required. However, today the number of controlled paths is limited since the control systems cannot cope with potential adverse interaction of these controlled paths. This problem can be overcome by either overall network controllers, which would desire a complete new high-speed network control system. Even in this case the adverse interaction cannot definitely be avoided.

A second approach is to design a controller working for fast actions on local input variables, but achieves coordination through exchange of information with selected parts of the entire system. This reflects the basic requirement for the NISC-architecture. For the realization of such a controller design the following specifications are defined:

- New controller design does not require a redesign of already installed network controllers
- Several network controllers work together with the same control approach
- Robustness according to requirements of power system operation (change of operational points during time periods of days and years)
- Modular controller design for system control and ancillary services; scalable for different control ranges
- No misbehavior in contingency situations

## 9.2 Architecture

Generally, one has to distinguish between predefined robust controllers for regular operation and contingency situations. In the following the controller for regular operation is referred to the function $\Im_1(\underline{u}_1)$. This function comprises several control algorithms for controlling the transmission path, e.g. active power flow control, reactive power flow control, voltage control, etc. The contingency case is covered by function $\Im_2(\underline{x},\underline{u}_2)$. This function affects the regular device control in order to adapt its behavior according to changing network conditions, in particular during contingencies. The overall structure of a NISC controller is shown in Figure 9.2.

In the simplest case the contingency controller does not affect the regular control function. For the initial design of the controller the function of the regular controller can be separated:

$$\Im_2(\Im_1(\underline{u}_1),\underline{u}_2) \equiv \Im_1(\underline{u}_1) \tag{9.1}$$

The design of the regular control function is traditionally based on a thorough network analysis where conventional robust controller design methodologies are applied e.g. $H_\infty$ [6]-[8]. For practical applications it is hard to get the dynamic system model to design the controller. The effort for this procedure is one reason for the limited use of network controllers in practice. Therefore the controller should be designed more or less independently from detailed system studies for each application. But at first the stability for such designs, independent from their special desired control characteristics, must be ensured.

If the controller has a certain desired characteristic for all operational points, the design can be done once without applying neither structural nor parameter changes during online operation. If not, the controller performance has be to checked in regular intervals and control parameters have to be updated accordingly. Therefore, the connection $D_2$ (see Figure 9.2) serves as a data channel used for downloading the updated control parameters.

## 9.2 Architecture

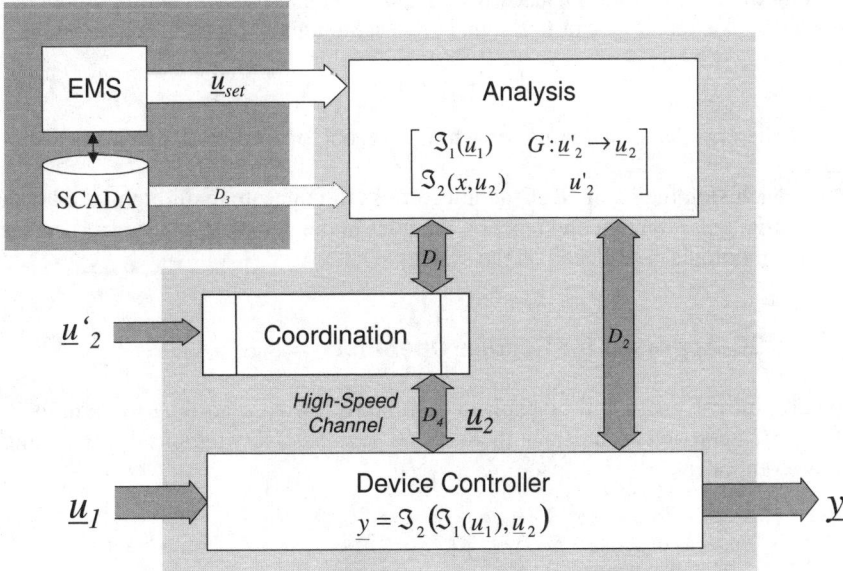

**Fig. 9.2.** Structure of NISC-Architecture

However from the theoretical point of view, the overall objective of this controller design methodology is to get rid of the connection between controller and SCADA-EMS-System $D_3$. The information exchange shall be reduced ideally to the set points $\underline{u}_{set}$ for the network controllers.

The contingency controller supervises the regular controller to prevent it from mal-functioning. This means coordination between the considered controlled transmission path and the entire system. One possible realization is a coordination instance, which derives (from measurement values $\underline{u}_2'$) the contingency case e.g. short circuit, line tripping, outages, overloading, under-voltage, etc. The result is an additional input $\underline{u}_2$ for the device controller upon which the regular control system structure is adapted to the contingency situation.

The coordination is time variant and depends on the actual network parameters and topology. Therefore the proposed NISC-architecture is despite its functional similarity not directly belonging to the class of adaptive controllers. The major difference lies in the mapping $G : \underline{u}_2' \rightarrow \underline{u}_2$, which defines what kind of measurement quantities are mapped on which additional input quantity for the device controller (see Figure 9.2). In particular in comparison to centralized real time network control systems, within this approach the amount of high speed data transmission is drastically reduced. No additional broadband SCADA-system is needed for the realization. However a certain exchange of date for online coordination of contingency cases cannot be avoided. Future optimization potential of the NISC-architecture lies in totally reducing the high speed data channel by substituting the coordination instance with a special signal processing unit on the device level. The major task of this signal processing unit is to establish a mapping

$$H : \underline{u}_1 \rightarrow \underline{u}_2 \tag{9.2}$$

and thereby deriving the contingency case out of locally available measurements.

In conclusion the ideal NISC-architecture shall concentrate all high-speed data processing, measurement and reaction schemes at the device level. Slow processes and methodologies are placed on the system level.

### 9.2.1 NISC-Approach for Regular Operation

The non-intrusiveness will be explained in the following according to Figure 9.3. The NISC-approach ensures that there are no new instability regions due to adding a new component.

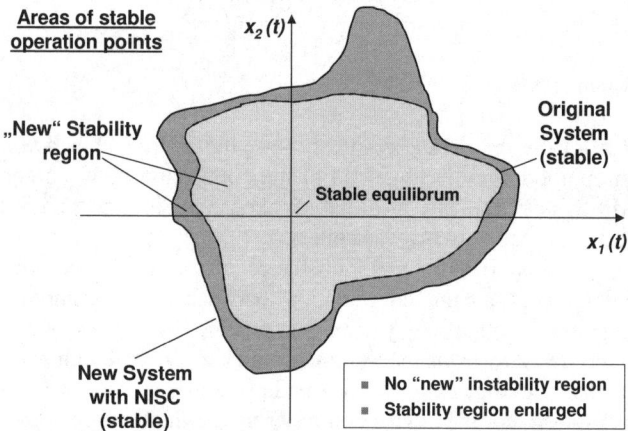

**Fig. 1.3.** Areas of stable operation points enlarged by adding new controllers with NISC-approach

The ideal goal of the NISC control design is to avoid the frequent update of the controller while ensuring certain robustness. There are several approaches possible to realize such a controller for the standard function of controlling the power flow or the voltage with the additional network element.

The first approach is coming from the theory of passivity. If a stable power system without the new controllable device is assumed, the system is passive if an energy function $V(T)$ exists for time points $T \geq 0$ [9].

$$V(T) \leq V(0) + \int_0^T y(t)u(t)dt \quad \forall \; u(.), \; T \geq 0 \tag{9.3}$$

If the additional network controller fulfills the same requirement and is also passive, then both systems in parallel or in a feedback loop are also passive and therefore stable. This means, that the additional component does not affect the stability itself if there is no energy input from this system. For the normal operation of fixing an operational point this is sufficient, but this approach does not tell anything about the damping of the resulting system. Also for additional components with storage characteristic this is not applicable.

Another nearly similar approach is the Controlled Lyapunov Function (CLF) for a system with the structure:

$$\dot{x} = f(x,u) = f_0(x) + \sum_{i=1}^{m} u_i f_i(x) \tag{9.4}$$

If the power system without control input is stable, it can be shown that there exists a positive energy function $V_{PS}(x)$ with $\dot{V}_{PS} \leq 0$. The system with the network controller is stable if, when $V_{PS}$ is combined with the energy function of the controllable element $V_{CO}$, the resulting function $V$ is a Lyapunov function for the new system. This holds if:

$$\dot{V} = \dot{V}_{PS} + \dot{V}_{CO} \leq \dot{V}_{CO} \leq 0 \tag{9.5}$$

In [10] this is shown with the example of a controllable series device. It is shown that the stability area of the resulting system is enlarged by adding the new component. To get an improved damping characteristic is a question of the controller design. The resulting controller must be checked to fulfill the above requirements for CLF. The results so far are adaptable for the basic control function. The robustness of the controller depends on the model of the device and is independent from the system's model so far the system can be assumed to be stable. Therefore a robust control design is desired.

To design a robust controller for specific characteristics it is desired to make the design based on a typical structural environment and not with a detailed system study. An approach for such a design is shown in [10] where the structure of the system is known, but not the exact parameter values.

With these approaches within the NISC-architecture a redesign of the controller can be avoided and the stable operation together with other controllers can be guaranteed. The stability is guaranteed and the robustness depends only on the device model. As a result the area of stable operation points remains the same after integrating a new controlled transmission path, which adds stable operation points.

### 9.2.2 NISC-Approach for Contingency Operation

The major difficulty for the application of network controllers is that it must be assured that they behave correctly during abnormal operation situations or contingency cases. In particular this is required for all kinds of fast controlling devices and therefore especially FACTS-devices. Many application studies have shown that the technical advantages of e.g. power flow controllers can only be profitably utilized in connection with a purposeful extension of the control and protection system. The critical factor is the dynamic behavior of the power system. This gets worsened and furthermore an overall endangering of the steady-state and dynamical system security is expected if the operation of network controllers like FACTS-devices are not coordinated properly.

The coordination has to be done according to changing operating situations or critical events in the power system. The NISC-architecture solved this problem due to its preventive coordination mechanism. This control is activated by a trigger signal reflecting a contingency event in the entire system. This broadcast activates the according local contingency control method within the device controllers (see Figure 9.4). After the contingency has been cleared the device controllers request a new planning and download cycle since the network topology or operation condition could have changed.

The analysis of the contingency cycle-time and the regular cycle-time shows that an online coordination of several network controllers cannot be achieved.

$$\Delta T_{CC} \ll \Delta T_{CR} \tag{9.6}$$

To implement a full dynamic system analysis online is not possible due to the centralized databases and analysis time effort. Therefore the underlying concept of coordination is referred to as preventive coordination since the coordination is done before execution starts.

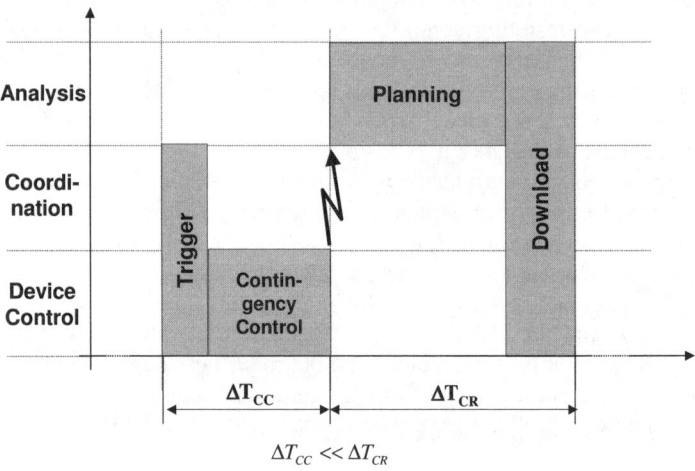

**Fig. 9.4.** Typical contingency control cycle within the NISC-architecture

This chapter has specified the requirements for fast network controllers especially FACTS-controllers. In particular power flow controllers require a coordinated approach, because of their interaction with wide parts of a power system. Adding FACTS-controllers shall always improve the stability of a system for all expected operations. Designs for regular and contingency operations can be separated. To be prepared for a contingency operation a analysis and planning phase has to be performed in cycles. The action schemes needs to be downloaded into the local controller. The controller is not prepared to act in contingency situations according to the pre-defined schemes. The required data are ideally locally available or need to be transmitted from pre-selected source in the system. The following chapters will show implementation examples for specific applications of this basic NISC-architecture.

# References

[1] Larsen EV, Sunchez-Gasca JJ (1995) Concepts for design of FACTS controllers to damp power swings. IEEE Transactions on Power Systems, vol 10, no 2
[2] Povh D, Haubrich H (1996) Global settings of FACTS controllers in power systems. CIGRE Session Paper 14-305
[3] Glavitsch H, Rahmani M (1998) Increased transmission capacity by forced symmetrization. IEEE Transactions on Power Systems, vol 13, no 1
[4] Esmeraldo PCV, Gabaglia CPR, Aleksandrov GN, Gerasimov IA, Evdokunin GN (1999) A proposed design for the new Furnas 500 kV transmission lines-the High Surge Impedance Loading Line. IEEE Transactions on Power Delivery, vol 14, no 1, pp 278-286
[5] Brochu J (1999) Interphase Power Controllers. Polytechnic International Press, Montreal
[6] Ngamroo I, Mitani Y, Tsuji K (1999) Robust load frequency control by solid-state phase shifter based on H∞ control design. IEEE PES Winter Meeting, vol 1, pp 725 - 730
[7] Taranto GN, Shiau JK, Chow JH, Othman HA (1997) Robust decentralized design for multiple FACTS damping controllers. IEE Proceedings Generation, Transmission and Distribution, vol 144, no 1, pp 61-67
[8] Wang L, Tsai MH (1998) Design of a H∞ static VAr controller for the damping of generator oscillations. International Conference on Power System Technology, Proceedings. POWERCON '98., vol 2, pp 785-789
[9] Ortega R, Loria A, Nicklasson PJ, Sira-Ramirez H (1998) Passivity-based Control of Euler-Lagrange Systems. Springer, Netherlands
[10] Andersson, G, Ghandari M, Hiskens IA (2000) Control Lyapunov Functions for controlled series devices. VII SEPOPE, Curitiba, Brazil
[11] Bulliger E, Allgöwer F (2000) Adaptive λ-tracking for nonlinear systems with higher relative degree. Proceedings of the Conference on Decision and Control 2000, Sydney, Australia

# 10 Autonomous Systems for Emergency and Stability Control of FACTS

The requirement specification in chapter 9 has clearly shown, that the uncoordinated use of FACTS-devices involves some negative effects and interactions with other devices, which leads to an endangerment of the steady-state and dynamical system security. This chapter shows one approach to overcome these difficulties and provides a solution for a coordinated control system fulfilling the specified requirements.

An autonomous control system for electrical power systems with embedded FACTS-devices is developed that provides the necessary preventive coordination. With methods of computational intelligence the system automatically generates specific coordinating measures from specified abstract coordinating rules for every operating condition of the power system without human intervention or control. This guarantees an optimal utilization of the technical advantages of FACTS-devices as well as the steady-state and dynamical system security. Interactions between the autonomous system and other existing controllers in electrical power systems are taken into consideration so that the autonomous system can completely be integrated into an existing conventional network control system.

## 10.1 Autonomous System Structure

The response time of FACTS-devices is in the range of some ten milliseconds. In case of critical events within the power system, e.g. faults or overloadings, FACTS-devices react immediately to these events due to their short response time. If the FACTS-devices are not adapted to the situation in and after such a critical event, this can lead to an endangerment of the steady-state and dynamical system security. The Non-Intrusive System Control (NISC) approach in chapter 9 defined the necessary interactions for regular and emergency control of FACTS-devices.

As a consequence, the application of FACTS-devices requires both a fast coordination of their controllers among one another and with power plants, loads, and conventional controlling devices within the power system. This coordination must guarantee the steady-state and dynamical system security in the case of critical events and has to be automatic, quick, intelligent, and preventive.

The NISC approach has separated the planning phase for coordinating actions from their local execution. One step further goes the autonomous system approach, where clearly separated autonomously acting components provide specific tasks. These tasks are in this case system analysis, coordination and execution of the specified control task. Autonomous systems generally represent an abstract informational-technological framework, which is specified in detail in [1]. Generally its architecture can be subdivided into several intelligent autonomous components communicating with each other.

The autonomous components themselves consist of different authorities called 'management', 'coordination', and 'execution'. Depending on the control level on which an intelligent autonomous component is placed, one of the three authorities dominates compared to the other two authorities. In order to specify the components on each control level every necessary local controller of the process must be determined concerning its structure. An autonomous component can be a control station, a process computer or a simple controller.

According to the hierarchical model of a control system for complex technical processes, e.g. electric power systems, the different control levels are called:

- network control level,
- substation control level,
- bay control level.

**Bay Control Level.** The physical coupling of the autonomous components on the bay control level is realized by sensors and actuators. The main task at the bay control level is 'execution', i.e. in this context mainly the application of control and adaptation algorithms.

**Substation Control Level.** Autonomous components on the substation control level mainly act as coordinators. They determine and plan the functionality of other components and delegate distinct special tasks.

**Network Control Level.** On the network control level autonomous components are working with information being generated from a model of the whole process, which can be implemented on this control level. The most important task of these components is the decomposition of global aims being generated here or prescribed by a human operator through the human-machine-interface.

The main capability of an autonomous control system is to act automatically without manual interactions. The autonomous system shall provide the following features:

- perform self-learning, self-organization, and can plan and optimize control actions,
- decentralized artificial intelligence enables quick autonomous actions.
- automatically adaptation to changes of the technical process in structure and parameters,
- operation of the process without human intervention.

To achieve this, some kind of knowledge about the required coordinating actions and adaptations must be embedded into the system on the specific levels. As a solution coordinating generic rules can be defined which are valid in any power system. These rules have to be adapted by a system analyses to the specific operational conditions.

## 10.2 Autonomous Security and Emergency Control

### 10.2.1 Model and Control Structure

In the following the autonomous system control will be demonstrated by the means of UPFC. The reason is that the UPFC provides fast power flow, voltage and damping control and therefore requires especially the coordinating control scheme. Other simpler FACTS-device controls can be derived from this general structure.

As shown in Fig. 10.1 the dynamic behavior of a UPFC can be modeled by a current source injecting the shunt current $\bar{I}_q$ and a voltage source inserting the longitudinal voltage $\bar{V}_l$. The dynamics of the two VSC are modeled by first order time delay elements ($PT_1$-Elements) with a time constant in the range between 15 and 30 ms [2].

In the model, the outputs of the operating point controllers are directly used by the converter control model for the calculation of $\bar{V}_l$ and $\bar{I}_q$. Furthermore a controller for improving the small signal stability of the system (damping controller) is implemented which will be dealt with in section 10.3. The outputs being fed back by the controller are the deviations from the setpoint values of active-power ($\Delta P_{ij}$), reactive-power ($\Delta Q_{ij}$), nodal voltage ($\Delta V_i$) and the corresponding serial current ($\Delta I_l$). The controller function is defined in equation 10.1. Its input and output vectors are defined in equations 10.2 and 10.3.

$$\Delta u = - F \Delta y \tag{10.1}$$

$$\Delta y = \begin{pmatrix} \Delta V_i & \Delta Q_{ij} & \Delta P_{ij} & \Delta I_l \end{pmatrix}^T \tag{10.2}$$

$$\Delta u = \begin{pmatrix} \Delta u_{V,D} & \Delta u_{Q,D} & \Delta u_{P,D} \end{pmatrix}^T \tag{10.3}$$

### 10.2.2 Generic Rules for Coordination

Coordination for the steady-state operation can e.g. be performed using optimal power flow techniques [3]. Concerning the dynamical operation, an adaptation of the control operations by FACTS-devices to changing operating situations or critical events in the power system has to be performed.

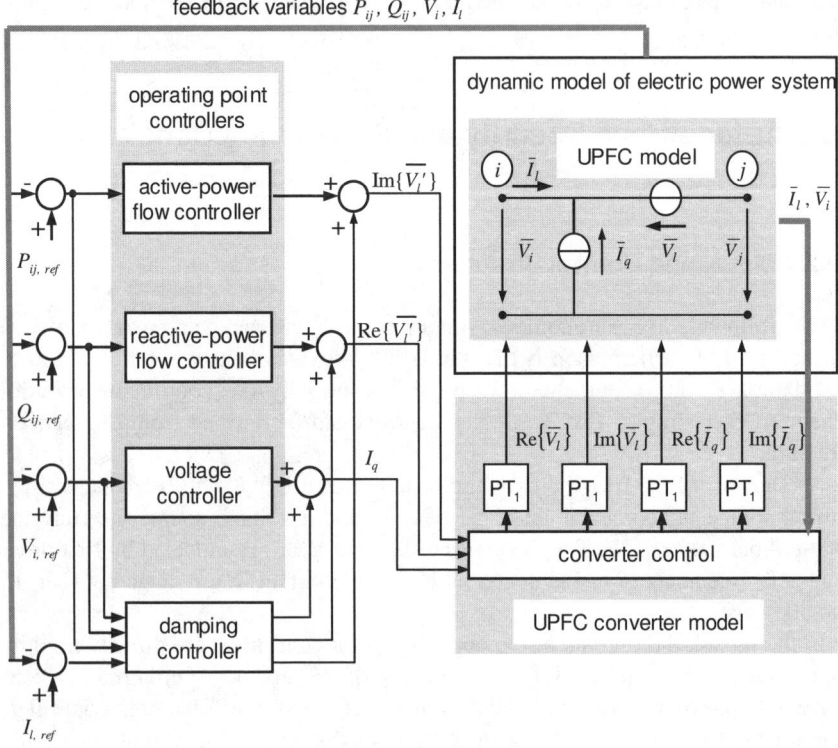

**Fig. 10.1.** UPFC modeling and control

Critical events, which require coordinating control measures to be applied to the embedded FACTS-devices, are:

- overloading of electrical devices,
- failure of electrical devices,
- short circuits in transmission elements,
- changes of the system's state.

Necessary coordinating control measures have to be applied in short term range after the occurrence of one of the above-mentioned events. The first three events are emergency cases requiring fast actions. The forth one concerns the damping control and will be analyzed in section 10.3.

The coordinating control measures can be formulated in a knowledge-based form as so-called generic rules [4]. Before they will be listed and explained, the definition of the terms 'control path' and 'parallel path', which concern the network topology, has to be given (see Table 10.1). For illustration, the topology of a simple example power system including one UPFC is shown in Figure 10.2.

## 10.2 Autonomous Security and Emergency Control

**Table 10.1.** Definition of terms

| Term | Definition |
|---|---|
| control path | transmission path in which a power flow controlling device (e.g. UPFC) is implemented and which only has junctions at its end-nodes |
| parallel path | transmission path which starts and ends at the same nodes as a control path and in which no power flow controlling device is implemented |

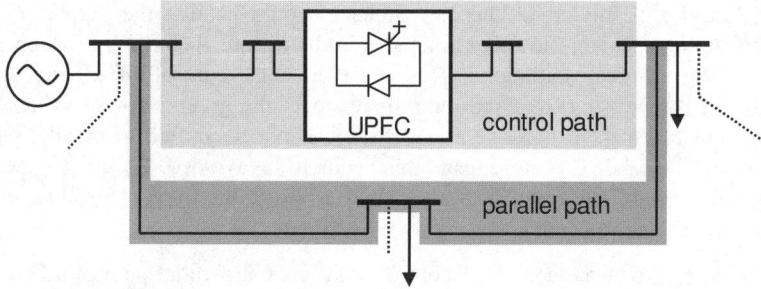

**Fig. 10.2.** Simple example power system used for definition of control and parallel paths

The existence of a parallel path is an essential necessity for a power flow controlling FACTS-device. Controlling the power flow over its control path a FACTS-device shifts the power flow from its control path to parallel paths and vice versa.

A system theoretical analysis shows the following four coordinating control rules:

1. **IF** a device on a parallel path of a FACTS-device is overloaded,
   **THEN** modify the $P$-setpoint-values of the FACTS-device

A power flow controlling FACTS-device can directly influence the active and reactive power flow over its control path. This leads to the above-mentioned shift of the power flow from the control path to parallel paths or vice versa. Consequently, power flows over parallel paths can be specifically influenced by changing the setpoint values for the active- and reactive-power flow of the control path. In this way overloadings of devices on parallel paths can be suppressed by changing the setpoint values of a power-flow controlling FACTS-device. The control path takes over the surplus of power flow which otherwise leads to the overloading of the device(s) on a parallel path.

This rule recommends modifying only the $P$-setpoint-values of FACTS-devices to suppress overloadings because these are mainly caused by active power flows. The reactive power-flow controlling functions of a FACTS-device can then be used for voltage control.

2. **IF** there is a failure of a device on a parallel path **AND** no further parallel path exists for a FACTS-device
**THEN** deactivate the power flow controllers of the FACTS-device

The existence of at least one parallel path to a control path is an important condition for the reasonable application of the power flow control function of a FACTS-device. As already described above, power flow control causes a shift of the power flow between control path and parallel path(s). Hence, if a failure of a device causes an opening of all parallel paths, the power flow control of a FACTS-device is hindered. The consequence would be that the outputs of the FACTS-device's power flow controllers would run into their limits, which may cause strong system oscillations. This is called 'false controlling effect', which means that the power flow controllers try to meet the given setpoint values, but they cannot reach them because power flow can not be shifted to parallel paths. According to the NISC requirements this needs to be avoided. By quickly deactivating the power flow controllers after such a failure the false controlling effect can effectively be prevented.

3. **IF** a short circuit happens on a control path or on a parallel path of a FACTS-device,
**THEN** slow down the operating point controllers of the FACTS-device

This coordinating measure prevents excessive power oscillations after a short circuit followed by automatic reclosing. The reason for this is that the power flow changes drastically during the short circuit. Mainly, a high reactive current flows over every line into the direction of the short circuit location. Because of the short response time of the FACTS-devices the power flow controllers respond immediately to the short circuit and try to meet the preset setpoint values. Also the voltage controller tries to fix the setpoint-voltage. Hence, the outputs of the operating point controllers will strongly increase within a short period of time and reach their limits even before the fault is clarified and the automatic reclosing is started. When the fault is removed after an automatic reclosing these large values of the manipulated variables of the operating controllers lead to strong oscillations. This is another kind of false controlling effect and has to be suppressed by suitable measures. Through slowing down the power flow controllers and the voltage controller during the short circuit and the automatic reclosing (decreasing of the PI controller parameters) this false controlling effect can be prevented.

The correct application of these three coordinating measures to FACTS-devices and their control enables the network operators to exploit the advantages being offered by FACTS for their steady-state and dynamical secure operation. The autonomous control system is designed to execute them automatically.

### 10.2.3 Synthesis of the Autonomous Control System

Due to the continuous changes of the operating states and the topology during the daily operation through varying loads, generations and switching operations, the specific coordinating control measures must be followed up automatically to these changes. Only under this condition the controller is able to react adequately on critical events in the changed system. This guarantees a dynamical and stationary secure behavior of the whole system. To ensure a quick reaction of the autonomous system, the specific coordinating measures have to be derived, before a critical event occurs. Hence, topology-changes of the network have to be analyzed continuously. This continuous adaptation of the specific coordinating measures for changing topologies is called 'preventive coordination' being performed by the autonomous control system. The three coordinating generic rules, which have been explained in the previous section, are the elementary tasks, which have to be fulfilled by the autonomous control system.

These first three rules mainly concern setpoint values for the operating point controllers and the operating controller's parameters. The development of the autonomous system is performed successively starting at the bay control level. Some elementary autonomous components are chosen and designed to be acting on this control level. After that, additional autonomous components on the other control levels are added. They provide the components on the bay control level with necessary specific information, which is generated automatically in dependence on the actual network topology.

### 10.2.3.1 Bay Control Level

Figure 10.3 shows the operating point controllers of a UPFC, which are extended by the additional controllers as autonomous components on the bay control level. They perform the basic measures, which are required by the first three generic rules and are explained in the following.

The coordinating measure given by the first generic rule requires a modification of the $P$-setpoint value of the UPFC in order to prevent overloadings on lines on its parallel paths. A simple but effective autonomous component performing this can be an integral-action controller forming an outer control loop. The actual active-power flows over all lines on parallel paths have to be observed by the autonomous component. As the degree of freedom for influencing power flows over parallel paths of one FACTS-device is equal to one. A UPFC can at the same time specifically prevent only one overloaded line. If several overloadings are detected, the line with the biggest overloading is chosen.

The actual deviation from the maximum allowed active power flow ($P\text{-}P_{max}$), which has a positive value in case of an overloading, is taken as the input of the integral-action controller. This way it adjusts the setpoint of the active-power flow controller of the UPFC until the active-power flow of the overloaded line is reduced to its maximum allowed value $P_{max}$. This is the basic idea of how the first generic rule is implemented on the bay control level. It guarantees the steady-state security of the power system.

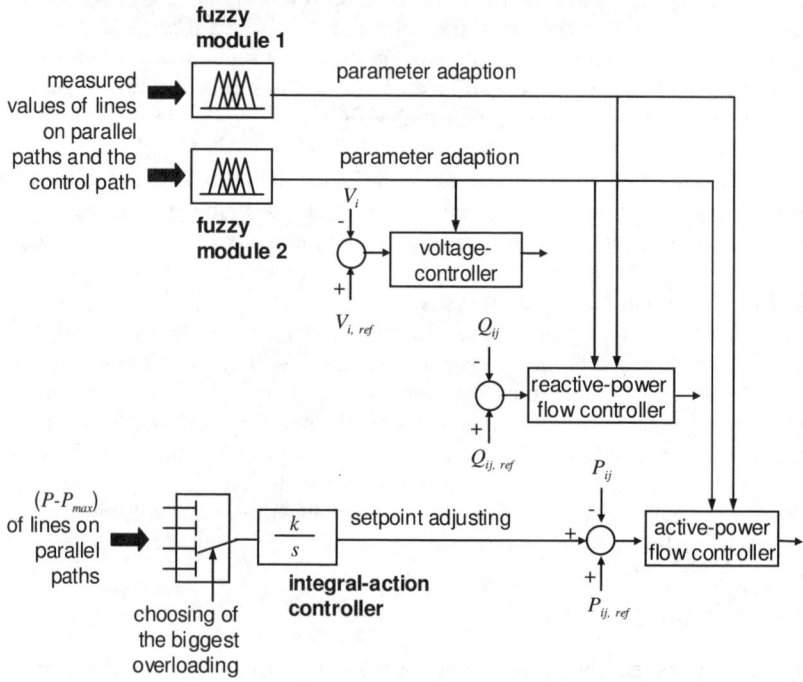

**Fig. 10.3.** Operating point controllers of a UPFC with autonomous components on the bay control level

When using this method in practice several additional measures have to be implemented. This comprises e.g. the detection if the reason of an overloading has disappeared after the overloading has been removed by the integral-action controller. In this case the setpoint adjusting by the integral-action controller has to be reset. Another important issue is the detection if an overloading is permanent or only temporary. Temporary overloadings can appear in case that the active power flow over a line oscillates around a value, which is directly below the maximum capacity. Those temporary overloadings are usually uncritical because they do not cause thermal problems. Hence they do not have to be treated by the autonomous control system. Additionally, it has to be respected that not all overloadings of lines on parallel paths can be removed by the *P*-setpoint adjusting. It strongly depends on the impact of a FACTS-device on the power flow of parallel paths, which can be high or very low. In case the impact is very low, usually a very big change of the *P*-setpoint is required for removing the overloading. As the UPFC has only limited control power, the setpoint adjusting will probably not be successful when trying to remove the overloading. These and further specific aspects are very important for the implementation of the method.

The second generic rule requires a deactivation of the power flow controllers in case of failures of distinct devices on parallel paths. The deactivation of the controllers shall be performed by quickly setting the controller parameters of the active and reactive power flow controller to zero. Adaptive control is chosen to be suitable for this. Since fuzzy adaptation provides a transparent knowledge based implementation of adaptation rules, a fuzzy module is chosen to be the autonomous component on the bay control level performing this task (fuzzy module 1).

In addition, such a fuzzy adaptation produces soft transitions between the activation and deactivation of the controllers. The knowledge bases are derived from the generic rule 2. This is performed by autonomous components on higher control levels and will be described in a later section. The input quantities of the fuzzy controller must be measured values of lines on parallel paths. From these input quantities the fuzzy controller must be able to clearly recognize failures of relevant transmission elements. Measurements of the currents or complex power flows over the concerning transmission elements can be taken as input quantities. Membership functions for the input quantities have to be chosen once and remain valid for all operating cases.

The implementation of the third generic rule on the bay control level is also done by a fuzzy controller performing an adaptation of the parameters of the operating controller (fuzzy module 2). It decreases the operating point controller's parameters in cases of short circuits on lines of the control path or on parallel paths so that the controllers are slowed down strongly, as it is required according to generic rule 3.

Short circuits (faults) must be reliably recognized by the input quantities of the fuzzy controller. Hence, the currents over those lines can be taken as input quantities for the fuzzy controller. Also here the membership functions have to be chosen only once.

### 10.2.3.2 Substation and Network Control Level

Autonomous components on the substation and the network control level have to generate specific additional information for the autonomous components on the bay control level (fuzzy modules, integral-action controller and damping controller). This must also be based on the generic rules.

The generic rules strongly depend on the network topology. They use the terms 'control path' and 'parallel path' as they have been defined above. For this reason, autonomous components on the network control level have at first to analyze automatically the network's topology. This is done recursively with the known backtracking technique. The result is an assignment of all parallel paths to each control path. For large and complex networks these calculations can take long computation time because theoretically a large number of parallel paths may exist. However, since the impact on parallel paths that are far away from the control path may be very small, the user can define a reasonable area of impact for each FACTS-device, in which it has sufficient impact on its parallel paths. These areas should be chosen such that the influence of the power flow over lines within the areas can be performed with a realistic amount of control power. The analysis of the network's topology for finding control and parallel paths can then be limited to these areas of impact.

With the result of the topology analysis the three generic rules can be brought to a set of concrete coordinating rules, which are valid for the actual network topology. To illustrate this, one example of a concrete rule for each generic rule shall be given:

1. **IF** line 11-19 is overloaded **THEN** modify the $P$-setpoint-values of UPFC 2
2. **IF** there is a failure of line 17-18 **THEN** deactivate the power flow controllers of UPFC 1
3. **IF** a short circuit happens on line 11-19 **THEN** slow down the operating point controllers of UPFC 2

This is how the rules may look like for an example real power system containing UPFCs. The complete sets of concrete coordinating rules may contain a large number of rules.

For the generic rules 2 and 3 the concrete rules are then translated by autonomous components into fuzzy rule bases for the fuzzy modules 1 and 2 on the bay control level for each FACTS-device. The rule bases are downloaded into the fuzzy modules 1 and 2.

Concerning generic rule 1 the result of the topology analysis is used by a further autonomous component to compute the impact of the FACTS-devices on lines on parallel paths. It computes the GSDF (generation shift distribution factors, [5]) in order to quantify the impacts of FACTS-devices on all lines of the parallel paths. Only if the impact of a FACTS-device on a line is big enough, it is sensible to include this line into the autonomous control in terms of preventing overloadings. If more than one FACTS-device has a certain impact on a line, the FACTS-device with the biggest impact on that line is determined to remove a possibly occurring overloading. This way the GSDF determine the lines, which have to be monitored by which FACTS-device with regard to overloadings. They also determine the parameters $k$ of the integral-action controllers. This mainly concerns the sign of the control action, which means if the $P$-setpoint has to be increased or decreased to remove a specific overloading of a transmission element.

In this way it can be guaranteed that the integral-action controllers perform their control actions to remove overloadings with the correct direction and the necessary intensity.

Figure 10.4 finally shows the autonomous components, which are necessary on the substation and the network control level in order to generate specific information for the fuzzy modules and the integral-action controllers as autonomous components on the bay control level.

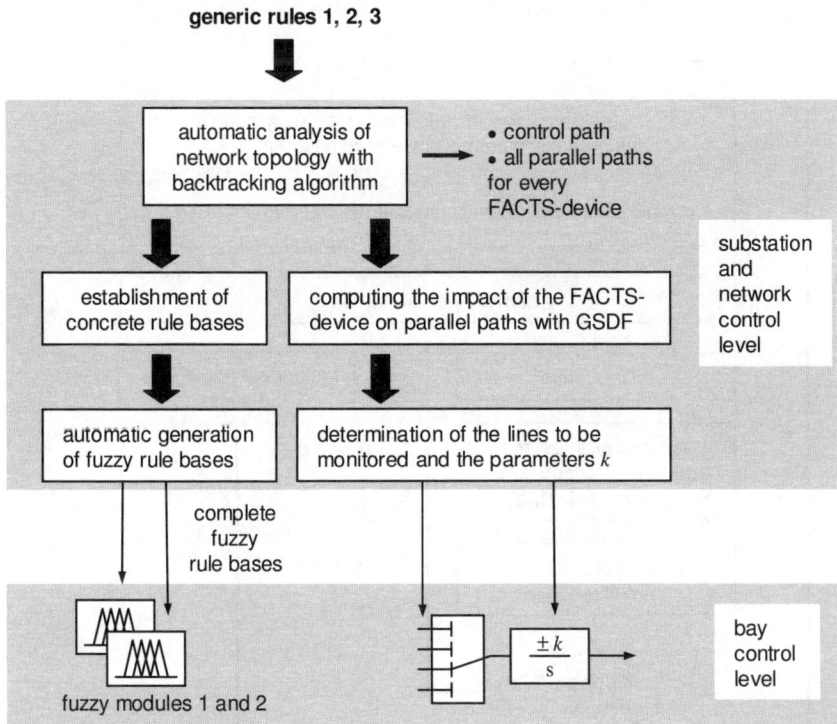

**Fig. 10.4.** Autonomous components on the substation and network control level for generic rules 1, 2 and 3

### 10.2.3.3 Preventive Coordination

As already mentioned before, the specific information for fuzzy and integral-action controllers (fuzzy rule bases etc.), which represent the coordinating measures in the case of overloadings, faults, and failures, is only valid for the network topology for which they have been generated. Since the topology of the system changes in the daily operation of the power system by switching operations, the fuzzy rule bases and additional information for the integral controllers must be followed up automatically to these modifications. Only under this condition it is guaranteed that the autonomous system can react correctly to critical events according to the above-mentioned generic rules. Such planned changes of the network topology are named in Fig. 10.5 with 'intended topological changes'. In addition, the occurrence of critical events, to which the autonomous system reacts by means of fuzzy parameter adaptation or setpoint adjusting, itself may lead to a changed topology, for instance through the unintentional failure of a transmission line.

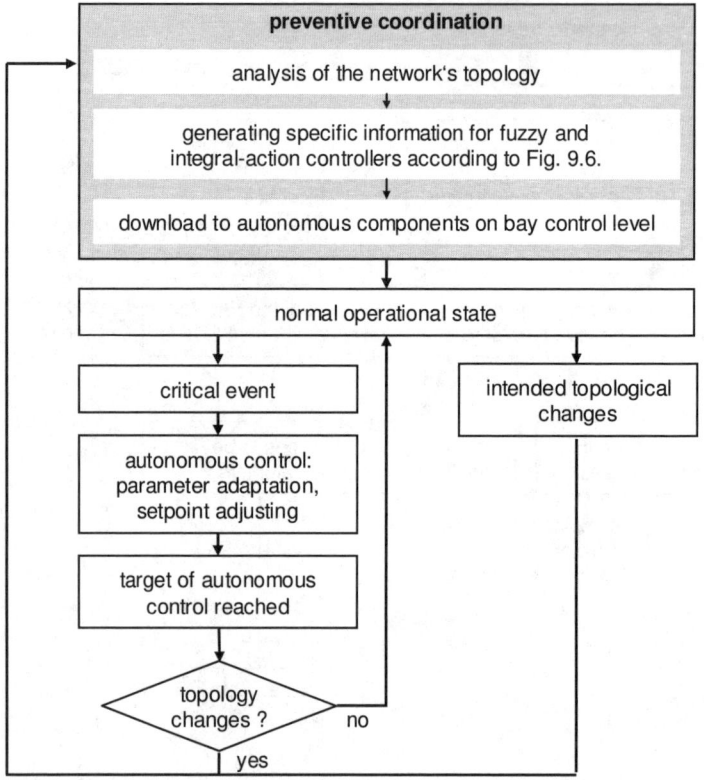

**Fig. 10.5.** Procedure for preventive coordination of FACTS-devices

For both cases of topology changes the previously described procedures for the generation of specific information for fuzzy and integral-action controllers being performed by autonomous components on the substation and the network control level have to be activated automatically.

Hence, new specific information is generated for the autonomous components on the bay control level. This is called 'preventive coordination'. The term 'normal operational state' means that at the moment no certain coordinating action of the autonomous system is required so that only the operating point controllers are in normal operation. 'Target of autonomous control reached' indicates that an overloading has been successfully removed or that operating point controllers have successfully been slowed down or deactivated in order to prevent false controlling effects respectively.

## 10.3 Adaptive Small Signal Stability Control

From a control systems theory point of view, an electric power system behaves like a non-linear, time-variant controlled system. Due to permanent changes of power generation, loads, and the networks topology, the dynamical state of the system varies strongly and continuously. FACTS-devices, which are equipped with simple damping controllers like a linear output feedback controller, shall improve the small signal stability of a power system during all its operating conditions. Therefore the controller has to be continuously adapted according to the changes of the dynamical states, which happen to the entire system.

The damping controller in Figure 10.1 is usually implemented to work in parallel to the operating point controllers, which means that the outputs of the damping-controller are added to the outputs of the operating point controllers. For example an output feedback controller with adaptable parameters can be used to improve the system's small signal stability. This controller is linear and it is usually parameterized for a power system model being linearized around an operating point. Constant parameter settings can usually only guarantee good control performance for the system operating around this point and not within the whole range of states in which it can operate. Hence, the damping controller parameters are to be adapted to changes of the system's state.

A fourth generic rule can be formulated to take this requirement into account.

4. **IF** a change of the dynamical state of the entire system happens,
   **THEN** adapt the parameters of the FACTS-device's damping controller.

### 10.3.1 Autonomous Components for Damping Control

The damping controller is designed as an output feedback controller whose feedback matrix $F$ has to be adaptable to changing conditions of the power system according to generic rule 4.

The bay control level does not contain any specific autonomous components for the adaptation of the damping controller since information about the whole system's state can only be provided from the entire power system point of view, i.e. from the network control level.

Autonomous components on the network and substation control level have to determine the damping controller parameters, i.e. the elements of the output feedback matrices $F$ of each FACTS-device being fitted with such a damping controller. As already done in the previous sections of this chapter, the ideas and concepts of how this is performed shall be revealed instead of presenting all details about their implementation.

As mentioned above, loads, generations and the network's topology determine the dynamical state of the power system as a controlled system, its input variables and its equivalent transmission function. The non-linear system equations for a current operating point can be linearized around this operating point, such that a set of linear coupled differential equations is received. Hence, the power system can be described as a first-order state space model, which is valid in a certain environment around the chosen operating point.

The computation of the eigenvalues of the system matrix $A$ gives information about its oscillatory characteristics, e.g. critical modes. Critical oscillation modes are modes with a small or even negative damping ratio. Furthermore, the eigenvalues have to be computed by an autonomous component in order to determine the modal transformation of the system. This is also done on the network control level. Regarding the input matrix $\boldsymbol{B_m}$ of the modal transformed system it can easily be analyzed, which FACTS damping controller has got a strong influence on which of the critical oscillation modes of the system.

The autonomous components assign to the critical mode with the lowest damping ratio one FACTS-device, whose damping controller has the biggest influence on that mode. The remaining FACTS-devices are then one by one assigned to other critical modes with higher damping ratios.

Using this selection and some further information, like the damping sensitivity factors (DSF) [6], a cost function is formulated which expresses the effectiveness of a chosen parameter set for the FACTS damping controllers, concerning the resulting damping ratios of the critical modes in the closed-loop operation. This cost function is then minimized using the well known Simulated Annealing algorithm as a numerical optimization technique [7] in order to determine the optimal output feedback control matrices $\boldsymbol{F_i}$ for each existing FACTS damping controller and the present system's state. Fig. 10.6 illustrates the whole described procedure being performed by autonomous components on the network control level in order to compute optimal FACTS damping controller parameters after a change of the dynamical state of the entire system.

## 10.4 Verification

In the following, two simulation examples are shown in order to illustrate the performance of the autonomous control system. For the investigations the example network according to Fig. 10.7 has been analyzed. The fuzzy rule bases and global information for integral controllers were generated with the described autonomous system, which has also been implemented into a simulation environment.

A failure of a transmission line and the scenario of a line overloading caused by a rapid increase of a load is simulated. In this way the effectiveness of the coordinating measures through generic rules 1 and 2 is shown. The per-unit quantities of the used example power system are: $S_b = 1250$ MVA and $U_b = 400$ kV.

The example system is derived from the extra-high voltage level of large power system, which has been reduced to the essential transmission elements, generators and loads.

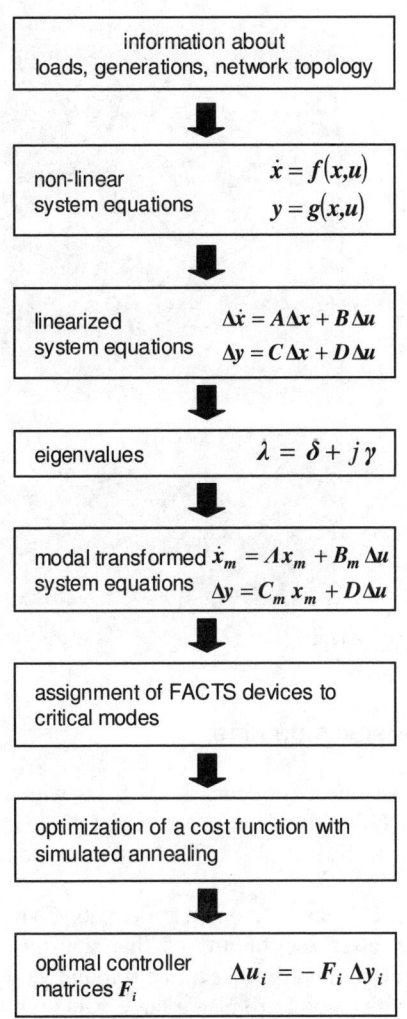

**Fig. 10.6.** Procedures being performed by autonomous components on the network control level for the automatic adaptation of FACTS damping controller parameters after changes of the dynamical system state (index $i$ denotes the $i$-th FACTS-device, if several FACTS-devices are installed in the power system)

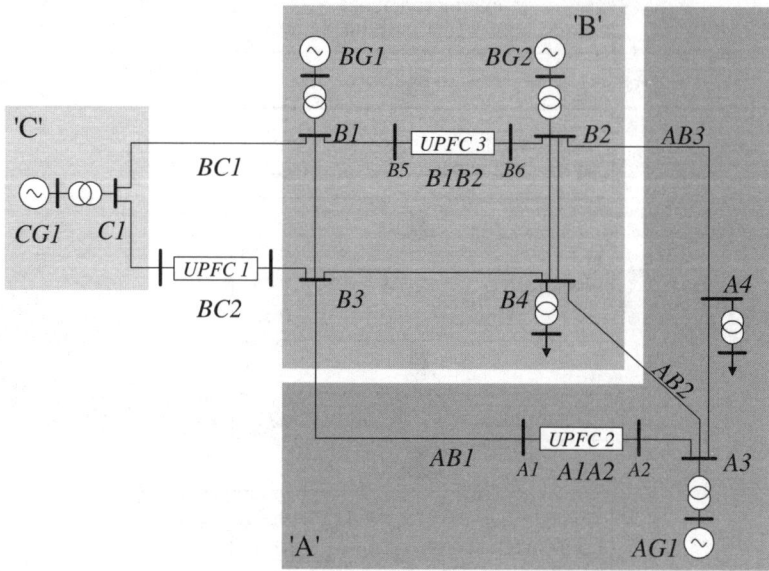

**Fig. 10.7.** Topology of the test system

### 10.4.1 Failure of a Transmission Line

A failure of line *BC1* is assumed. It occurs at $t = 0.1$ s with a duration of 4.9 s. Line *BC1* has before been correctly identified as a part of a parallel path of UPFC 1 by the topology analysis. The results of the automatic topology analysis are listed in Table 10.2.

The control path of UPFC 1 has only one parallel path. This means that there is no parallel path existing after the failure of this transmission line. Without autonomous control the controllers try to keep the setpoint value for active and reactive power flow over UPFC 1 and produce a large value for $V_l$ up to its limit of 0.15 pu (see Fig. 10.8). This is due to the false controlling effect. However, the setpoint values cannot be kept because of the missing parallel path. This large value of $V_l$ produces strong power oscillations during the failure of the line. They can be seen in Fig. 10.9.

**Table 10.2.** Result of the automatic topology analysis

| FACTS-device | UPFC 1 | UPFC 2 | UPFC 3 |
|---|---|---|---|
| control path (node numbers) | C1-B3 | B3-A3 | B1-B2 |
| parallel paths (node numbers) | C1-B1-B3 | B3-B4-B2-A4-A3<br>B3-B4-A3 | B1-B3-B4-B2<br>B1-B3-B4-A3-A4-B2 |

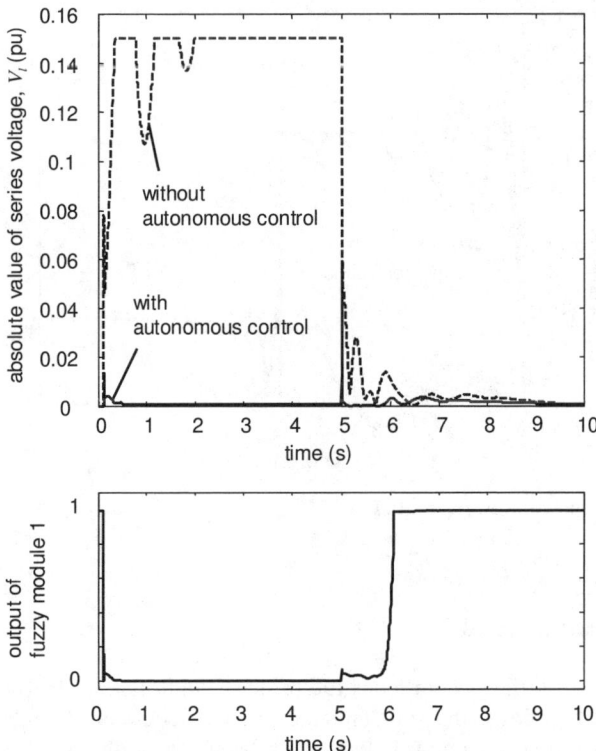

**Fig. 10.8.** Series voltage of UPFC 1 (above) and output of fuzzy module 1 of UPFC 1

When the autonomous control system is in use, fuzzy module 1 for the generic rule 2 in UPFC 1 reacts immediately by deactivating the power flow controllers. The effect is visible in Fig. 10.8 with the output of the fuzzy module and the resulting outputs of the power flow controllers. The two power flow PI-controllers of UPFC 1 only cause a small increase of the manipulated variables during the failure. Consequently, the oscillations in the system are calmer during the failure than without the application of the autonomous control system.

It has to be mentioned that the shown effect only results from the slowing down of the controllers in order to prevent the false controlling effect. No FACTS damping controllers are present within that system. The damping could even be further improved if FACTS damping controllers were used.

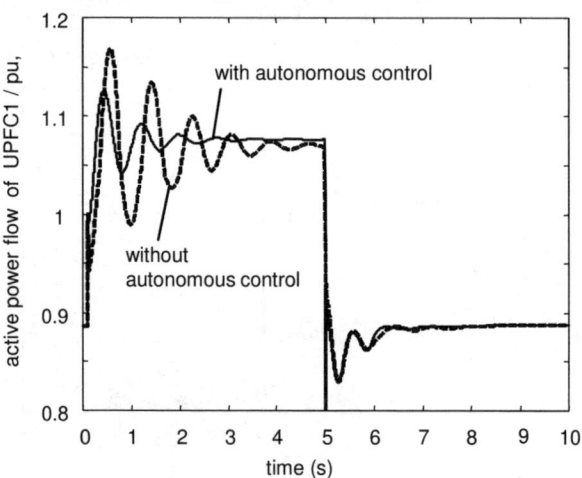

**Fig. 10.9.** Active power flow over UPFC 1

### 10.4.2 Increase of Load

After an increase of the system loads the primary controllers of the power plants operate in order to cover the supplementary power requirement. Independent of this the three UPFC fix the power flows over the control paths constant with their fast power flow controllers. Consequently, they can for this moment not be used for the transmission of primary control power.

The capacity of line *AB2* is used to approximately 94 % before the load increase. A sloping increase of the system loads of 14 % happens at $t = 1$s. The primary control power generated by *AG1* has to be transmitted to the load at node *B4* e.g. over line *AB2*, since the control of UPFC 2 first keeps the unchanged setpoint values. Without the autonomous control system a non-permissible overloading of *AB2* occurs (see Fig. 10.10). If this condition continued, a tripping of the transmission line would be inevitable.

In the topology analysis, which has been executed before by the activated autonomous control system, this line is recognized as an element of a parallel path to UPFC 2, so that its integral-action controller counteracts directly on the line overloading and increases the setpoint value for the active power flow (see Fig. 10.10, below). This causes a shift of the power flow and a relief of *AB2*, so that its maximum loading limit of 0.62 pu (active power) can be kept (see Fig. 10.10, above). At $t = 1000$ s the loads are decreased to their original values, which means that the reason for the overloading has now disappeared. This has been simulated to show that the autonomous control system is able to reset itself when its coordinating control actions are not needed any more. The integral-action controller reduces its output back to zero.

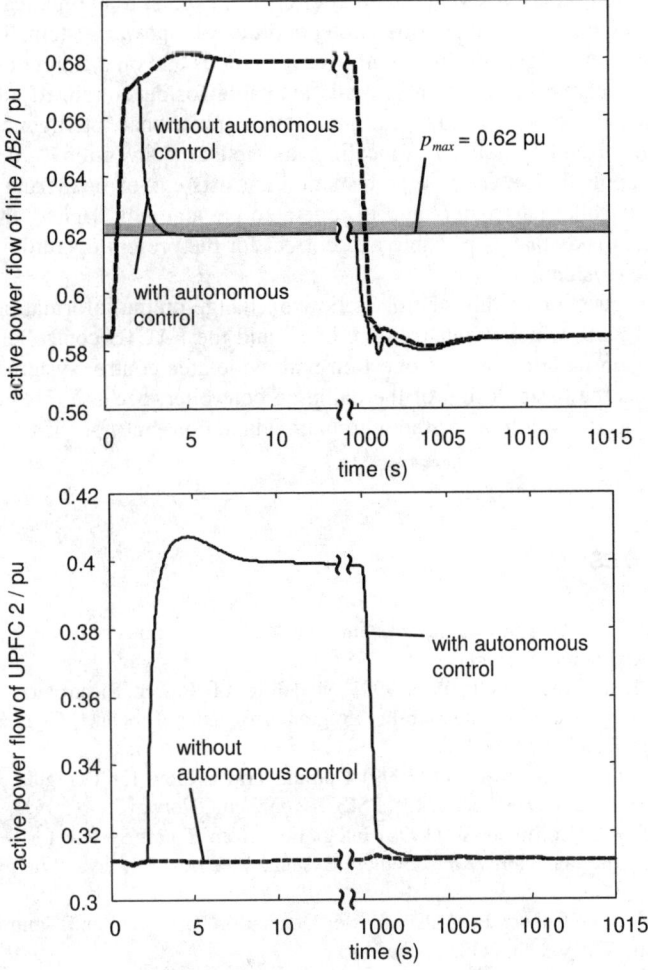

**Fig. 10.10.** Active power flow over line *AB2* (above) and active power flow over UPFC 2 (below)

In conclusion, the use of FACTS-devices offers a flexible management of network-operation from a technical and economical point of view. Beyond this they may cause many negative effects, which occur due to their short response time after critical events. Hence their advantages can only be used if automatic, quick, intelligent, and preventive coordinating measures are performed to eliminate those negative effects. The theory of autonomous systems offers a structural approach for a coordinating control system for FACTS-devices. The necessary coordinating measures, which have to be executed by an autonomous control system, can be formulated as four generic rules. An autonomous control system has been developed and implemented for such a coordination so that the steady-state and dynamical system security is guaranteed after critical events. It automatically applies the four generic rules for every operation condition of a power system. The control system specifies the generic rules within several steps and on different control levels so that concrete information is made available for decentralized autonomous components. For this, several techniques of Computational Intelligence, such as Fuzzy Control and Simulated Annealing, as well as conventional control techniques are applied. The concrete information consists e.g. of fuzzy rule bases and damping controller parameters and is generated preventively. Hence, the reaction of the system is as fast as possible and correct for the present operating condition of the power system.

An open question in this section is how exchange online information between the parallel path, which means remote lines, and the FACTS controller. The following section will answer this question with wide area control systems. Another open issue is the basic design of the damping controller, because this chapter has only introduced a solution for the automatic adaptation, but not for the design itself.

## References

[1] Rehtanz C (2003) Autonomous Systems and Intelligent Agents in Power System Contol and Operation, Springer
[2] Cigré Task Force 38.01.08 (1999) Modeling of Power Electronics Equipment (FACTS) in Load Flow and Stability Programs. Technical Brochure, Cigré SC 38, WG 01.08, Ref. No. 145
[3] Handschin E, Lehmköster C (1999) Optimal Power Flow for Deregulated Systems with FACTS-devices. Proc. of 13$^{th}$ PSCC, Trondheim, Norway
[4] Handschin E, Hoffmann W (1992) Integration of an Expert System for Security Assessment into an Energy Management System. Electrical Power & Energy Systems, vol 14
[5] Wood AJ, Wollenberg BF (1996) Power Generation, Operation and Control, 2$^{nd}$ Edition. John Wiley & Sons Inc., New York
[6] Chen XR, Pahalawaththa NC, Annakkage UD (1997) Design of Multiple FACTS Damping controllers. Proc. of International Power Engineering Conference IPEC 1997, Singapur, pp 331-336
[7] King, RE (1999) Computational Intelligence in Control Engineering. Control Engineering Series, Marcel Dekker, Inc., New York, Basel

# 11 Wide Area Control of FACTS

FACTS-control has always to cope with speed and in the case of power flow control with exchange of system wide information. The high speed exchange of data to react on contingencies needs to be ensured to fulfill the requirements of the NISC-architecture according to the specifications in chapter 9. Online monitoring of the system status is needed for the optimization of the FACTS-device applications. Especially for power flow control and power system oscillations a dynamic performance evaluation supports an optimized transmission capability and an adaptive damping control.

Although pioneered already in the 80s, it is not until now phasor measurement units (PMU) have become widely available in power systems [1]. However, since recently wide-area measurement systems based on PMUs are becoming proven technology and are seen by many utilities as one of the most promising ways to gain more detailed information to operate the networks closer to the limits. Typically a wide-area measurement system based on phasor measurements provides access to system-wide data with a time resolution of tens of Hertz. The amount of gathered data becomes large, and the data need proper processing to be used either for the operator support or as part of the control system especially for FACTS. This chapter discusses wide area measurement and control systems as part of the coordinating FACTS-control.

## 11.1 Wide Area Monitoring and Control System

A wide-area measurement system (WAMS) can provide streaming measurements at update rates of 10-20 Hz, which enables monitoring not only of slow phenomena such as voltage and load evolution dynamics, but also faster phenomena such as oscillatory, transient and frequency dynamics. However, because of the high time-resolution of the measurements a WAMS will deliver huge amounts of data that need specific algorithms to use the provided information. The WAMS serves as the infrastructure necessary to implement wide-area stability control or system protection schemes [2].

Figure 11.1 shows the basic setup of a WAMS system. PMUs are placed within a critical area of the power system. This area could be for example a specific corridor. The PMUs are placed to derive a model of the specific area out of their measurements.

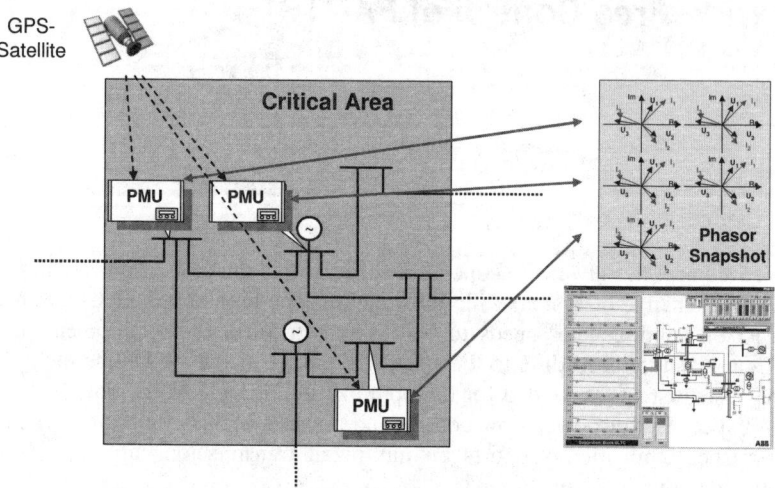

**Fig. 11.1.** Basic setup of a wide area measurement system

All PMUs are time-synchronized via a GPS-satellite time signal. Therefore the date from different PMUs can be directly compared which allows to directly measure voltage and current angles. The measured data are transmitted via communication channels to a central computer running the applications. Figure 11.2 shows the more detailed architecture of a WAMS system including the interfaces to SCADA/EMS and substation automation as well as the closed control loops back to network controllers like FACTS-devices.

The central computer contains services for preprocessing the incoming phasor measurements and basic services. The incoming measurement data must be sorted according to their time stamps and missing information must be detected. If the number of PMU data allows a full observability of the system a PMU based state estimation can be calculated. With PMUs in every second substation even topology detection can be performed. For most applications an estimation of a model of a specific area like a corridor is sufficient and limits the complexity of the WAMS.

An interface to the SCADA/EMS system allows receiving topology information and device parameters, like line inductance and the switching status. On the other hand, PMU information can be integrated into the conventional State Estimation to improve the accuracy. The Graphic User Interface (GUI) of the WAMS system can be kept separately or integrated into the SCADA/EMS screens.

The WAMS system runs various applications for wide area monitoring, control and protection. The monitoring performs for example stability assessments. Based on this information, control or protection actions, like the FACTS-control or for example load shedding schemes can be executed. The control signals are going back either directly to local controllers for specific devices or to substation automation systems.

## 11.1 Wide Area Monitoring and Control System

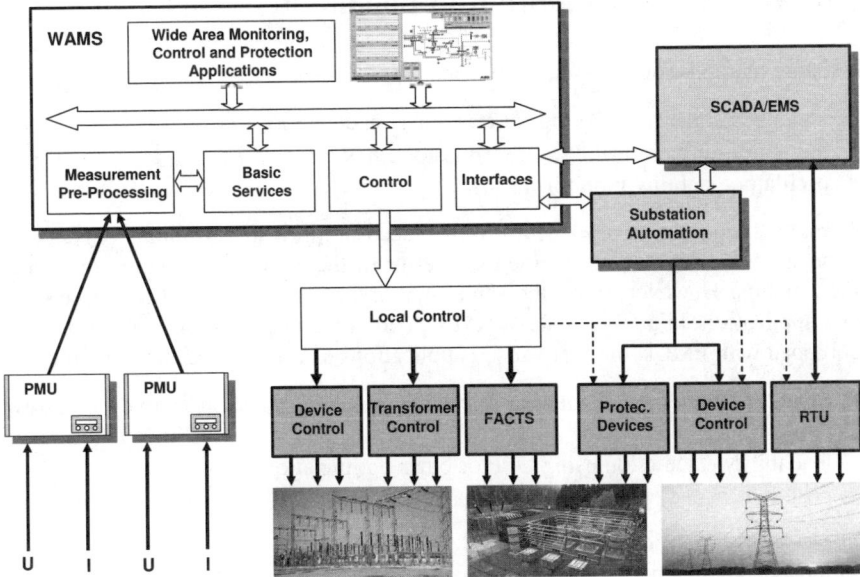

**Fig. 11.2.** Architecture of a wide area monitoring, control and protection system

The maximum performance of the applications in terms of speed is mainly limited by the communication channels. The data transmission from PMU to the central system and back to a device controller can be assumed to be between 50 and 200 ms for each direction.

In general a WAMS is structurally placed between SCADA/EMS and local control and protection systems. Figure 11.3 shows the basic characteristics.

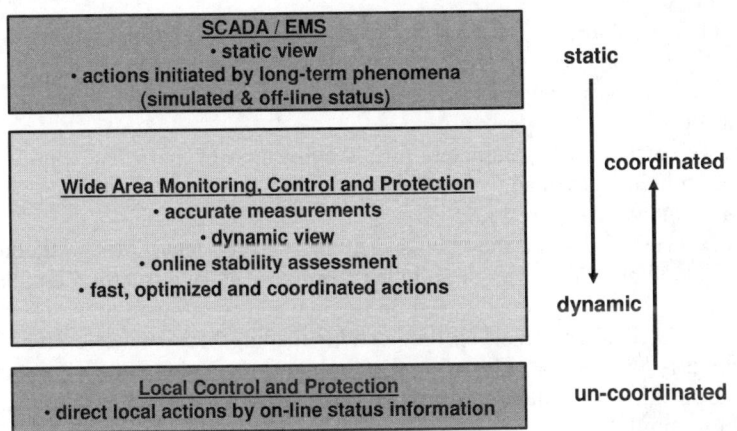

**Fig. 11.3.** Wide area monitoring, control and protection system capabilities in comparison to SCADA/EMS and local control and protection

## 11.2 Wide Area Monitoring Applications

Existing methods for using the PMU data are:

- Voltage stability monitoring for transmission corridors,
- thermal limit monitoring for transmission lines,
- oscillatory stability monitoring.

These methods can be used with PMUs placed in a few key locations only. Each application has its own requirements in terms of the number of required measurement points. However, often the same measurements can be used for more than one application. In a large-scale WAMS where a major part of the substations are equipped with PMUs, more advanced applications can be utilized, for example:

- State- and topology calculation providing dynamic snapshots of the power system,
- loadability calculation using OPF or other optimization techniques,
- post-contingency prediction of system state, especially for voltage stability.

These second applications are based on a completely observed network from which a detailed network model is derived.

### 11.2.1 Corridor Voltage Stability Monitoring

In real power systems main limitations are typically caused by transmission corridors between generation and load areas or for trading purposes between regions. If these transmission corridors extend a certain length, voltage stability is the limiting factor, which needs to be carefully supervised to utilize the corridor to a maximum extend.

The main principle of the corridor voltage stability monitoring is to use the measurements from both ends of a transmission corridor, reduce them to lump currents and voltages, and to compute a reduced equivalent model of the transmission corridor.

First we calculate the parameters of a T-equivalent of the actual transmission corridor, including any load or generation that may be present in the transmission corridor as shown in Figure 11.4.

This reduced model can then be used to analytically determine the theoretical maximum loading of the corridor and the margin to voltage instability. Optionally, load shedding can be activated based on the loadability estimate to avoid voltage collapse in the load region when the corridor loading becomes excessive. Since the method is based on a reduced equivalent network model, which is estimated on-line from the PMU measurements, no parameter input is required to estimate the stability limit.

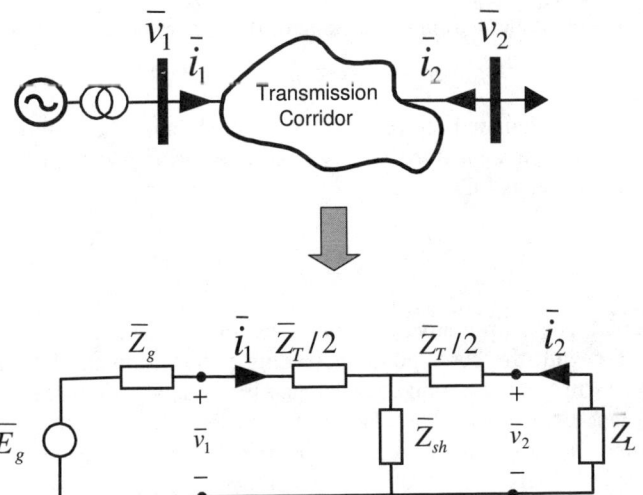

**Fig. 11.4.** T- and Thevenin-equivalents of a transmission corridor fed by a generation area

Applying Ohm's and Kirchoff's laws, with the known complex quantities (measured phasors) $\bar{v}_1$, $\bar{i}_1$, $\bar{v}_2$ and $\bar{i}_2$ we can calculate the complex impedances $\bar{Z}_T$, $\bar{Z}_{sh}$ and $\bar{Z}_L$ as follows

$$\bar{Z}_T = 2\frac{\bar{v}_1 - \bar{v}_2}{\bar{i}_1 - \bar{i}_2} \tag{11.1}$$

$$\bar{Z}_{sh} = \frac{\bar{v}_1\bar{i}_2 - \bar{v}_2\bar{i}_1}{\bar{i}_1^{\,2} - \bar{i}_2^{\,2}} \tag{11.2}$$

$$\bar{Z}_L = \frac{\bar{v}_2}{-\bar{i}_2} \tag{11.3}$$

The complex voltage $\bar{E}_g$ and impedance of the equivalent voltage source $\bar{Z}_g$ cannot be simultaneously calculated in the same straightforward way, so one of them must be assumed to be known to avoid the time delay of an estimation procedure like the one in [3] and [4]. If the generators have voltage controllers and can be assumed to stay within their capability limits, $\bar{E}_g$ can assumed to be constant and $\bar{Z}_g$ could then be calculated using:

$$\bar{Z}_g = \frac{\bar{E}_g - \bar{v}_1}{\bar{i}_1} \tag{11.4}$$

However, in most practical cases it is more realistic to assume that $\bar{Z}_g$ is known since it typically comprises of the step-up transformers and short transmission lines to the beginning of the transmission corridor. It is therefore preferential to calculate the equivalent complex voltage of the generators as follows:

$$\overline{E}_g = \overline{v}_1 + \overline{Z}_g \overline{i}_1 \tag{11.5}$$

Once we have calculated the parameters of the T- and Thevenin equivalent, a second Thevenin equivalent for the combined generation and transmission corridor can be calculated as follows:

$$\overline{Z}_{th} = \frac{\overline{Z}_T}{2} + \frac{1}{\frac{1}{\overline{Z}_{sh}} + \frac{1}{\overline{Z}_T/2 + \overline{Z}_g}} \tag{11.6}$$

This second Thevenin equivalent comprises of the impedance from equation (11.6) together with the corresponding feeding voltage and the load impedance from above. With this very simple model stability analysis can be performed analytically in a straightforward way. However, practical corridors usually comprise of several lines not always connected to the same sending and receiving node. In this case a reduced network model must be calculated.

Consider the example network diagram in Figure 11.5. To apply the network reduction procedure, first the main load and generation centers must be identified. In this case, a distinct generation center can be found in the area above cut 1, which contains three major generators and some shunt compensation but only a few minor loads. Between cuts 1 and 2 is an area with no generation equipment and only a few minor loads. This is the transmission corridor, whose stability is of interest. In the equivalencing procedure described above, these loads will be implicitly included in the shunt impedance. Below cut 2 is an area with predominantly load character. There are some minor generators, but in cases where the voltage stability is endangered, these generators would have exceeded their capability limits and thus no longer contribute to stabilization. It is therefore reasonable to include them in the shunt impedance modeling of the load.

After identifying the region boundaries, which are given by the two transfer cuts we can define two virtual buses, one for each end of the transmission corridor. These are the buses directly adjacent to a cut. Buses 6, 13 and 14 of the original system are grouped into virtual bus 1, and buses 24, 15 and 16 into virtual bus 2. The part of the system between cuts 1 and 2 becomes the virtual transmission corridor. At least one complex voltage in the area of each virtual bus and the complex currents on each line crossing a cut must be measured by PMUs.

We can then compute the currents at either end of the virtual transmission corridor using:

$$\overline{i}_i = \left( \frac{p_{cut,i} + jq_{cut,i}}{\overline{v}_i} \right)^* \quad i \in 1, 2 \tag{11.7}$$

Here $p_{cut,i}$ and $q_{cut,j}$ refer to the sum of the power transfers through cut $i$, and $\overline{v}_i$ as the average of the voltages included in virtual bus $i$.

## 11.2 Wide Area Monitoring Applications

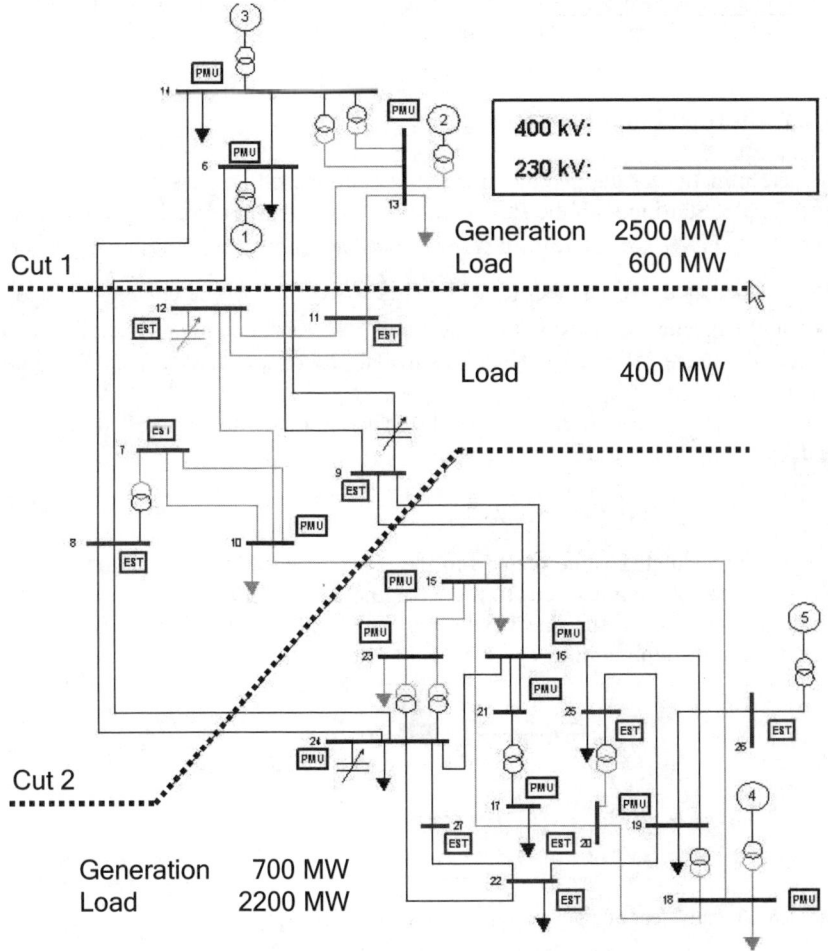

**Fig. 11.5.** Diagram of a real power system with a corridor situation

Computations of stability margins have to be carried out based on this virtual transmission corridor model. The stability analysis can be performed analytically with the second Thevenin equivalent. The point of maximum power transfer $p_{L\,max}$ can be calculated for an assumed load increase with constant power factor.

$$p_{L\max} = \Re\left[\bar{Z}_{th}\left|\frac{\bar{E}_{th}}{2\bar{Z}_{th}}\right|^2\right] \tag{11.8}$$

Normally at least a part of the load has constant power characteristics, and the point of maximum power transfer as given by equation (11.8) then also becomes a loadability limit. Past this limit there is a loss of equilibrium and a voltage collapse will occur. Therefore it becomes a stability limit.

## 11.2.2 Thermal Limit Monitoring

The determination of the average line temperature based on phasor information is quite simple. Starting with the PI-equivalent of the line in Figure 11.6, the line parameters $R$, $X_L$, $X_C$ are determined from the voltage and current phasors $\bar{v}_1$, $\bar{i}_1$, $\bar{v}_2$ and $\bar{i}_2$, whereas the resistance $R$ has the largest variability. The changes of inductance and capacitance are small during operation. As an example, the change of the line resistance $\Delta R$ of 10 % leads to a loadability change of $\Delta s_{max} = 6.5$ % for a typical 400-kV-line.

If the actual value of $R$ is determined, the actual line temperature can be calculated according to the following formula:

$$\frac{R_1}{R_2} = \frac{T_1 + T_0}{T_2 + T_0} \qquad (11.9)$$

$R_1$ is the calculated value of $R$ from the phasor measurements. $R_2$ and $T_2$ are a pair of values, which are given from the original design of the line. $T_0$ in eq. (1) is a material constant for the line wires. (e.g. $T_0 = 228$ °C for aluminum). With the given values, the temperature $T_1$ can be calculated.

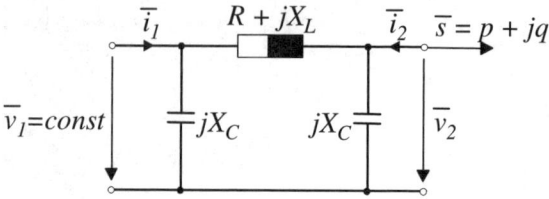

**Fig. 11.6.** PI-equivalent of a power transmission line

This calculated temperature is the average temperature of the entire line between the two measurement points. This temperature includes the actual situation of ambient conditions like wind speed, sun and line current. Consequently, these data offer much more information than the line current as a loadability limit only. The drawback is that this kind of information cannot identify hot spots and therefore sometimes not replace local temperature measurements.

## 11.2.3 Oscillatory Stability Monitoring

Initiated by the normal small changes in the system load and disturbances such as generator or line trips, oscillations are characteristics of a power system. However, a small load increase in a line flow, for instance a couple of MWs, may make the difference between stable oscillations, which are acceptable, and unstable oscillations, which have the potential to cause a system collapse. It is another matter of fact that increasing long-distance power transfers cause the inter-area modes to become lightly damped or even unstable. FACTS-devices like the TCSC are damping these oscillations. However, there is even no warning to the transmission operator so far, if a new operating condition causes an unstable oscillation or not and if the controller works well.

The objective has been to develop an algorithm for a real-time monitoring of oscillations from on-line measured signals; in other words, to estimate the parameters characterizing the electromechanical oscillations such as frequency and damping, and to present this information to the operator in a user-friendly environment of the operator station [5]. This kind of information can hardly be obtained only by watching the measured signals displayed in the time-domain. The on-line collected measured data from the WAMS are subject to a further evaluation with the objective to estimate dominant frequency and damping of the electro-mechanical oscillatory modes during normal operation of the power system. The power system is assumed being driven by small disturbances around a nominal operating point. Methods considered here are parametric, model-based ones. Evaluation of the estimated model parameters enables quantitative detection of oscillations and other properties of the system, such as actual system stability. Moreover, similar models obtained using the same identification techniques may be used for a stabilizing controller design or controller adaptation according to the autonomous scheme introduced in chapter 9. Taking into account the trade-off between model complexity and suitability to represent narrow spectra, linear autoregressive models have been focused on.

The basic scheme is outlined in Figure 11.7. The power system is assumed being driven by white noise disturbances $e(k)$ around a nominal operating point. The system is modeled by a linear autoregressive model with adjustable time-varying coefficients. The system outputs $y(k)$ are the measurements provided by PMUs. The ever-present measurement error is represented by $d(k)$.

An adaptive Kalman filter is used to evaluate the parameters of a reduced-order linear equivalent dynamic model of the power system based on a selection of the measurement inputs. Later the damping and frequency of the dominant modes are extracted through eigenvalue analysis of the equivalent model.

The model-based estimation method chosen here is based on an auto-regressive (AR) model with adjustable time-varying coefficients

$$y(k) = \sum_{i=1}^{n} a_i y(k-i) - \varepsilon(k) \qquad (11.10)$$

with $\varepsilon$ given by

$$\varepsilon(k) = \hat{y}(k|k-1) - y(k) \qquad (11.11)$$

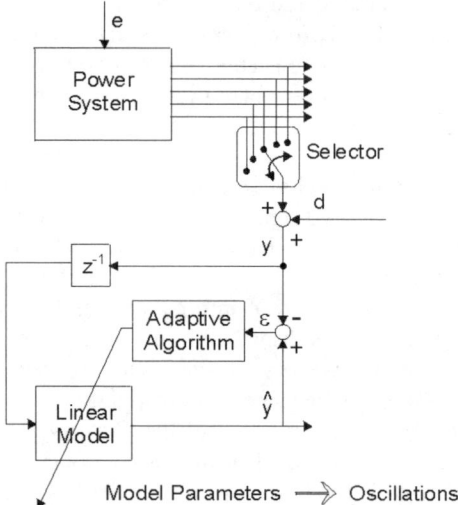

**Fig. 11.7.** Basic scheme for the proposed detection of power system oscillations

The measured signal $y$ may contain some measurement noise $d$. An adaptive algorithm recursively optimizes the criterion (11.12) and yields the optimal parameters of the AR-model, generating possibly the same sequence of data $\hat{y}$ as the measured $y$. The goal is to obtain the parameters of oscillations characterized by their frequency $f_i$ and damping $\xi_i$. They are obtained repeatedly once per given period with the so called refresh time $T_r$ from the AR-model for the set of its $n$ parameters $a_i(k)$. The refresh time defines how often the dominant oscillations are to be calculated from the estimated model parameters and displayed to the operator. This is a trade-off between the computational power of the computer on which the application is running, taking also into account how rapidly the power system varies with time.

Therefore, the first step of the presented approach is to estimate recursively these coefficients $a_i(k)$ that minimize the sum of squared prediction errors

$$J = \min_{a_i} \sum \varepsilon^T \varepsilon = \min_{a_i} \sum (\hat{y}(k|k-1) - y(k))^2 \qquad (11.12)$$

where $\hat{y}(k|k-1)$ denotes the prediction value of $y(k)$ for measurements given up to time $(k-1)$. Recall that the poles of this model contain the required information about the time-varying system dynamics, which depends on the operating point of the power system. The poles can be calculated solving the characteristic equation (11.13) for a set of actual values of $a_i(k)$ frozen at time $k$.

$$z^n - a_1 z^{n-1} \ldots - a_{n-1} z - a_n = 0 \qquad (11.13)$$

The assumption here is that the power system is operated at the same operating point for a certain period of time that enables the estimated coefficients to converge. Indeed, this is no constraint in practice since the estimated model parameters converge to their new values fast enough compared to the dynamics of the power system, if e.g. the algorithm proposed here is employed.

For our purpose - estimation of the damping and frequency of the dominant oscillations - the most suitable conversion of the estimated discrete-time model to a continuous-time model is the Tustin's approximation. This one has the advantage of mapping the left half-s-plane into the unit-disc in z-plane and vice versa [6]. Hence, stable discrete-time systems are transferred into stable continuous-time systems whose eigenvalues are then the basis of calculation of the parameters of oscillations. The relationship between $z$ and $s$ to obtain continuous-time poles resp. eigenvalues $\lambda_i = \alpha_i = i\omega_i$ is for the Tustin's approximation given by:

$$z = \frac{1 + sT_s/2}{1 - sT_s/2} \tag{11.14}$$

A normal operating power system is stable. This means $\alpha_i < 0$ for all $i=1...n$. Oscillations are characterized by complex eigenvalues. The real part $\alpha_i$ gives the information about the decay rate and the imaginary part

$$\omega_i = 2\pi f_i \tag{11.15}$$

about the frequency $f_i$ [Hz] of the oscillatory mode. A practical measure for damping is the relative damping:

$$\xi_i = 100 \frac{-\alpha_i}{\sqrt{\alpha_i^2 + \omega_i^2}} \% \tag{11.16}$$

It has turned out that the adaptive Kalman filtering technique [7] is an appropriate tool to identify the optimal model parameters. This approach has shown the smallest prediction error and the shortest estimation time; i.e. the number of iterations necessary for the parameters to converge. At the same time, applying this method to different measured signals, the results were not very sensitive with regard to the set of the tuning parameters. The standard set of recursive equations to be solved on-line is for the Kalman filter known to be (11.17), see e.g. [7].

$$\begin{aligned} g(k) &= \frac{K(k-1)u(k)}{u^T(k)K(k-1)u(k) + Q_m} \\ \varepsilon(k) &= u^T(k)p(k-1) - y(k) \\ p(k) &= p(k-1) + \varepsilon(k)g(k) \\ K(k) &= K(k-1) - g(k)u^T(k)K(k-1) + Q_p \end{aligned} \tag{11.17}$$

To ensure good numerical robustness of the standard estimation algorithm, the equations (11.17) have been extended by some additional ones (11.18). The covariance matrix $K(k)$ is enforced here to remain symmetrical, and for a better parameter tracking, a regularized constant trace algorithm is used with $c_1/c_2 \cong 10^4$.

$$K(k) = \frac{K(k) + K^T(k)}{2}$$
$$K(k) = \frac{c_1 K(k)}{tr(K(k))} + c_2 Q_p \tag{11.18}$$

All the variables can be initialized with zeros, except for the covariance matrix $K(0)$, which should be initialized with a unity matrix multiplied by a large constant. The most important parameter for tuning is the model order $n$. Its selection is one of the most important aspects for the use of AR models. If one selects a model with too low order, the obtained spectrum will be highly smoothed. On the other hand, if the order is too high, faked low-level peaks in the spectrum will be introduced. The measured signal $y(k)$ is filtered through a digital band-pass filter (with the cut-off frequencies 0.1Hz and 2Hz).

Besides the described real-time estimation of frequency and damping, the running mean value and the amplitude of oscillations can be calculated. This is performed by two self-tuning digital low-pass filters placed before and after the input band-pass filtering. The time constants of these two filters are simply taken over from the estimated dominant frequency.

To show the performance of the algorithm it has been applied to real measurement data from existing WAMS installations. Figure 11.8 shows an example which contains a drastically change in the system structure during the measurement period.

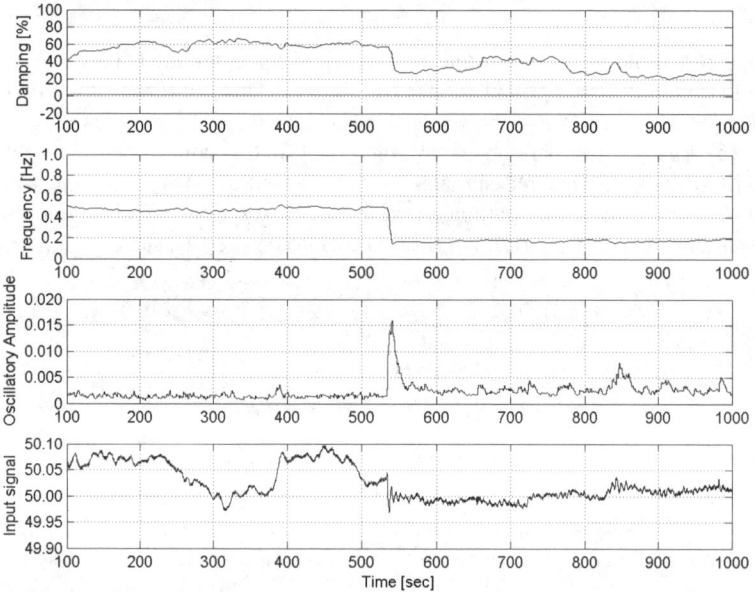

**Fig. 11.8.** Oscillation detection algorithm applied to real PMU measurement data

On the left hand side the system has well damped oscillatory modes and only variations by very slow control actions. After the change in the system structure at 530 s the dominant frequency is going down to 0.2 Hz with a reduced damping. The situation is still uncritical. The provision of this information to the operator during the switching in the system made them feel much more comfortable, because of the better observability.

### 11.2.4 Topology Detection and State Calculation

When enough measurements are available, it is possible to completely detect the status of each network element and to calculate each voltage and current of the network. Topology detection and state calculation is used to provide snapshots of the power system 10-20 times per second. The present topology of the network is inferred from the raw PMU measurements, and therefore the WAMS does not rely on other sources, such a SCADA system for topology information. Therefore the date rate of the PMU can be followed without time delays implied by other sources. The basic function is that the PMU measurement directly shows if a line or another network element is in service or switched off. With several measurements in a region a selection of assumed topology can be verified to fit to the actual one. Another part of the topology detection is an islanding identification to detect if the system has separated into smaller areas.

The next step after the topology analysis is the state-calculation, which is executed once for each island, and serves the purpose of computing the voltages and currents at each bus in the island, also those where no PMUs have been installed.

The classical iterative state estimation, which primarily serves to identify measurement errors, can be represented by the non-linear measurement model in equation (11.19).

$$\underline{z} = \underline{h}(\underline{x}) + \underline{v} \qquad (11.19)$$

The states $\underline{x}$ are complex bus voltages, with magnitude and phase angle. Traditionally, phase angles could not be measured due to impossibility to handle the synchronization of measurement devices. The classical state estimation derives these values from other measurements, such as voltage magnitudes or line active and reactive power flows and eliminated measurement errors $\underline{v}$.

The basic problems of the state estimation are coming from several sources. The network parameters are changing over time with ambient conditions (e.g. temperature, radiation, aging of devices etc.). The topology of the network needs to be updated automatically or manually dependent on the switching status of the devices (line in or out e.g. for service or after a fault). If the topology is not maintained carefully in the system, the state estimation results are wrong. Furthermore the state estimation assumes that the system is in a steady-state situation. In transient situations e.g. after a series of faults, the topology status and the measurement values are not necessarily fitting together which gives bad results for the state estimation. In fringe areas of a power system the redundancy of measurements is usually not given or weak. This means that the state estimation is not able to compensate either bad measurement values or inaccuracies in network parameters.

These drawbacks to the classical state estimation can be eliminated by a WAMS based one. A PMU based state is introduced in [8][9]. If only PMUs are used, the angles of voltage and currents are directly measured. The measurement model in this case is linear according to (11.20).

$$\underline{z} = \underline{\underline{H}} \cdot \underline{x} + \underline{v} \tag{11.20}$$

In this case the influence of the network parameters remains until PMUs are used to determine as well the actual parameters like in the thermal line monitoring algorithm in section 11.2.2. Due to the linearity of the equation, the PMU based state-estimation is a non-iterative process and therefore a solution in predictable time can be guaranteed.

Setting up a PMU based state estimation to the system in Figure 11.5 leads to a selected number of buses with PMU from which the missing network states can be determined. The figure shows the buses where the PMUs have been placed (marked PMU) and where voltages and currents are estimated (marked EST).

## 11.2.5 Loadability Calculation based on OPF Techniques

A detailed voltage stability assessment can be made based on the network model and state information that is received from the state- and topology calculation. Based on the network and state information, a load increase can be simulated on a number of selected buses until the point of maximum loadability is reached. The procedure employs nonlinear optimization techniques to compute the maximum transfer capacity for the topology with which the power system is currently operating but also for various contingency scenarios. Such techniques have been proposed for off-line application for example by [10], but similar ideas can be applied to on-line applications. In mathematical terms the general formulation is:

$$\begin{aligned} \text{maximise} \quad & f(p,x) \\ \text{subject to} \quad & g(p,x) = 0 \\ & h(p,x) \le 0 \end{aligned} \tag{11.21}$$

The function to maximize $f(p,x)$ can be arbitrarily chosen based on the criteria to be optimized. In this case it is chosen as a fictitious active power transfer to a set of load buses known *a-priori* to be critical for the voltage stability or the transfer to a predefined critical area or through a corridor. The optimization variable $p$ can be scalar or vector valued and is the parameter that is varied to simulate a load increase. The function $g(p,x)$ represents the constraints given by the network equations as well as the steady state response of the FACTS-control systems and other controllers. The function $h(p,x)$ contains various operational constraints such as voltage or current limits and actuator limits of the FACTS-devices. The vector $x$ contains the (static) state variables of the network equations, and are implicitly determined by the equality constraints.

From the solution of (11.21), the maximum allowable transfer to the region can be computed and a power margin taken as the difference between the maximum transfer and the transfer at the current operating point.

Before the calculation an admittance matrix is constructed based on the snapshot from the wide-area measurement system. This admittance matrix is required to evaluate the function $g(p,x)$. If desired, N-1 contingency screening can be done by repeatedly solving (11.21). Different contingencies are modeled by modifying the admittance matrix in case of simulated line trips or by changing the load-flow input data in case of generator trips, prior to the solution of (11.21). In this optimization method the setpoints of FACTS-devices can be included as variables. More details on that will be shown in a later section 11.3.

The drawback of this method is the high demand of on-line data and measurements. It requires a WAMS installation with full observability which means both state and topology estimation, and consequently a relatively large number of PMUs. The advantage is that accurate power margins can be calculated and optimal setpoints can be generated. This method is also applicable for general network topologies.

## 11.2.6 Voltage Stability Prediction

For the problem of emergency voltage stability control the two phenomena of short and long term voltage instability must be addressed. If a system is in normal operation, only cascaded or combined outages lead to instability. In most of the practical system collapses long term unfolding instabilities occurred. The reason is that after the initial contingencies the weak situation was not detected well, following events were not foreseen and no appropriate remedial actions were taken. Therefore, either long-term voltage instability or a following event caused by protection mismatch occurred. In both cases the complexity of the problem is beyond that can be foreseen with pre-calculations on N-x base. Therefore any algorithm must be triggered and run after the first events.

After a contingency occurs, the system is in a dynamic phase, which is in the case of long-term voltage instability determined by control actions for instance from tap changers (ULTC), overload capacity of generators and load recovery [11][12]. This characteristic leads to a retarded behaviour, which may lead to a collapse. The idea is to predict just after an event if a collapse might occur.

After a contingency, a sliding data window of PMU measurements is used to determine the actual system and especially load characteristic. A dynamic model is fed with this information. The equilibrium of this model is determined without a time domain simulation. If there is an equilibrium, the system is predicted to be stable, otherwise the system will collapse. Figure 11.9 shows the principle of this approach. Phase 1 is the normal operation where N-1 calculations can be performed. Phase 2 is stable under the assumption that phase 1 was N-1 stable. After the second contingency it is not obvious at the beginning if the system is stable or not.

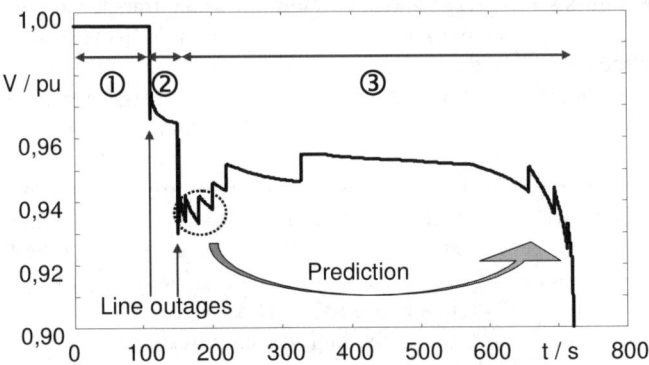

**Fig. 11.9.** Simulated voltage collapse of the power system in Figure 11.5

From the conventional viewpoint of the operator, the voltage is coming back and the system seems to be stabilized during the phase 3 of the simulation. But the underlying dynamics lead the system finally into a collapse. The information from the prediction algorithm can also be used for the determination of stabilising actions.

The steady state equilibrium of the full dynamic system model (11.22) is determined using a model reduction. $G$ are the power flow equations and $z$ the bus voltage magnitudes and angles. $F$ are the remaining equations and $x$ the remaining state variables.

$$\dot{x} = F(x,z)$$
$$0 = G(x,z)$$
(11.22)

At first, all short-term transients in the model are neglected. ULTC, voltage controllers, reactive power limiters and load characteristics can be approximated by their steady-state behavior. To find the equilibrium of the remaining equation system (11.23) a Newton-Raphson algorithm is applied. In (11.23) $F_s$ are the simplified equations with the reduced state vector $x_s$.

$$0 = F_s(x_s, z)$$
$$0 = G(x_s, z)$$
(11.18)

With this model simplification the transient characteristics are separated from the interesting steady-state ones.

To set up the full algorithm the following steps has to be performed. While the system is running in a steady-state situation, the steady-state values of bus voltages $V_0$ and load powers $P_0$ and $Q_0$ must be traced and contingencies such as changes in the topology must be detected. After a contingency is detected the parameters of an applied load model, which describes the voltage dependency of the power, must be determined.

A general load model is shown in (11.24). $P_0$ and $V_0$ are the base power and voltage before the contingency and $\dot{P}$ and $\dot{V}$ are the power and voltage gradients at a certain time step $t$. $\boldsymbol{p}$ is a vector containing all unknown load parameters.

$$P(t) = f(P_0, V_0, \dot{P}(t), \dot{V}(t), V(t), \boldsymbol{p})  \qquad (11.24)$$

An example of a typical load model is the Hill and Karlsson model in (11.25), which shows the typical load recovery characteristic after voltage steps [13]. But also any other model, like e.g. composite ones, can be used.

$$P = -T_p \dot{P} + P_0 \left(\frac{V}{V_0}\right)^{\alpha_s} + \frac{P_0}{V_0} \dot{V} \; T_p \; \alpha_t \left(\frac{V}{V_0}\right)^{\alpha_t - 1}  \qquad (11.25)$$

To determine the load parameters, a sliding window of voltages $V$ at each bus and feeder loads $P$, $Q$ are collected. $\dot{P}$ and $\dot{V}$ are the mean values of the gradients between two timely neighboring measurement points. A set of load equations (11.24) for different time steps within this window builds a nonlinear equation system, which has to be solved for the unknown parameters $\boldsymbol{p}$. This equation system can be solved with a non-linear solver algorithm (e.g. Nelder-Mead). When the number of equations is greater than the load parameters the equation system is over-determined, which increases the accuracy and robustness of the results. Alternatively, a simplified linear solving algorithm is proposed in [14].

The algorithm must be calculated for each relevant load in the system. If it can be seen that the loads behave similarly in a certain area of the system, the number of calculations can be reduced to single examples for each area.

The determined load parameters are fed into the simplified system model to be solved for the equilibrium as described above. This equilibrium point is the predicted state of the system, which might be tens of seconds in the future. If no equilibrium is found, the transient phase will end in a collapse. In both cases a positive or negative power margin can be determined with a continuation power flow [15][16] or optimization technique from the previous section.

Figure 11.10 shows the loadability as a power margin *PM* for the predicted and the non-predicted case for the power system from Figure 11.5 after two contingencies, which are not leading to a collapse. The sensitivity after the first contingency is still low. Therefore the effect of the prediction is also low. After the second contingency the system is operating in a more sensitive resp. non-linear operational point showing a more significant difference between the algorithm with and without prediction. From the beginning of the prediction after the second contingency it needs about 50 s until the non-predicted power margin *PM* is the same as the predicted one. Therefore the forecast of the proposed algorithm is about 50 s in this example. As a result, the criticality of the system will be predicted earlier and also remedial actions can be taken without delay.

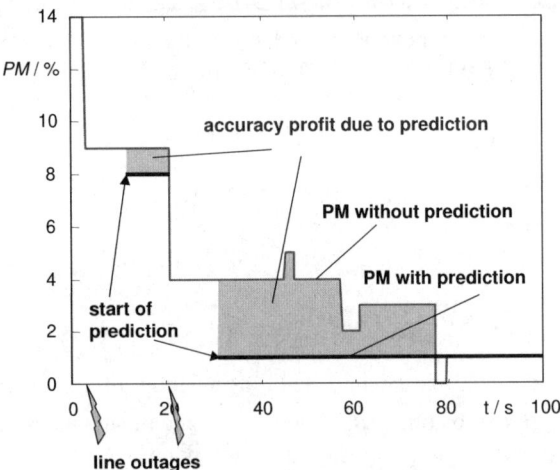

**Fig. 11.10.** Power Margin calculated with and without prediction during two contingencies in the power system in Figure 11.5

## 11.3 Wide Area Control Applications

As shown in the previous chapters a significant added value of FACTS-devices can be gained by introducing secondary control that coordinates the setpoints of FACTS-controllers. In particular contingency cases are important, since they can have dramatic influence on the network flows. The speed and continuous control capability of FACTS-devices make these devices especially useful for improving transfer capacity in these contingency cases, since they can adapt to new flow situations much faster than traditional devices.

All the algorithms in the previous section serve as the basic information to identify critical system situations and instabilities. The algorithms are designed based on the PMU information and therefore provide the information dynamically and in very short time intervals. Both are mandatory to use the speed of the FACTS-devices. This section introduces a selection of algorithms and examples using the basic monitoring methods and apply them in FACTS control schemes. The criteria from the NISC architecture in chapter 9 and the autonomous system in chapter 10 are considered.

In the following a general method based on full network supervision is introduced. This method is a predictive voltage stability control with FACTS setpoint determination using optimization techniques. A simplified method based on limited PMU information can be derived from this first one. A coordination of FACTS control is achieved by using feedback from selected remote PMU measurements. This method enhance transmission capacity restricted for instance by voltage stability in well defined network situations, for instance like corridors.

## 11.3.1 Predictive Control with Setpoint Optimization

Setpoints for the FACTS-devices have to be determined in three basic cases. The first one is the optimization of an actual operational situation. This is equal to the classic Optimized Power Flow (OPF). A second application is the pre-contingency calculation to be prepared for the next event, according to the autonomous system approach in chapter 10. The system has to determine set values for a selection of critical contingencies which might occur. The third application is the post-contingency case, where either directly or after executing the prediction algorithm of section 11.2.6 an optimal set of FACTS setpoints has to be determined.

By a slight modification of equation (11.21) in section 11.2.5 we can extend the method so that it also generates optimal setpoints for FACTS-devices. The resulting setpoints are optimal in the sense that they maximize the loadability criterion $f$ $(p,x,u)$. Now, a vector of the FACTS-device setpoints ($u$) is included together with $p$ as optimization variables. The modified optimization problem becomes:

$$\begin{aligned} &\text{maximise} &&f(p,x,u) \\ &\text{subject to} &&g(p,x,u) = 0 \\ & &&h(p,x,u) \le 0 \end{aligned} \quad (11.26)$$

The solution of (11.26) yields the optimal FACTS setpoints as well as the maximum loadability when these setpoints are applied. If the contingency screening is applied using this method, also FACTS setpoints that maximize the loadability can be pre-computed for a list of credible disturbances. Based on this method the real corridor situation of Figure 11.5 is discussed in the following.

For power transfer capability increase, shunt connected devices such as SVCs have been proven cost-effective, especially when fast or continuously controllable compensation is necessary due to stability or voltage quality concerns. Typically, an SVC also has a voltage controller that controls the terminal voltage so that it is close to a (fixed) reference value. To control power flow, series connected devices such as phase-shifting transformer, TCSC or DFC can be used.

The controllers that are typically embedded in the FACTS-devices are here referred to as primary controllers, and are typically of P- or PI- type, with special supplementary controllers like damping controllers. Normally, the setpoints for FACTS-devices are kept constant or changed manually on a slow timescale based on market activities or optimal power-flow calculations for the base case. Typical FACTS-device controllers operate purely based on local criteria with the objective to control a single local quantity such as voltage or power-flow. The performance objectives of the controllers do not consider their effect on the power system as a whole.

The optimization basically introduces a secondary control loop that generates the setpoints for the primary FACTS-controllers as discussed in the autonomous system in chapter 10. Additionally to the basic rules in the autonomous system the optimization algorithm provides concrete numbers for the set value adaptation, which have to be pre-calculated. The approach avoids the conflict of the secondary control with the objectives of the local primary control loops. For example, using a power-flow control device to reduce power flow to a load area can jeopardize system stability since it would introduce additional (apparent) reactance which could cause or contribute to voltage instability. When the system is operating close to or possibly even beyond stability limits, it would be wise to relax the primary control objectives in favor of the objective to improve stability margins. The task of the secondary controllers is thus to detect when stability margins are small and to carry out appropriate setpoint corrections to improve stability margins.

Figure 11.11 shows the results of a loadability analysis and secondary control actions of the system in Figure 11.5 for the base case and different contingency cases. The system has three FACTS-devices, one Power Flow Control Device (PFD) between bus 6 and 9 and two SVCs at buses 12 and 24. Usually the SVCs would try to keep the voltage according to their reference value independent from the corridor situation. Since one of the aims is to demonstrate the benefits of wide-area FACTS-control, the analysis is made three times. Once with the FACTS-devices deactivated, and once with the FACTS-devices using traditional local controllers and once with the FACTS-devices using the optimal setpoints to maximize transfer capability to the load region. The dashed line illustrates the actual loading, and the bars the maximum possible loading for a particular contingency and with a particular configuration of the FACTS-devices. One observation that can be made in the figure is that the added capacity by coordinated optimal FACTS-control can stabilize a system that would otherwise be unstable as in the case for contingency number 1 in the Figure 11.11.

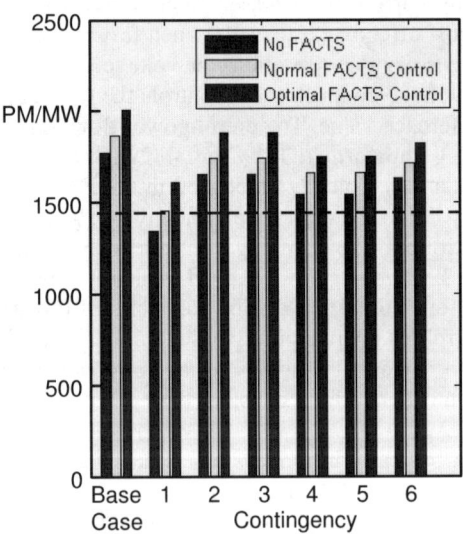

**Fig. 11.11.** Loadability analysis of the power system from Figure 11.5 showing the power margin (PM) for base case and contingency cases

For the shown contingency cases the set-values are pre-calculated. New predefined actions have to be determined continuously according to slightly changing operational conditions and especially after a contingency to be prepared for at least the following one like described in chapter 10. If there is no time to perform the pre-calculation because of multiple contingencies, a post calculation approach has to be applied like the prediction method in section 11.2.6.

But if the contingency is more severe and there is not time to perform the prediction, which means that the system for instance suffers a short-term voltage instability, either predefined actions on a system base or conventional protections like under-voltage protection schemes have to act. In conclusion the control scheme is as follows:

- Steady state or predicted stable operation: The set-values follow the continuous change of the system operation. Predefined actions are determined continuously to act after next contingency
- One contingency occurs (long-term instability): If available and necessary, predefined actions are taken. The prediction process is started and stabilising actions are taken or predefined for next contingency.
- Cascaded contingencies occur (long-term instability): The prediction process is started and stabilising actions are taken or predefined for next contingency.
- Short-term instability: Either predefined system wide actions are taken or conventional protection schemes are acting.

### 11.3.2 Remote Feedback Control

Instead of using full system observability like in the previous section, the remote feedback control schemes use selected remote measurements to detect stability problems in a power system and to determine adapted FACTS setpoints. For simple network topologies, guidelines for the design of feedback controllers to coordinate multiple FACTS-devices can be described. The main advantage of these schemes is that they are simple to implement and stand-alone, but they must be customized for each particular network.

This approach is applicable when the system situation is simple enough and the problems which might occur are limited in number and are well predictable. The remote feedback controller can be derived from a rule base defining a set of actions to be taken in a number of situations. Therefore this kind of control is part of the autonomous system description in chapter 10.

For the situation in Figure 11.5 a remote feedback controller can be designed. The Power Flow Control Device (PFD) has a nominal setpoint for the active power transfer equal to the power transfer through the line before the device is activated. However, as disturbances are applied or the load level changes, the PFD will keep the power transfer through itself close to the reference value until it has saturated. Applying this control the PFD is limiting the maximum transfer over the corridor, because it is not coordinated with the SVCs. It even may reduce the benefits of the SVCs. This is due to the detrimental behavior of the PFD, which introduces additional reactance to reduce the flow through line 6–9 when the load in the southern area is increased. Applying the optimal control, however makes it possible to use the PFD to its full potential to increase transfer capacity. The transfer capacity increase becomes significant about 11 % over the corridor. The capacity increase is achieved through a setpoint adaptation for the PFD as well as the SVC at bus 12.

Figure 11.12 shows how a secondary control loop based on feedback control can be used to achieve near-optimal control. This new secondary loop uses the voltage of bus 24 as feedback signal and operates on the PFD power reference. The PFD secondary loop is using a PI controller with deadband ±0.08 pu, output limits of ±4 pu, the gain $K_r = 100$ and the rise time $T_r = 1$ s.

Figure 11.13 shows the PV-curves obtained through dynamic simulation of the system with the PFD and two SVCs. The three curves show results with FACTS-devices disabled, with the FACTS-devices using conventional local controllers and with the feedback based secondary control scheme.

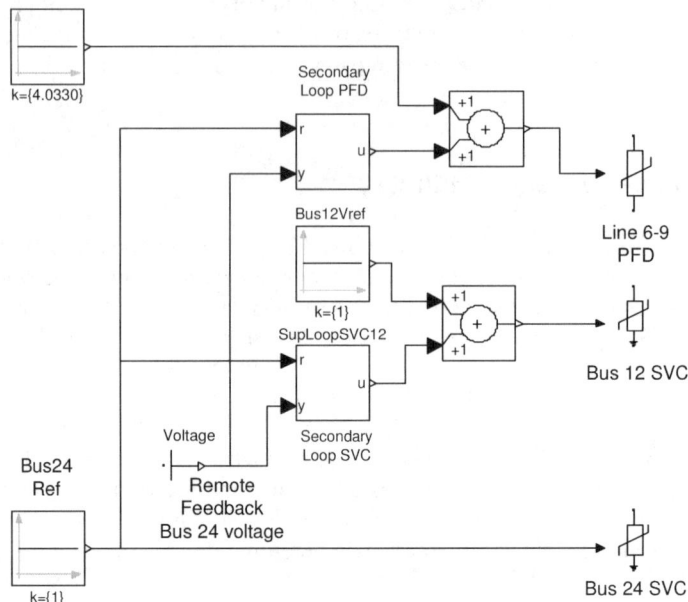

**Fig. 11.12.** Secondary control loop for two SVCs and one power flow control device (PFD)

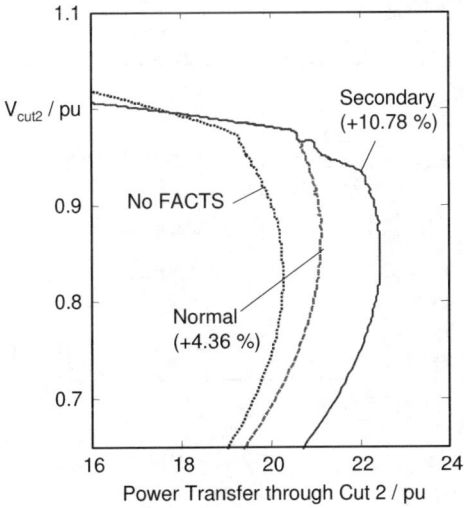

**Fig. 11.13.** PV curves for the system in Figure 11.5 without FACTS, for conventional local FACTS control (normal) and a secondary control scheme

In a second case study it will be shown how wide-area control can be used also in the case of active power flow control when the network topology is more complex than a simple radial corridor as in the first case study. The case design has been inspired by a collapse situation in the central part of the European interconnected network represented by a simplified system in Figure 11.14. The wide-area control scheme is designed only from the point of view of one area (Area 2). It is clear however, that the best solution for the system as a whole would be a global controller for all FACTS-devices in the system. However, since different TSO's operate the system, it would not be easy to implement such a control scheme because of organizational reasons. We therefore consider only regional wide-area controls.

The scheme here has been designed to avoid corridor overloads and not to maximize NTC (loadability) as in the other case. For a control system with this complexity, the optimization based scheme from the previous section 11.3.1 could also be considered, but a remote feedback control based on pre-defined rules is applicable as well.

The corridor between Area 1 and Area 2 (L1) has a transfer limit of 1.2 pu (on a 1000 MVA base). The corridor between Area 1 and Area 4 (L2) normally has a higher transfer limit, but after a line trip the transfer limit will be decreased to about 1 pu. This corridor is equipped with a phase shifting transformer (T1) equipped with a local power-flow control loop. The low voltage part of the corridor within Area 2 (L5) has a transfer limit of 0.65 pu. This corridor is actually composed of many parallel lines and cannot easily be equipped with compensation or power flow controllers. Area 1 is primarily an exporting area, Area 2 is import-export neutral, Area 3 is exporting and Area 4 is an importing area.

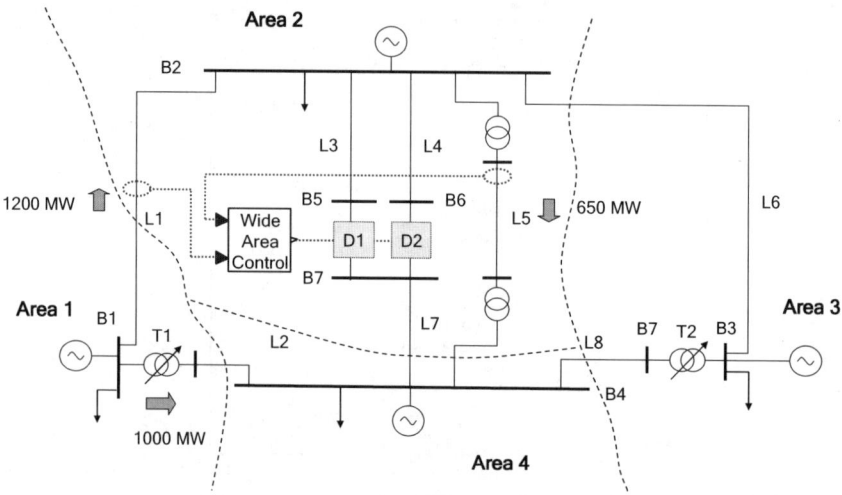

**Fig. 11.14.** Four area system representing a part of the European interconnected network. Transfer limits are marked with block arrows. Power Flow Controller Devices D1 and D2 with Wide Area Control Scheme.

For a first scenario the elements D1 and D2 are neglected. A critical scenario for this configuration is that corridor L2 experiences a line trip which decreases the transfer limit to 1 pu. Therefore, the active power reference value for T1 is reduced to the same value. Figure 11.15 shows the power flows on the corridors following this setpoint decrease.

The flow through corridor L2 decreases to a value close to its reference, however the transformer saturates at its maximum tap step. As the flow decreases through L2, the flow is shifted to the corridor L1 which is overloaded as well as the lines L3 and L4. At this stage the loading of corridor L5 is still below the transfer limit. This case illustrates that the central Area 2 is vulnerable to overload, when the PSTs in Area 1 or Area 3 are used to redirect power-flows away from corridors L2 or L8. This is a natural consequence of Area 2 having the only uncontrolled path leading to the import Area 4.

In order to provide better controllability on the north-south corridor consisting of line L3 and L4, which is the main path, power flow control devices are considered for installation in this corridor. We first consider two phase-shifting transformers (PST), identical to T1 installed at the locations D1 and D2. The two new PSTs are using constant active power references. Figure 11.16 shows simulation results for this case. The two new devices successfully keep the flow on corridor L1 well below the limit, but as a consequence more power is forced through the low voltage corridor L5, which is overloaded instead.

The previously described scenario clearly demonstrates that the standard constant power reference controllers for the PSTs at D1 and D2 are too egoistic. They successfully keep the flow at L3 and L4 close to the reference value, but their control leads to overload in the low voltage corridor.

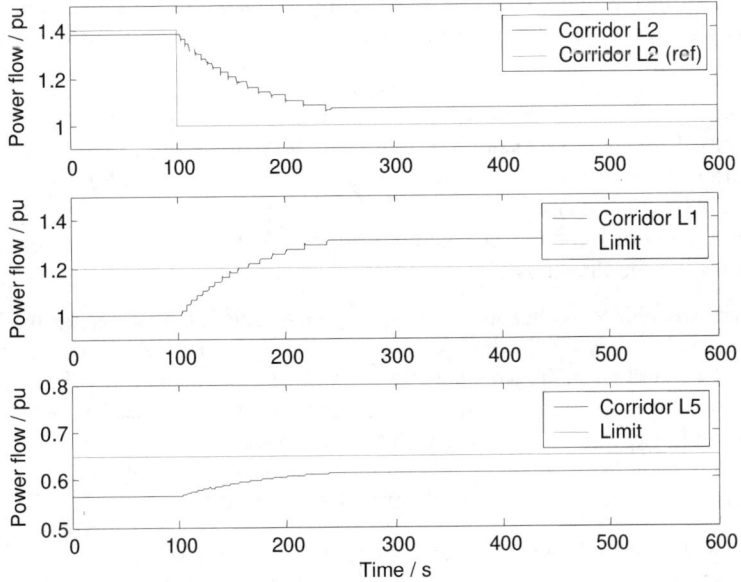

**Fig. 11.15.** Power transfers in the corridors following reference change to 1 pu for T1

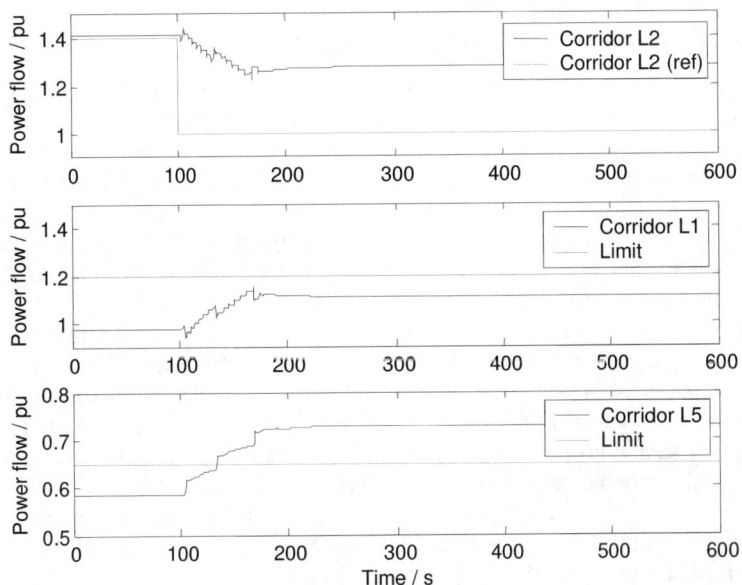

**Fig. 11.16.** Power transfers in the corridors with PSTs at D1 and D2

In the following, a wide-area controller based on feedback is designed using the power transfer through L1 and L5 as measurements and the power references for D1 and D2 as actuator. The controller has the following priorities as rules for its actions as follows:

- Priority 1 - Avoid overload of low voltage corridor L5 (0.65 pu),
- Priority 2 - Avoid overload of corridor L1 (1.00 pu),
- Priority 3 - Control power flow through L3 and L4 to predefined reference (2.74 pu in this case),
- Alarm operator if the two conflicting Priority 1 and Priority 2 objectives cannot be fulfilled simultaneously.

The first two objectives become conflicting since reducing overload on the lower voltage level comes at the cost of lowering the apparent impedance (using the PSTs) of L3 and L4. This will increase the flow on the uncontrolled L1 corridor. In case of alarm, the operators must request the operators of Area 1 to reduce production until the overload situation in Area 2 is solved.

By inspection of the network topology, we can design the following controller logic that would address the above described objectives:

- In case of overload on L5, increase setpoints for D1 and D2
- In case of overload on L1 and if there is sufficient margin on L5 corridor, decrease power reference for D1 and D2
- If there is an overload on L1 and not sufficient margin on L5 corridor, the operator should be alarmed, so that a request for relieving the corridor by generation reduction in Area 1 can be made.
- If there is no overload on either L1 or L5 keep reference constant on D1 and D2

Figure 11.17 shows the responses to the initial power reference change at L2 when the wide-area controller is used as a secondary controller for the PSTs at D1 and D2. As decided by the prioritization of the objectives, the secondary controller relieves the overload on L5 but allows overload on L1. The benefit is that L3, L4 and L5 are together used to its maximum. Since both overload constraints could not be met simultaneously, an alarm signal is given at about 190 seconds that shows that further actions are necessary to relieve overload. Because of the response time of the PSTs D1 and D2, there is an overshoot in the limit for L5.

In conclusion the wide-area control scheme allows using the installed transmission capacity between Area 2 and 4 to its maximum, independent of what is happening in the areas around. The scheme acts fully automatic and is transparent for the operator. In case of limitations a warning is generated asking for manual interactions. These actions could be included as well in the wide-area scheme requiring setting up the scheme between different TSO operation areas.

The same approach which was discussed here with PSTs can be applied to fast controllable power flow controllers like the Dynamic Flow Controller (DFC) as well. Figure 11.18 shows the respective results for the same scenario.

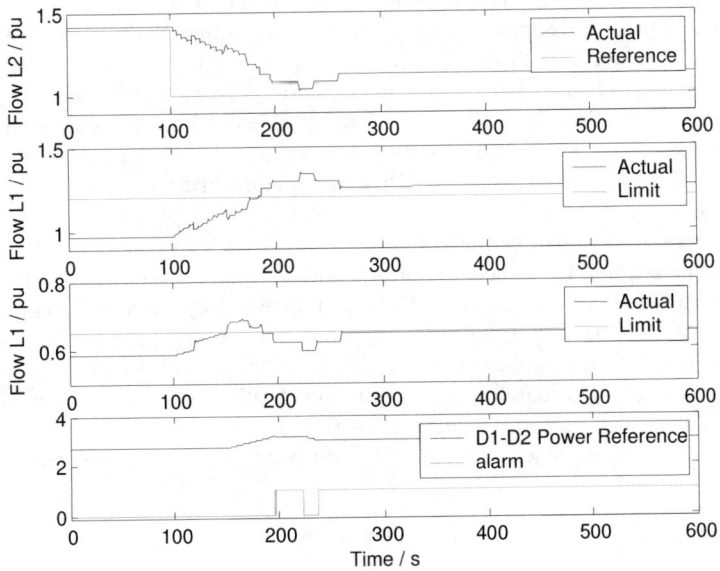

**Fig. 11.17.** Power transfers in the corridors with secondary control for PSTs at D1 and D2

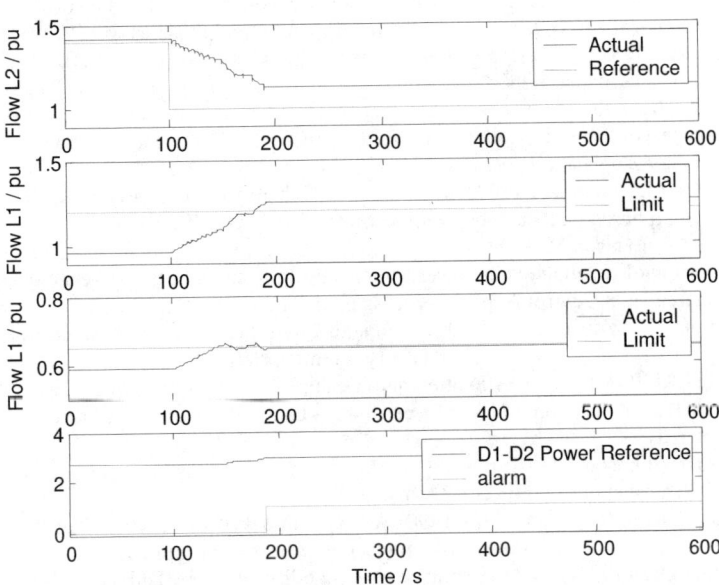

**Fig. 11.18.** Power transfers in the corridors with secondary control for Dynamic Flow Controllers DFC instead of PSTs at D1 and D2

The approach which was discussed here is fulfilling the requirements of the NISC-architecture in chapter 9. The solution shown is one practical representation of the rule base according to the autonomous system approach from chapter 10. In case of topology changes within the corridor between Area 2 and 4 the rules might be adapted. Other functionalities of the FACTS-devices like damping control or a coordination of the power flow control devices with reactive compensation like SVCs to maximize the transfer capability (see section 11.3.1) can be implemented additionally.

The case studies in this chapter have shown that significant added value of FACTS-devices can be gained by introducing wide-area control that coordinates the setpoints of FACTS-controllers. Only with the consideration of the approaches presented in the chapters 9, 10 and 11 a beneficial use of power flow controlling FACTS-devices can be achieved. Normal operations as well as emergency situations have been considered. The speed and continuous control capability of FACTS-devices make them especially useful for improving transfer capacity in emergency situations, since they can adapt to new flow situations much faster than traditional devices.

## References

[1] Phadke AG, Thorpe J, Adamiak MG (1983) A New Measurement Technique of Tracking Voltage Phasors, Local System Frequency and Rate of Change of Frequency. IEEE Transactions on Power Apparatus and Systems, vol PAS-102, no 5
[2] CIGRE (2000) System protection schemes in power net-works, CIGRE Task Force 38.02.19, Technical Report
[3] Vu K, Begovic M, Novosel D, Saha MM (1997) Use of local measurements to estimate voltage-stability margin, 20th International Conference on Power Industry Computer Applications IEEE, pp 318–323
[4] Warland L, Holen AT (2002) Estimation of distance to voltage collapse: Testing an algorithm based on local measurements. 14th PSCC, Sevilla, Spain
[5] Korba P, Larsson M, Rehtanz C (2003) Detection of Oscillations in Power Systems Detection of Oscillations in Power Systems using Kalman Filtering Techniques. IEEE Conference on Control Applications, Istanbul, Turkey
[6] Astrom KJ, Wittenmark B (1996) Computer Controlled Systems. Prentice-Hall
[7] Haykin S (1996) Adaptive Filter Theory, Prentice Hall
[8] Nuqui RF (2001) State Estimation and Voltage Security Monitoring Using Synchronized Phasor Measurements. Dissertation, Virginia Polytechnic Institute and State University
[9] Phadke AG, Thorp JS, Karimi KJ (1986) State estimation with phasor measurements. IEEE Transactions on Power Systems, no 1
[10] Van Cutsem T, Vournas C (1990) Voltage Stability of Electric Power Systems. Power Electronics and Power Systems Series, Kluwer Academic Publishers
[11] Taylor CW (1994) Power System Voltage Stability. McGraw Hill, New York
[12] Daalder J, Gustafsson MN, Krantz NU (1997) Voltage Stability: Significance of load characteristics and current limiters. IEE Proc. Generation, Transmission and Distribution, vol 144, no 3, pp 257-262

[13] Hill D J, Karlsson D (1994) Modeling and identification of nonlinear dynamic loads in power systems. IEEE Trans. on Power Systems, vol 9, no 1, pp 157- 163
[14] Rehtanz C (2001) Wide area protection and online stability assessment based on Phasor Measurement Units. IREP - Bulk Power Systems Dynamics and Control V, Onomichi, Japan
[15] Ajjarapu V, Christy C (1991) The continuation power flow: A tool for steady state voltage stability analysis. IEEE PICA '91 Baltimore, pp 304-311
[16] Flueck EH, Dondeti, JR (2000) A new continuation power flow tool for investigating the nonlinear effects of transmission branch parameter variations. IEEE Transactions on Power Systems, vol 15, no 1, pp 223-227

# 12 Modeling of Power Systems for Small Signal Stability Analysis with FACTS

Small signal stability in a power system is the ability of the system to ascertain a stable operating condition following a small perturbation around its operating equilibrium. Power system disturbances can be broadly classified into two categories; large and small. Disturbances such as generation tripping, load outage, faults etc have severe influences on the system operation. These are large disturbances and the dynamic response and the stability conditions of the system are assessed within the standard framework of transient stability analysis and control. The system is modeled as a non-linear dynamic process. A large number of references dealing with this problem exist in power engineering literature [1]-[3]. Essentially the researchers have applied non-linear system theories and simulations to establish a clear understanding of the dynamic behavior of power system under such conditions. Effective tools to analyze and devise various non-linear control strategies are now in place.

The power system largely operates under quasi-equilibrium state except when undergoing large disturbance situations. The disturbances of small magnitude are very common. Such disturbances can come from the random fluctuation in loads induced by weather conditions etc. These small and gradual disturbances do not lead to severe excursion of system operating variables such as machine angle and speed from their operating equilibrium values. It is observed that the electromechanical oscillations observed in the post-fault recovery stage of the system are usually linear in nature [4]. The theory of linear system analysis has provided a deep insight into the operating behavior of an interconnected power system under such situations. The assumption of a linear system model around an operating equilibrium has revealed many interesting conclusions. Most often these conclusions are not consistent with what have been observed in the field under similar set of operating circumstances. A better understanding of the nature of the system dynamics helps to plan control strategies for secure operation.

This chapter will focus on modelling and analysis of power system dynamic behavior under small disturbances. A brief description of the modelling of various components in power systems including FACTS-devices is given. This chapter will focus on the dynamic model of FACTS-devices as their steady state power flow models have already been discussed in the earlier chapters. The overall system model is linearized for small signal stability analysis through eigenvalue approach. The small signal analysis will be applied in an interconnected power system model with FACTS-devices. The approach of modal controllability [4] will be described and applied to examine control capability of FACTS-devices from various locations in the systems. We will also describe the methods of modal observability [4] to identify the most effective feedback signals for the control design of the FACTS-devices to produce greater stability margin. The aim of this chapter is

- to develop a clear understanding of how linear system theory can provide enhanced insight into power system dynamic behavior under various operating situations,
- to develop a better understanding of the control needs and specifications.

## 12.1 Small Signal Modeling

### 12.1.1 Synchronous Generators

The primary sources of electrical energy are the synchronous generators. They are electromechanical energy conversion devices that are driven at synchronous speed by steam, hydro and gas turbines, depending on the source of mechanical energy. The rotor houses a field winding which is excited through direct current to produce flux. The flux produces a rotational voltage in the stator windings which are connected to the grid. The magnitude of the voltage is controlled through an automatic voltage regulator (AVR). Power output is varied through controlled admission of steam or water or gas using a governor. The general approach to synchronous machine modelling is quite mature. The high frequency stator transient is usually ignored. Besides the field winding, the rotor might have a closed physical winding. The solid rotor body provides a closed rotor winding effect. At a speed other than synchronous, voltage is induced and currents circulate. They provide damping action against rotor speed deviation. Consequently the windings are known as damper winding. The rotor damping effect is modeled by closed windings of suitable inductances and time constants. The number of damper windings used to represent rotor damping effect depends on the nature of study. For small signal stability, two damper windings in the q-axis and one damper in the field axis are adequate. One can neglect the damper winding for model simplification at the cost of introducing some degree of conservatism in small signal stability results.

Let us assume an interconnected power system with $m$ machine and $n$ bus. We consider four windings on the rotor (one field and one damper in d-axis and two dampers in q-axis). For $i = 1$ to $m$, the following equations represent machine dynamics [5]. We have considered a d-q axis modeling of machine with the q-axis leading the d-axis and taken generator current as positive, i.e. IEEE convention [6] and [7].

$$\frac{d\delta_i}{dt} = \omega_i - \omega_s \tag{12.1}$$

## 12.1 Small Signal Modeling

$$\frac{d\omega_i}{dt} = \frac{\omega_i}{2H_i}\left[T_{mi} - T_{eleci} - T_{Di}\right] \quad (12.2)$$

$$\frac{dE'_{di}}{dt} = -\frac{1}{T'_{qoi}}\left[+E_{di}' + \left(X_{qi} - X'_{qi}\right)\left[I_{qi}\left[-\frac{\left(X'_{qi} - X''_{qi}\right)}{\left(X'_{qi} - X_{lsi}\right)^2}\left(-\Psi_{2qi}\right.\right.\right.\right.$$
$$\left.\left.\left.+ \left(X'_{qi} - X_{lsi}\right)I_{qi} - E'_{di}\right)\right]\right] \quad (12.3)$$

$$\frac{d\Psi_{1di}}{dt} = \frac{1}{T''_{doi}}\left[-\Psi_{1di} + E'_{qi} + \left(X'_{di} - X_{lsi}\right)I_{di}\right] \quad (12.4)$$

$$\frac{d\Psi_{2qi}}{dt} = -\frac{1}{T''_{qoi}}\left[\Psi_{2qi} + E'_{di} - \left(X'_{qi} - X_{lsi}\right)I_{qi}\right] \quad (12.5)$$

$$T_{eleci} = \frac{(X''_{di} - X_{lsi})}{(X'_{di} - X_{lsi})}E'_{qi}I_{qi} + \frac{(X'_{di} - X''_{di})}{(X'_{di} - X_{lsi})}\Psi_{1di}I_{qi}$$
$$+ \frac{(X''_{qi} - X_{lsi})}{(X'_{qi} - X_{lsi})}E'_{di}I_{di} - \frac{(X'_{qi} - X''_{qi})}{(X'_{qi} - X_{lsi})}\Psi_{2qi}I_{di} - (X''_{qi} - X''_{di})I_{qi}I_d \quad (12.6)$$

$$T_D = D(\omega_s - \omega) \quad (12.7)$$

The stator current equations are algebraic in nature because of the assumption made earlier. They are:

$$V_i \cos(\delta_i - \theta_i) - \frac{\left(X''_{di} - X_{lsi}\right)}{\left(X'_{di} - X_{lsi}\right)}E'_{qi} - \frac{\left(X'_{di} - X''_{di}\right)}{\left(X'_{di} - X_{lsi}\right)}\Psi_{1di} + R_{si}I_{qi} - X''_d I_{di} = 0 \quad (12.8)$$

$$V_i \sin(\delta_i - \theta_i) + \frac{\left(X''_{qi} - X_{lsi}\right)}{\left(X'_{qi} - X_{lsi}\right)}E'_{di} - \frac{\left(X'_{qi} - X''_{qi}\right)}{\left(X'_{qi} - X_{lsi}\right)}\Psi_{2qi} - X''_{qi}I_{qi} - R_{si}I_{di} = 0 \quad (12.9)$$

where, for the $i^{th}$ machine

$\delta_i$: rotor angle (radian)
$\omega_i$: rotor speed (radian per second)
$E_{fdi}$: exciter voltage on stator base (p.u.)
$E_{qi}'$ ($E_{di}'$): quadrature (direct) axis transient voltage (p.u.).
$\psi_{1di}$ ($\psi_{1qi}$): flux linkage in the direct (inner quadrature) axis damper (p.u.)
$I_{qi}$ ($I_{di}$): stator q-axis(d-axis) component of currents (p.u.)

$V_i$, $\theta_i$:            bus voltage magnitude and angle respectively.
$X_{lsi}$:            stator leakage reactance (p.u.)
$R_{si}$:            stator resistance (p.u.)
$X_{di}$, $X_{di}'$, $X_{di}''$:        direct axis synchronous, transient and sub-transient reactance (p.u.) respectively.
$X_{qi}$, $X_{qi}'$, $X_{qi}''$:        quadrature axis synchronous, transient and sub-transient reactance (p.u.) respectively.
$T_{do}'$, $T_{do}''$:        direct axis open circuit transient and sub-transient time constants (seconds) respectively.
$T_{qo}'$, $T_{qo}''$:        quadrature axis open circuit transient and sub-transient time constants (seconds) respectively

### 12.1.2 Excitation Systems

The excitation system provides the necessary rotor flux to induce a voltage in the stator. The excitation voltage $E_{fdi}$ is never manipulated directly but is changed through the action of the exciter. Excitation systems are broadly classified into two types: slow DC excitation and fast static excitation [4].

A typical slow excitation system (termed DC1A exciter) [4] consists of four basic blocks as shown in Fig 12.1. They are the exciter, amplifier, excitation-stabilizer and terminal voltage sensor. A basic model for an exciter is given by equation (12.10) where $S_e$ is the saturation in the exciter. It is approximated as an exponential function. The constants $K_e$ and $T_e$ relate to exciter gain and time constant respectively. The $K_e$ varies with the operating conditions. For each operating condition, it is assumed that $K_e$ is such as to make the voltage regulator output zero in the steady state. In order to automatically control the terminal voltage a measured voltage signal must be compared to a reference voltage and amplified to produce the exciter input, $V_r$. The amplifier can be a pilot exciter or a solid state amplifier. In either case, the amplifier is modeled as a first order differential equation as shown in equation (12.11). The regulator is often equipped with a stabilizing transformer that is modeled by equation (12.12).

The symbols $K_a$, $T_a$ and $K_f$, $T_f$ are the gain and time constant of the amplifier and stabilizer circuit respectively. The terminal voltage sensor is modeled as a first order block with a filter time constant $T_r$ and shown in equation (12.13).

$$T_e \frac{dE_{fd}}{dt} = -\left[K_e E_{fd} + S_e(E_{fd})E_{fd}\right] + V_r \quad (12.10)$$

$$T_a \frac{dV_r}{dt} = -V_r + K_a V_i \quad ; \quad V_{r\,min} \leq V_r \leq V_{r\,max} \quad (12.11)$$

$$T_f \frac{dR_f}{dt} = -R_f + \frac{K_f}{T_f} E_{fd} \quad (12.12)$$

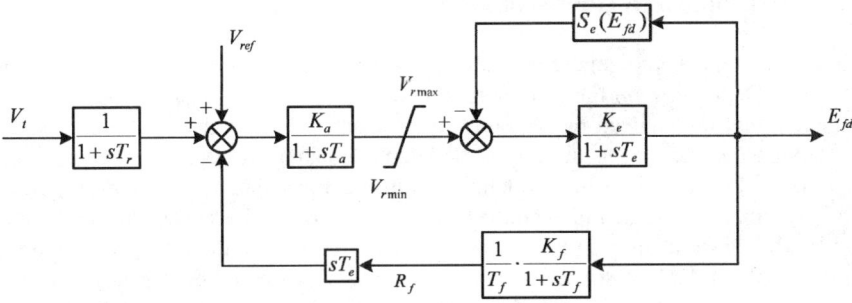

**Fig. 12.1.** Block Diagram of a DC1A-type Excitation System

$$\frac{dV_{tr}}{dt} = -\frac{1}{Tr}\left[V_{tr} - V_t\right] \tag{12.13}$$

Modern large machines are equipped with fast acting type Thyristor based excitation systems. The exciter power is drawn from the generator bus through an exciter transformer. Such a type of excitation system (ST1A) is often modeled as a single time-constant block [4]. The error signal is used as input and $E_{fd}$ as output. Figure 12.2 shows a small signal representation for a high gain (of the order 200 to 400) and fast (of the order of a few milliseconds) exciter. Normally, $T_a$ is neglected. When $T_a$ is ignored, the dynamics are described by the following two equations:

$$\frac{dV_{tr}}{dt} = -\frac{1}{Tr}\left[V_{tr} - V_t\right] \tag{12.14}$$

$$E_{fd} = K_A(V_{ref} - V_{tr}) \tag{12.15}$$

In the case where the voltage regulator gain $K_a$ is too large for better transient stability performance, the damping torque introduced by the exciter becomes negative. In order to ensure a well-damped post-fault response of the system, the regulator block is preceded by a transient gain reduction (TGR) block. However, with properly designed power system stabilizer (PSS) this block is not necessary.

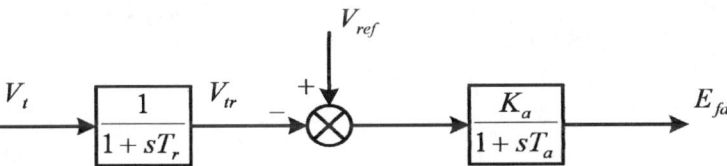

**Fig. 12.2.** Block Diagram of a Fast ST1A-type Excitation System

### 12.1.3 Turbine and Governor Model

In a power plant, the generator is driven either by a steam-turbine or a hydraulic turbine. Depending on the size, construction and principle of operation, different small signal models can be derived. For a steam turbine with tandem compound structure, various stages should be modeled adequately to represent the torsional dynamics [4]. The dynamic modeling of turbine and governor plays an important role in small signal stability studies. Interesting conclusions were drawn from a utility based field study in turbine and governor model validation. It was reported [8] that about 40% of simulated response could be observed during large generation trips in Western Electric Co-ordination Council (WECC) system. This allowed a response based modeling for the turbine governor system. The validated model produced a system response that matched closely the measured response of the system. Normally, for the electromechanical modes in the frequency range of 0.2 to 2.0 Hz, the dynamic interaction of these turbine masses can be an important consideration, if the associated governor is not properly tuned. All present day speed-governing systems are expected to be properly tuned, making the turbine less interactive. The inclusion of a small signal model of the turbine can increase the frequency of the low frequency electromechanical modes very slightly. As long as governors are properly tuned with adequate dead-band they will not have any adverse effect on power system damping. In view of this and for the sake of a simple model, the mechanical input to generator is assumed constant. However, for mid-term and long-term stability studies, which address system recovery from severe upsets with time, accurate modeling of turbine, governor is essential.

### 12.1.4 Load Model

In power system stability and power flow studies, the loads are modeled as seen from the bulk delivery point at transmission voltage level. Based on the way voltage and frequency influence loads at the delivery point, they are classified into two broad categories: static and dynamic.

In the static approach, both real and reactive loads are modeled as a non linear function of voltage magnitude. It also includes average frequency deviation $(\Delta f)$. A static load model expresses the characteristics of the load at any instant of time as algebraic functions of the bus voltage magnitude and frequency at that instant. The active power component, $P$, and reactive power component, $Q$, are considered separately. The voltage dependency of the load characteristics is represented by the exponential model [4] as given in the following two equations.

$$P = P_0 \left( \overline{V} \right)^a \qquad (12.16)$$

$$Q = Q_0 \left( \overline{V} \right)^b \qquad (12.17)$$

$$\overline{V} = \left(\frac{V}{V_0}\right) \tag{12.18}$$

$P_0$, $Q_0$ and $V_0$ are the values at the initial operating condition. The parameters of this model are the exponents 'a' and 'b'. With these exponents equal to 0, 1, 2, the model represents load of a constant power (CP), constant current (CC) or constant impedance (CI) type respectively. The exponent 'a' (or 'b') are sensitivities of power to voltage at $V = V_0$. For composite system loads, the exponent 'a' usually lies in the range between 0.5 and 1.8. Exponent 'b' varies as a non-linear function of the voltage. For $Q$ at higher voltages, 'b' tends to be significantly higher than 'a'. An alternative model that has been widely used to represent the voltage dependency of loads is the polynomial model.

$$P = P_0 \left[ p_1 (\overline{V})^2 + p_2 \overline{V} + p_3 \right] \tag{12.19}$$

$$Q = Q_0 \left[ q_1 (\overline{V})^2 + q_2 \overline{V} + q_3 \right] \tag{12.20}$$

This model is commonly referred to as the ZIP model as it is composed of constant impedance Z, constant current I and constant power P components. The parameters of the model are the coefficients '$p_1$' to '$p_3$' and '$q_1$' to '$q_2$' that denote the proportion of each component. The frequency dependency of the load characteristic is usually represented in the exponential and polynomial models by a factor as follows:

$$P = P_0 (\overline{V})^a \left[ 1 + K_{pf} \Delta f \right] \tag{12.21}$$

$$Q = Q_0 (\overline{V})^b \left[ 1 + K_{qf} \Delta f \right] \tag{12.22}$$

$$P = P_0 \left[ p_1 (\overline{V})^2 + p_2 \overline{V} + p_3 \right] \left( 1 + K_{pf} \Delta f \right) \tag{12.23}$$

$$Q = Q_0 \left[ q_1 (\overline{V})^2 + q_2 \overline{V} + q_3 \right] \left( 1 + K_{qf} \Delta f \right) \tag{12.24}$$

Typically, $K_{pf}$ ranges from 0 to 3.0 and $K_{qf}$ ranges from -2.0 to 0.0.

Power system loads during a disturbance behave dynamically. However, because of the distributed nature of loads, it is difficult to get an equivalent dynamic representation of them. A large single induction motor load is modeled in the d-q reference frame almost in the same way as the synchronous generator. Some researchers represent loads through differential equations involving load voltage magnitude and angle as state variables. A power recovery model has been suggested in [9] for analyzing voltage stability related problems. It is shown that such models can capture voltage instability events more realistically.

The response of most of the composite loads to voltage and frequency changes is fast and the steady state condition for the response is reached very quickly. This is true at least for modest changes in the voltage/frequency. The use of the static models described in the previous sections is justified in such cases.

There are, however, many cases where it is necessary to account for the dynamics of the load components. Studies of inter-area oscillations, voltage instability and long term stability often require load dynamics to be modeled. A study of systems with large concentrations of motors also requires the representation of load dynamics. Reference [4] discusses various models in use for stability studies and proposes a general model that encompasses a large variety of models with suitable modification of the coefficients. A CIGRE task force, formed to investigate the causes of the Swedish system blackouts in 1983, produced the following recommendations [10] on the effect of load models in stability studies in stressed power systems.

$$P_L = P_0 + P + K_{pw}\frac{d\theta_L}{dt} + K_{pv}\left(V_L + T\frac{dV_L}{dt}\right) \quad (12.25)$$

$$Q_L = Q_0 + Q + K_{qw}\frac{d\theta_L}{dt} + K_{qv1}V_L + K_{qv2}V_L^2 \quad (12.26)$$

Where, $P$ and $Q$ are static power loads, $P_0$ and $Q_0$ are the constant power portion of the induction motor load and the rest depends on bus voltage and frequency deviation. The symbols $V_L$ and $\theta_L$ are the bus voltage magnitude and phase angle respectively. In our research we have simplified the load model shown in equation (12.25) and (12.26) according to:

$$P_L = P_{L0}\left(\frac{V_L}{V_{L0}}\right)^{np} + K_{pw}\frac{d\theta_L}{dt} + K_{pv}T\frac{dV_L}{dt} \quad (12.27)$$

$$Q_L = Q_{L0}\left(\frac{V_L}{V_{L0}}\right)^{nq} + K_{qw}\frac{d\theta_L}{dt} \quad (12.28)$$

### 12.1.5 Network and Power Flow Model

Power is transmitted over long distance through overhead lines of high voltage ranging from 230 kV to 1,100 kV. These overhead lines are classified according to length, based on the approximations used in their modeling:

- Short line: Lines shorter than 50 miles (80 km) are represented as equivalent series impedance. The shunt capacitance is neglected.
- Medium line: Lines, with length in the range of 80 km to about 200 km, are represented by nominal π equivalent circuits.
- Long Line: Lines longer than about 200 km fall in this category. For such lines the distributed effects of the parameters are significant. They need to be

represented by equivalent π circuits or alternatively as cascaded sections of shorter lengths, with each section represented by a nominal π equivalent.

For stability studies involving low frequency oscillations it is reasonable to assume a lumped parameter model. The approximation introduces a bit of conservatism in the margin of stability. However for simulation of lightning or switching transients, the distributed parameter model is used. High voltage transmission cables are also modeled in a similar way to overhead lines but they have much larger shunt capacitance than that of EHV lines of similar length and voltage rating.

In the steady state power frequency network model, the power flow equations at each node can be expressed as:

$$P_{G,k} - P_{L,k} = \sum_{m=1}^{n} V_k V_m (G_{km} \cos(\theta_k - \theta_m) + B_{km} \sin(\theta_k - \theta_m)) \quad (12.29)$$

$$Q_{G,k} - Q_{L,k} = \sum_{m=1}^{n} V_k V_m (G_{km} \sin(\theta_k - \theta_m) - B_{km} \cos(\theta_k - \theta_m)) \quad (12.30)$$

The symbols $P_{G,k}$ and $Q_{G,k}$ are real and reactive power generated respectively at the $k^{th}$ bus. They are expressed as functions of bus voltage magnitude, angle and armature current as:

$$P_{G,k} = V_k \cos(\delta_k - \theta_k) I_{qk} - V_k \sin(\delta_k - \theta_k) I_{dk} \quad (12.31)$$

$$Q_{G,k} = -V_k \sin(\delta_k - \theta_k) I_{qk} - V_k \cos(\delta_k - \theta_k) I_{dk} \quad (12.32)$$

### 12.1.6 FACTS-Models

In this section, the steady-state and small-signal dynamic models of three most commonly used FACTS-devices are described. They are Static VAr Compensator (SVC), Controllable Phase Shifter (CPS) and Thyristor Controlled Series Capacitors (TCSC). We will mainly describe their steady state power flow characteristic and small signal dynamic characteristics. The power injection model has been used for steady-state representation of these devices as it is most suitable for incorporation into an existing power flow algorithm without altering the bus admittance matrix. The power injection equations governing this type of model are described for each of the devices. The small-signal dynamic models of the series connected devices are presented considering a single time constant block representing the response time of the power electronics based converters. For the shunt voltage control devices, a separate voltage control loop is involved with suitable response time of the voltage sensing hardware and time constants of the voltage regulator block.

#### 12.1.6.1 SVC-Model

A commonly used topology of a Static VAr compensator (SVC), shown in Fig 12.3, comprises a parallel combination of a Thyristor Controlled Reactor and a fixed capacitor. It is basically a shunt connected static var generator/absorber whose output is adjusted to exchange capacitive or inductive current so as to maintain or control specific parameters of the electrical power system, typically bus voltage.

The reactive power injection of a SVC connected to bus k is given by

$$Q_k = V_k^2 B_{svc} \tag{12.33}$$

$B_{svc} = B_C - B_L$; the symbols $B_C$ and $B_L$ are the respective susceptances of the fixed capacitor and the Thyristor Controlled Reactor. It is also important to note that a SVC does not exchange real power with the system.

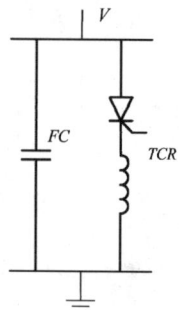

**Fig. 12.3.** SVC block diagram

The small-signal dynamic model of an SVC is given in Fig 12.4 [7]. $\Delta B_{svc}$ is defined as $\Delta B_C - \Delta B_L$. The differential equations from this block diagram can easily be derived as

$$\frac{d}{dt}\Delta B_{svc} = \frac{1}{T_{svc}}\left[-\Delta B_{svc} + \left(1 - \frac{T_{v1}}{T_{v2}}\right)\Delta V_{r-svc} - \frac{K_v T_{v1}}{T_{v2}}\Delta V_{t-svc}\right]$$
$$+ \frac{K_v T_{v1}}{T_{v2} T_{svc}}\left[\Delta V_{ss-svc} + \Delta V_{ref}\right] \tag{12.34}$$

$$\frac{d}{dt}\Delta V_{r-svc} = \frac{1}{T_{v2}}\left[-\Delta V_{r-svc} - K_v \Delta V_{t-svc} + K_v V_{ref} + K_v V_{ss-svc}\right] \tag{12.35}$$

$$\frac{d}{dt}\Delta V_{t-svc} = \frac{1}{T_m}\left[\Delta V_t - \Delta V_{t-svc}\right] \tag{12.36}$$

$K_v$, $T_{v1}$, $T_{v2}$ are the gain and time constants of the voltage controller respectively; $T_{svc}$ is the time constant associated with SVC response while $T_m$ is the voltage sensing circuit time constant. The SVC can work either in voltage control mode or in susceptance control mode.

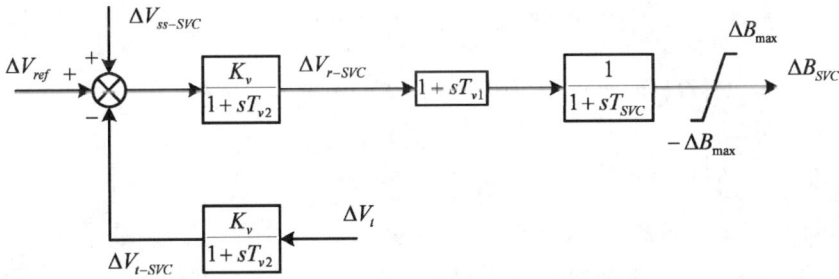

**Fig. 12.4.** SVC dynamic model

### 12.1.6.2 TCPS-Model

A Thyristor Controlled Phase Shifter (TCPS) can exert a continuous shift on the phase angle of voltage between the two ends of the line in which the TCPS is connected. A typical TCPS consists of an exciter and booster transformer pair. Fig 12.5 shows a typical TCPS connected in the line between bus k and m with its exciter transformer being fed by bus k. The injected voltage can be modeled as an ideal voltage source $V_{se}$ in series with the line impedance $Z_{km}$. The injection model [11] is obtained by replacing the voltage source by an equivalent current source $I_{se}$ in parallel with the line as shown in Fig 12.5 where $I_{se}$ and $I_{sh}$ are given by equation (12.37) and (12.38) and the injected power at both ends by (12.39) and (12.40).

$$\overline{I}_{se} = \frac{\overline{V}_{se}}{Z_{km}} \tag{12.37}$$

$$\overline{I}_{sh} = \overline{I}_k - \overline{I}_{se} \tag{12.38}$$

$$\overline{S}_k = \overline{V}_k \left( -\overline{I}_{sh} - \overline{I}_{se} \right)^* \tag{12.39}$$

$$\overline{S}_m = \overline{V}_m \left( \overline{I}_{se} \right)^* \tag{12.40}$$

The power injections $S_k$ and $S_m$ are given as

$$\overline{S}_k = \overline{V}_k \left( -\overline{I}_{sh} - \overline{I}_{se} \right)^* \tag{12.41}$$

$$\overline{S}_m = \overline{V}_m \left( \overline{I}_{se} \right)^* \tag{12.42}$$

The real and reactive components of the above two equations provide expression for nodal power injections which can be incorporated into network power flow equations. They are as follows:

$$P_{inj,k} = V_k V_m \left[ G_{km} \left\{ \cos\theta_{km} - \cos(\theta_{km} + \phi) \right\} + B_{km} \left\{ \sin\theta_{km} - \sin(\theta_{km} + \phi) \right\} \right] \tag{12.43}$$

$$Q_{inj,k} = V_k V_m \left[ G_{km} \{\sin\theta_{km} - \sin(\theta_{km} + \phi)\} - B_{km} \{\cos\theta_{km} - \cos(\theta_{km} + \phi)\} \right] \quad (12.44)$$

$$P_{inj,m} = V_m V_k \left[ G_{mk} \{\cos\theta_{mk} - \cos(\theta_{mk} - \phi)\} + B_{mk} \{\sin\theta_{mk} - \sin(\theta_{mk} - \phi)\} \right] \quad (12.45)$$

$$Q_{inj,m} = V_m V_k \left[ G_{mk} \{\sin\theta_{mk} - \sin(\theta_{mk} - \phi)\} - B_{mk} \{\cos\theta_{mk} - \cos(\theta_{mk} - \phi)\} \right] \quad (12.46)$$

where $\theta_{i,j} = \theta_i - \theta_j$.

The small-signal dynamic model of a controllable phase shifter is given in Fig 12.6. The small signal model is given by:

$$\frac{d}{dt}\Delta\phi = \frac{1}{T_{cps}} \left[ -\Delta\phi + \Delta\phi_{ref} + \Delta\phi_{cps} \right] \quad (12.47)$$

The symbol $T_{cps}$ represents the response time of the Thyristors.

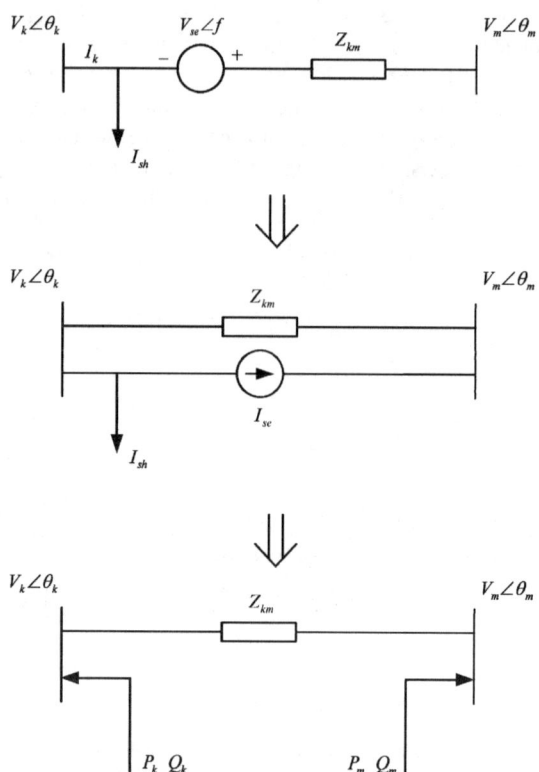

**Fig. 12.5.** TCPS block diagram

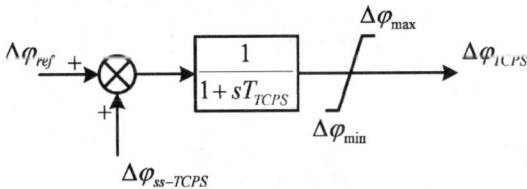

**Fig. 12.6.** TCPS dynamic model

### 12.1.6.3 TCSC-Model

A Thyristor Controlled Series Capacitor (TCSC) is a capacitive reactance compensator that consists of series capacitor banks shunted by Thyristor Controlled Reactors in order to provide a smoothly variable series capacitive reactance. Let us consider that the TCSC is connected in the line between bus k and m. In this case the resistance of the line is neglected for simplicity of the calculation. If $I$ is the current flowing through the line, the TCSC having capacitive reactance $X_c$ can be represented by a voltage source $V_{se}$ as shown in Fig 12.7, where $V_{se}$ is given by $V_{se} = jX_c I$. The injection model [11] is obtained by replacing the voltage source by an equivalent current source $I_s$ in parallel with the line as shown in Fig 12.7 where $I_s$ is given by $I_s = V_{se}/X_{km}$. The current source $I_s$ corresponds to the injection powers $S_k$ and $S_m$ which are given by

$\overline{S}_k = \overline{V}_k (-I_s)^*; \overline{S}_m = \overline{V}_m (I_s)^*$. These expressions are resolved into real and reactive components to produce the following nodal power injection expressions:

$$P_k = \frac{k_c}{(k_c - 1)} V_k V_m B_{km} \sin(\theta_k - \theta_m) \quad (12.48)$$

$$Q_k = \frac{k_c}{(k_c - 1)} B_{km} \left[ V_k^2 - V_k V_m \cos(\theta_k - \theta_m) \right] \quad (12.49)$$

$$P_m = \frac{k_c}{(k_c - 1)} V_m V_k B_{mk} \sin(\theta_m - \theta_k) \quad (12.50)$$

$$Q_m = \frac{k_c}{(k_c - 1)} B_{mk} \left[ V_m^2 - V_m V_k \cos(\theta_m - \theta_k) \right] \quad (12.51)$$

where, $k_c = \frac{X_C}{X_L}$.

The small-signal dynamic model [7] of a controllable series capacitor is given in Fig. 12.8. The symbol $T_{tcsc}$ represents the response time of the Thyristor. The model for various other FACTS-devices can be developed in similar manner.

332    12 Modeling of Power Systems for Small Signal Stability Analysis with FACTS

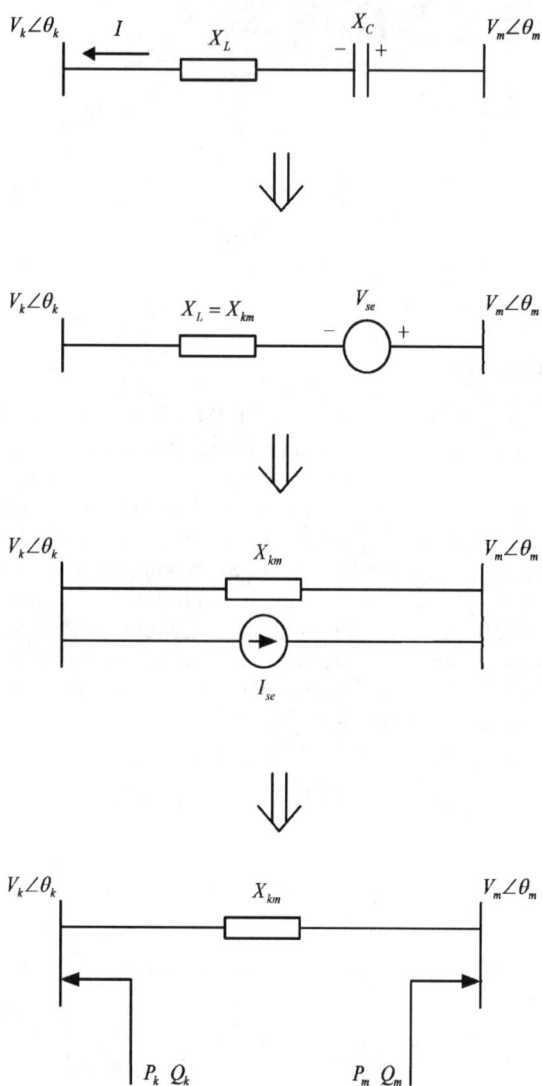

**Fig. 12.7.** TCPS block diagram

## 12.1 Small Signal Modeling 333

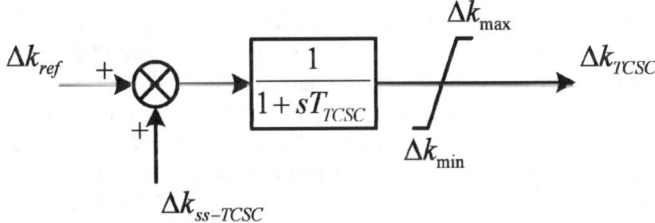

**Fig. 12.8.** TCPS dynamic model

### 12.1.7 Study System

The standard modeling approach described earlier will be applied to a study system. Fig. 12.9 shows a single line diagram of a 16-machine and 68-bus system model [7]. This is a reduced order equivalent of the New England Test System (NETS) and the New York Power System (NYPS) model. There are nine generators in NETS area and three in NYPS area. The three neighboring utilities are represented as three equivalent large generators #14, #15 and #16. The generators, loads and imports from other neighboring areas are representative of operating conditions in the early 1970s.

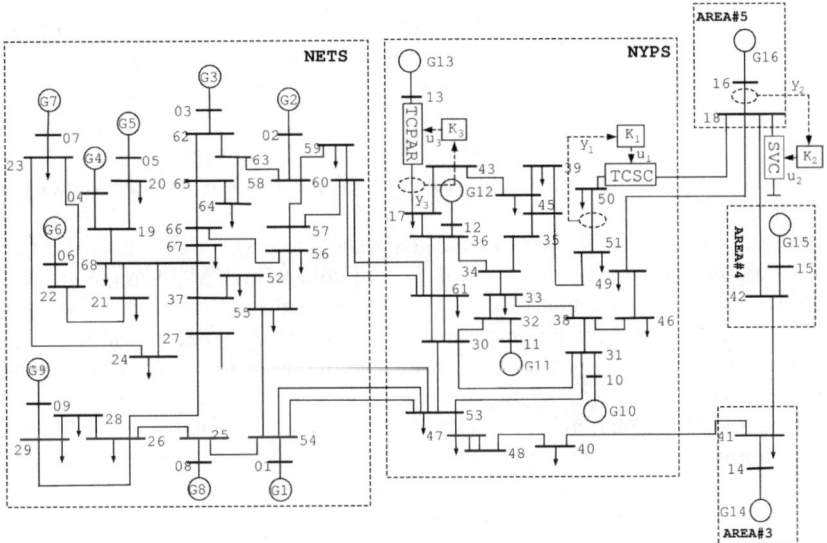

**Fig. 12.9.** 16-machine 68-bus study system

The first eight machines have slow excitation (IEEE type DC1A) whilst machine #9 is equipped with a fast acting static excitation system (IEEE ST1A). This machine is also assumed to have a speed input power system stabilizer (PSS) to ensure adequate damping of the electromechanical mode of this machine. The rest of the machines are under manual excitation control. We will analyze this system later for various modeling approximation and power flow and load characteristics to develop a better understanding of the system dynamics. We will also carry out the analysis with three FACTS-devices, SVC, TCSC and TCPS, in this system to see their influence on system dynamics.

## 12.2 Eigenvalue Analysis

### 12.2.1 Small Signal Stability Results of Study System

It is mentioned earlier that small signal stability is the ability of the system to maintain stable equilibrium when subjected to small disturbances. For small disturbances the response of the system will have linear behavior i.e. the equations that describe the resulting response of the system may be linearized for the purpose of analyses.

The behavior of a dynamic system such as a power system, as described in the earlier sections, can be expressed as a set of first order differential and algebraic (DAE) equations of the form:

$$\dot{x} = f(x,z,u) \tag{12.52}$$

$$0 = g(x,z,u) \tag{12.53}$$

$$y = h(x,z,u) \tag{12.54}$$

where, $x$ is a state vector, $z$ is an algebraic variable vector, $u$ is an input vector and $y$ is an output vector.

From a power system's perspective, the state vectors $x$ are generator angle, speed, transient voltage, flux; excitation system voltage and AVR output etc. The algebraic variables $z$ are bus voltage magnitude, angle, stator currents etc. The control variables $u$ are excitation control reference voltage, mechanical input etc. The choice of output $y$ variables depends on stabilizing signals such as line power, machine speed, bus voltage magnitude etc.

Let us linearize equations (12.52) to (12.54) around an initial $(x_0, z_0, u_0)$ operating equilibrium and express the result as:

$$\Delta \dot{x} = \frac{\partial f}{\partial x} \Delta x + \frac{\partial f}{\partial z} \Delta z + \frac{\partial f}{\partial u} \Delta u \tag{12.55}$$

$$0 = \frac{\partial g}{\partial x} \Delta x + \frac{\partial g}{\partial z} \Delta z + \frac{\partial g}{\partial u} \Delta u \tag{12.56}$$

$$\Delta y = \frac{\partial h}{\partial x}\Delta x + \frac{\partial h}{\partial z}\Delta z + \frac{\partial h}{\partial u}\Delta u \qquad (12.57)$$

The algebraic variable $\Delta z$ can be eliminated from the above to produce a state-space description of the system given by:

$$\Delta \dot{x} = A\Delta x + B\Delta u \qquad (12.58)$$

$$\Delta y = C\Delta x + D\Delta u \qquad (12.59)$$

where,

$$A = \frac{\partial f}{\partial x} - \frac{\partial f}{\partial z}\left(\frac{\partial g}{\partial z}\right)^{-1}\frac{\partial g}{\partial x}; \quad B = \frac{\partial f}{\partial u} - \frac{\partial f}{\partial z}\left(\frac{\partial g}{\partial z}\right)^{-1}\frac{\partial g}{\partial u} \qquad (12.60)$$

$$C = \frac{\partial h}{\partial x} - \frac{\partial h}{\partial z}\left(\frac{\partial g}{\partial z}\right)^{-1}\frac{\partial g}{\partial x}; \quad D = \frac{\partial h}{\partial u} - \frac{\partial h}{\partial z}\left(\frac{\partial g}{\partial z}\right)^{-1}\frac{\partial g}{\partial u} \qquad (12.61)$$

The symbol $\Delta$ from equations (12.58) and (12.59) will be dropped in an effort to follow the notation of standard state space description. Unless stated otherwise, henceforth all perturbed variables will mean incremental variables. This approach of representing power system behavior in DAE form is largely followed for small signal analysis.

The eigenvalues $\lambda$ of $A$ are the roots of the characteristic equation:

$$det(\lambda I - A) = 0 \qquad (12.62)$$

Symmetric or Hermitian matrices will have real eigenvalues. On the other hand, non symmetric or non Hermitian matrices will have a few complex eigenvalues occurring in conjugate pairs. The complex conjugate eigenvalues are due to the fact that the matrix $A$ is real and so the characteristic polynomial has real coefficients.

Let us take any complex conjugate eigenvalues $\lambda_{1,2} = \sigma \pm j\omega$. The real part $\sigma$ relates to damping and the imaginary part $\omega$ relates to the frequency of oscillation. In power system small signal stability literatures, usually damping ratio $\rho$ and linear frequency $f$ (Hz) are used. These are related to $\lambda_i$ as follows:

$$\lambda_i = \sigma_i \pm j\omega_i; \rho_i = -\frac{\sigma_i}{\sqrt{\sigma_i^2 + \omega_i^2}}; f_i = \frac{\omega_i}{2\pi} \qquad (12.63)$$

Let us use the classical model of all the generators with constant impedance load. In the classical model the airgap flux remains constant; so the effect of voltage regulator and damper circuit are absent. Only swing equations involving variables $\delta$ and $\omega$ are considered. The system will have thirty two state variables and so it will have as many eigenvalues. The eigenvalues are displayed in Table 12.1. There is one zero eigenvalue, one negative real and fifteen pairs of eigenvalues that occurs as complex conjugates. In this case the complex conjugate eigenvalues are known as electromechanical modes as they originate from the swing equations.

**Table 12.1.** Eigenvalues in classical model

| $\lambda_i$ | $\rho_i$ | $f_i$ (Hz) |
|---|---|---|
| 0.0 | | |
| -0.1328 | 1.000 | 0.00 |
| -0.0626±j2.4212 | 0.026 | 0.38 |
| -0.0697±j3.1469 | 0.022 | 0.50 |
| -0.0507±j3.9564 | 0.013 | 0.63 |
| -0.0810±j4.9710 | 0.016 | 0.79 |
| -0.0927±j6.1810 | 0.015 | 0.98 |
| -0.0684±j6.7719 | 0.005 | 1.15 |
| -0.0387±j7.2178 | 0.007 | 1.22 |
| -0.0568±j7.6901 | 0.009 | 1.26 |
| -0.0752±j7.9630 | 0.006 | 1.27 |
| -0.0542±j7.9915 | 0.006 | 1.34 |
| -0.0537±j8.4600 | 0.006 | 1.55 |
| -0.0617±j9.7802 | 0.006 | 1.55 |
| -0.0716±j9.7820 | 0.007 | 1.55 |
| -0.0701±j9.8349 | 0.007 | 1.56 |
| -0.1161±j11.467 | 0.010 | 1.82 |

We first explain the origin of the zero eigenvalue. The machine speeds and angles are expressed in absolute terms thereby introducing redundancies in the state variables and the resulting state matrix is singular. The zero eigenvalue is because of redundancy in angle but this can be removed by taking one machine angle as a reference and expressing all other angles with respect to it. This will result in reduction of angle state variable of the reference machine from the differential equations. Sometimes a second zero eigenvalue can exist when the generator torque is independent of machine speed deviations, i.e. mechanical damping is neglected and governor action is not represented. Because non-uniform damping is used, the second zero eigenvalue in this case does not exist. This situation can also arise when the ratios of inertia constant to damping coefficient in all the machines are uniform which can be avoided by setting the speed of one machine as reference (following the assumption of infinite inertia of a speed referenced machine) and expressing speed deviation of other machines with respect to the reference one. If one particular machine angle and speed are taken as reference, the dynamics of that particular machine will not affect the swing equation. In practice it is not done, as this introduces difficulties in indexing and manipulating various matrices and vectors in vector based computation. Usually the eigenvalues will not be exactly zero as initial conditions are not exact because of mismatches in power flow convergence, however small they might be.

The eigenvalues characterised by frequencies 0.38, 0.50, 0.63 and 0.79 Hz are known as inter-area modes [4] involving machines across a large portion of the system. The other eigenvalues in the table are local modes, involving one or two machines and hence the effect is localised.

We now show eigenvalue analysis results for detailed machine models, both with and without FACTS-devices, and for different network configuration and load situations. The eigenvalues for the system using full models are computed and displayed in Table 12.2. Each machine is modelled to have three damper windings, one field winding and an excitation control system. The first eight generators use DC excitation, while machine #9 is equipped with fast excitation. The load is constant impedance in nature. Other machines are placed on manual excitation control. One can see that one local mode is unstable, which is connected to machine #9. This is due to a fast excitation control system in that machine. This mode is stabilised with the help of a speed input PSS. The behaviour of the system with the PSS is also shown in Table 12.2. It can be observed readily that the inclusion of damper windings in the model has increased the damping of the electromechanical modes in general. The effect of excitation control is also seen to have improved the frequencies of oscillations.

**Table 12.2.** Electromechanical modes in detailed model

| Detailed model without PSS | | Detailed model with PSS | |
|---|---|---|---|
| $\rho_i$ | $f_i$ (Hz) | $\rho_i$ | $f_i$ (Hz) |
| 0.0165 | 0.3916 | 0.0643 | 0.3830 |
| 0.0436 | 0.5022 | 0.0436 | 0.5019 |
| 0.0345 | 0.6263 | 0.0560 | 0.6193 |
| 0.0498 | 0.7907 | 0.0499 | 0.7907 |
| 0.0627 | 1.0710 | 0.3061 | 0.8539 |
| 0.0578 | 1.1583 | 0.0630 | 1.0707 |
| -0.0043 | 1.1895 | 0.0589 | 1.1584 |
| 0.0793 | 1.2050 | 0.0798 | 1.2045 |
| 0.0743 | 1.2716 | 0.0574 | 1.2640 |
| 0.0070 | 1.2951 | 0.0745 | 1.2718 |
| 0.0349 | 1.3516 | 0.0502 | 1.3418 |
| 0.0976 | 1.5400 | 0.0977 | 1.5400 |
| 0.0690 | 1.5455 | 0.0681 | 1.5470 |
| 0.0906 | 1.5639 | 0.0907 | 1.5637 |
| 0.0615 | 1.8760 | 0.0616 | 1.8759 |

The effect of FACTS on the damping of electromechanical modes is also investigated. We assume TCPS, TCSC and SVC are located in the network to facilitate power flow and provide network voltage support. The TCPS is assumed to be installed in the line between bus 37 and bus 68, the TCSC between bus 69 and bus 50 and the SVC is located at bus 18. The effect of each of these devices is investigated separately. The results are displayed in Table 12.3. It can be seen that the steady state outputs from theses FACTS-devices do not improve the damping. We also include three devices and computed their combined effect in system damping. The observation was that the overall system damping did not improve appreciably. This means additional control known as supplementary power oscillation damping is necessary for improved system response. We now examine the effect of various levels of power flow on the damping of these electromechanical modes. We assume a 700 MW flow between NETS and NYPS as base and adjust the load and generation in both the areas to create a flow that varies from 100 MW to 900 MW.

**Table 12.3.** Effects of FACTS on electromechanical modes

| No FACTS | | SVC (only) | | TCSC (only) | | TCPS (only) | |
|---|---|---|---|---|---|---|---|
| $\rho_i$ | $f_i$ (Hz) | $\rho_i$ | $f_i$ (Hz) | $\rho_i$ | $f_i$ (Hz) | $\rho_i$ | $f_i$ (Hz) |
| 0.06 | 0.38 | 0.06 | 0.38 | 0.06 | 0.39 | 0.06 | 0.38 |
| 0.04 | 0.50 | 0.04 | 0.50 | 0.04 | 0.50 | 0.04 | 0.50 |
| 0.05 | 0.62 | 0.05 | 0.62 | 0.05 | 0.62 | 0.05 | 0.62 |
| 0.05 | 0.79 | 0.05 | 0.79 | 0.05 | 0.79 | 0.05 | 0.79 |
| 0.30 | 0.85 | 0.30 | 0.85 | 0.30 | 0.85 | 0.30 | 0.85 |
| 0.06 | 1.07 | 0.06 | 1.07 | 0.06 | 1.07 | 0.06 | 1.07 |
| 0.06 | 1.16 | 0.06 | 1.16 | 0.06 | 1.16 | 0.06 | 1.16 |
| 0.08 | 1.20 | 0.08 | 1.20 | 0.08 | 1.20 | 0.08 | 1.20 |
| 0.06 | 1.26 | 0.06 | 1.26 | 0.05 | 1.26 | 0.05 | 1.26 |
| 0.07 | 1.27 | 0.07 | 1.27 | 0.07 | 1.27 | 0.07 | 1.27 |
| 0.05 | 1.34 | 0.05 | 1.34 | 0.05 | 1.34 | 0.05 | 1.34 |
| 0.10 | 1.54 | 0.10 | 1.54 | 0.09 | 1.54 | 0.10 | 1.54 |
| 0.07 | 1.54 | 0.07 | 1.54 | 0.07 | 1.54 | 0.06 | 1.54 |
| 0.09 | 1.56 | 0.09 | 1.56 | 0.09 | 1.56 | 0.09 | 1.56 |
| 0.06 | 1.87 | 0.06 | 1.87 | 0.06 | 1.87 | 0.06 | 1.87 |

The results are shown in Table 12.4. It is seen that at higher level of power flow, the damping and frequencies of the first inter-area mode reduces. The other modes do not show much change because they are not affected by the power flow between theses two areas but instead are affected by flows between other areas.

This is owing to the reduced voltage at the two ends, which is picked up by the AVR in each area. The degradation of damping is not much because of the fact that only one area (NETS) has slow excitation control and the other area (NYPS) is on manual excitation control.

**Table 12.4.** Effects of power flow on electromechanical modes

| 100 MW | | 500 MW | | 700 MW | | 900 MW | |
|---|---|---|---|---|---|---|---|
| $\rho_i$ | $f_i$ (Hz) | $\rho_i$ | $f_i$ (Hz) | $\rho_i$ | $f_i$ (Hz) | $\rho_i$ | $f_i$ (Hz) |
| 0.07 | 0.39 | 0.07 | 0.38 | 0.06 | 0.38 | 0.06 | 0.37 |
| 0.04 | 0.50 | 0.04 | 0.50 | 0.04 | 0.50 | 0.04 | 0.50 |
| 0.06 | 0.64 | 0.06 | 0.63 | 0.05 | 0.62 | 0.05 | 0.60 |
| 0.05 | 0.79 | 0.05 | 0.79 | 0.05 | 0.79 | 0.04 | 0.79 |
| 0.30 | 0.85 | 0.30 | 0.85 | 0.30 | 0.85 | 0.30 | 0.85 |
| 0.06 | 1.07 | 0.06 | 1.07 | 0.06 | 1.07 | 0.06 | 1.06 |
| 0.06 | 1.15 | 0.06 | 1.15 | 0.06 | 1.16 | 0.06 | 1.16 |
| 0.08 | 1.20 | 0.08 | 1.20 | 0.08 | 1.20 | 0.08 | 1.20 |
| 0.06 | 1.26 | 0.06 | 1.26 | 0.05 | 1.26 | 0.05 | 1.26 |
| 0.07 | 1.28 | 0.07 | 1.27 | 0.07 | 1.27 | 0.07 | 1.26 |
| 0.05 | 1.34 | 0.05 | 1.34 | 0.05 | 1.34 | 0.05 | 1.34 |
| 0.10 | 1.54 | 0.10 | 1.54 | 0.09 | 1.54 | 0.10 | 1.54 |
| 0.07 | 1.54 | 0.07 | 1.54 | 0.07 | 1.54 | 0.06 | 1.54 |
| 0.09 | 1.56 | 0.09 | 1.56 | 0.09 | 1.56 | 0.09 | 1.56 |
| 0.06 | 1.87 | 0.06 | 1.87 | 0.06 | 1.87 | 0.06 | 1.87 |

Usually fast excitation control significantly reduces the damping of low frequency modes. The frequency also reduces slightly with power flow because of reduced synchronising power co-efficient at relatively high rotor angles.

We now investigate the effect of load characteristics on the damping. Table 12.5 displays the results. So far in all our calculations, constant impedance (CI) loads are assumed. We have investigated the effect of various load characteristics such as constant current (CC), constant power (CP) and dynamic load at a few buses. The results in Table 12.5 show that the damping action from the constant power type of load is least. This is why the voltage and angle stability margin involving constant power type of load is low. The induction motor type load on the other hand produces better damping because they are asynchronous in nature. It is very difficult to quantify the effect of loads on damping. It is system specific and depends on the relative locations of loads, generation and tie lines in the system. The effect of tie line strength on system damping is investigated next and the results are displayed in Table 12.6.

We assume 700 MW flows with different tie line strength. The base case is with all ties in operation. One line between bus 53 and 54 connecting NETS with NYPS is then taken out. The eigenvalue analysis shows that the damping of the first three inter-area modes is reduced. In addition, if one line between bus 60 and 61 is taken out, the damping is reduced further. This quantitatively confirms our simple understanding that the power system with weak tie-line strength experiences oscillations. The mechanism of reduction in damping can be attributed to higher angular separation between the two areas. The maximum power transfer capacity reduces because of high transmission impedance. This demands increase in angular separation between two areas for the same amount of power flow. As the voltages at different buses are reduced; the overall operating situation leads to reduced damping and frequency of oscillations.

**Table 12.5.** Effect of load characteristics on electromechanical modes

| CC | | CP | | CI | | Dynamic | |
| --- | --- | --- | --- | --- | --- | --- | --- |
| $\rho_i$ | $f_i$ (Hz) | $\rho_i$ | $f_i$ (Hz) | $\rho_i$ | $f_i$ (Hz) | $\rho_i$ | $f_i$ (Hz) |
| 0.05 | 0.38 | 0.05 | 0.36 | 0.06 | 0.38 | 0.07 | 0.38 |
| 0.04 | 0.51 | 0.04 | 0.53 | 0.04 | 0.50 | 0.04 | 0.50 |
| 0.06 | 0.62 | 0.06 | 0.64 | 0.05 | 0.62 | 0.07 | 0.62 |
| 0.05 | 0.79 | 0.05 | 0.79 | 0.05 | 0.79 | 0.04 | 0.79 |
| 0.30 | 0.88 | 0.29 | 0.91 | 0.30 | 0.85 | 0.30 | 0.85 |
| 0.06 | 1.07 | 0.06 | 1.07 | 0.06 | 1.07 | 0.06 | 1.07 |
| 0.06 | 1.15 | 0.06 | 1.15 | 0.06 | 1.16 | 0.06 | 1.16 |
| 0.08 | 1.20 | 0.08 | 1.20 | 0.08 | 1.20 | 0.08 | 1.20 |
| 0.06 | 1.26 | 0.06 | 1.26 | 0.05 | 1.26 | 0.06 | 1.26 |
| 0.07 | 1.27 | 0.07 | 1.27 | 0.07 | 1.27 | 0.07 | 1.27 |
| 0.05 | 1.34 | 0.05 | 1.34 | 0.05 | 1.34 | 0.05 | 1.34 |
| 0.10 | 1.54 | 0.10 | 1.54 | 0.09 | 1.54 | 0.10 | 1.54 |
| 0.07 | 1.54 | 0.07 | 1.54 | 0.07 | 1.54 | 0.07 | 1.54 |
| 0.09 | 1.56 | 0.09 | 1.56 | 0.09 | 1.56 | 0.09 | 1.56 |
| 0.06 | 1.87 | 0.06 | 1.87 | 0.06 | 1.87 | 0.06 | 1.87 |

**Table 12.6.** Effect of tie-line strength on electromechanical modes

| No outage | | Line 53-54 out | | Line 60-61, 53-54 out | |
|---|---|---|---|---|---|
| $\rho_i$ | $f_i$ (Hz) | $\rho_i$ | $f_i$ (Hz) | $\rho_i$ | $f_i$ (Hz) |
| 0.06 | 0.38 | 0.05 | 0.36 | 0.04 | 0.36 |
| 0.04 | 0.50 | 0.04 | 0.50 | 0.04 | 0.50 |
| 0.05 | 0.62 | 0.04 | 0.59 | 0.04 | 0.57 |
| 0.05 | 0.79 | 0.05 | 0.79 | 0.05 | 0.79 |
| 0.30 | 0.85 | 0.30 | 0.85 | 0.30 | 0.85 |
| 0.06 | 1.07 | 0.06 | 1.07 | 0.06 | 1.05 |
| 0.06 | 1.16 | 0.06 | 1.15 | 0.06 | 1.15 |
| 0.08 | 1.20 | 0.08 | 1.20 | 0.08 | 1.20 |
| 0.05 | 1.26 | 0.06 | 1.25 | 0.06 | 1.26 |
| 0.07 | 1.27 | 0.07 | 1.27 | 0.07 | 1.27 |
| 0.05 | 1.34 | 0.04 | 1.31 | 0.05 | 1.34 |
| 0.09 | 1.54 | 0.10 | 1.54 | 0.10 | 1.54 |
| 0.07 | 1.54 | 0.07 | 1.54 | 0.07 | 1.54 |
| 0.09 | 1.56 | 0.09 | 1.56 | 0.09 | 1.56 |
| 0.06 | 1.87 | 0.06 | 1.87 | 0.06 | 1.87 |

### 12.2.2 Eigenvector, Mode Shape and Participation Factor

If $\lambda_i$ is an eigenvalue of $A$, $v_i$ and $w_i$ are non zero column and row vectors respectively such that the following relations hold:

$$(A - \lambda_i I) v_i = 0 \tag{12.64}$$

$$w_i (A - \lambda_i I) = 0 \tag{12.65}$$

The vectors $v_i$ and $w_i$ are known as right and left eigenvectors of matrix $A$. In a matrix with all distinct eigenvalues (not strictly necessary) one can arrange all eigenvectors and eigenvalues through compact matrix notations such as:

$$AV = V\Lambda \tag{12.66}$$

$$WA = \Lambda W \tag{12.67}$$

where,

$$V = \begin{pmatrix} v_1 & v_2 & .. & v_{n-1} & v_n \end{pmatrix} \tag{12.68}$$

$$W = \begin{pmatrix} w_1^t & w_2^t & .. & w_{n-1}^t & w_n^t \end{pmatrix}^t \tag{12.69}$$

$$\Lambda = diag(\lambda_1 \quad \lambda_2 \quad .. \quad \lambda_{n-1} \quad \lambda_n) \tag{12.70}$$

The eigenvector matrices can be used as transformation matrices to transform the state variables $x$ into decoupled modal variables $z_m$. The advantage of this transformation is that these variables are decoupled. The time domain behaviour of each of them completely represents the contribution of a particular eigenvalue ($\lambda_i$)

to overall system response. Pre-multiplying (12.66) by $V^I$ and (12.67) by $W^I$ modal matrix $\Lambda$ can be obtained. One can transform physical state variables $x$ into modal variables $z$ with the help of eigenvector matrices $V$ and $W$ as follows:

$$x = Vz \qquad (12.71)$$

$$z = Wx \qquad (12.72)$$

The right eigenvector ($v_i$) is known as mode shape corresponding to $\lambda_i$. The mode shape is very useful in identifying a group of coherent generators in a multi-machine system. We compute right eigenvector corresponding to eigenvalue $0.0626 \pm j2.4212$ in Table 12.1. The entries in the eigenvector corresponding to machine speed are shown in Fig. 12.10. This shows two clusters of generators oscillating against each other. This is very important information for devising control strategies.

Of the fifteen complex conjugate eigenvalues, the first four when ranked in ascending order of frequency are known as the inter-area modes. In a large power system, it is important to quantify the role of any particular generator in one particular mode. This helps to simplify the dynamic characterization of the entire system by the reduced dynamic model to make the analysis and control synthesis much easier. It is natural to suggest that the significant state variables influencing a particular mode are those having large entries corresponding to the right eigenvector of $\lambda_i$. The problem of entries in an eigenvector is that they can not be compared with each other because they have different units and scaling i.e. entries in the eigenvector corresponding to state variables such as speed, angle, flux, voltage etc. can not be compared. Let us take a closer look at equations (12.71) and (12.72).

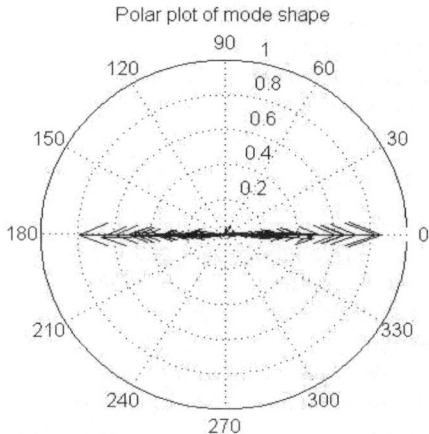

**Fig. 12.10.** Mode shape of first inter-area mode

The relation between the physical state and modal variables provides an important insight. Any arbitrary element $v_{ki}$ in $V$ can be seen as contribution of the $i^{th}$ mode in the $k^{th}$ state variable, i.e. activity of the $i^{th}$ mode in the $k^{th}$ state variable. On the other hand $w_{ik}$ corresponds to a weighted contribution of the $k^{th}$ state variable to $i^{th}$ mode. The product $w_{ki}v_{ik}$ is, however, a dimensionless measure known as participation factor [12]. Both $V$ and $W$ can be assumed to be orthogonal and they can be scaled suitably such that $w_k v_i = 1.0$. The conditions in (12.64) and (12.65) would still be satisfied. The more generic definition of participation factor [5] is given as:

$$p_{ki} = \frac{|v_{ik}||w_{ki}|}{\sum_{k=1}^{k=n}|v_{ik}||w_{ki}|}$$ (12.73)

The participation factors for all of the fifteen complex modes are computed and shown in Table 12.7

**Table 12.7.** Normalized Participation factors

| Eigenvalues ($\lambda_s$) | Normalized participation factors | Machine |
|---|---|---|
| -0.0626+j2.4212 | 0.10,0.06,0.05,0.03,0.03,0.03,0.026 | 13,15,16,9,14,6,3 |
| -0.0697+j3.1469 | 0.24,0.23,0.02 | 16,14,15 |
| -0.0507+j3.9564 | 0.25,0.04,0.03,0.03,0.03,0.027 | 13,9,6,12,5 |
| -0.0810+j4.9710 | 0.32,0.13,0.04 | 15,14,16 |
| -0.0927+j6.1810 | 0.38,0.03,0.02 | 9,5,6 |
| -0.0684+j6.7719 | 0.16,0.15,0.06,0.034,0.03,0.02 | 2,3,5,6,4,7 |
| -0.0387+j7.2178 | 0.38,0.07 | 12,13 |
| -0.0568+j7.6901 | 0.20,0.14,0.09,0.05 | 5,6,7,4 |
| -0.0752+j7.9630 | 0.25,0.24 | 2,3 |
| -0.0542+j7.9915 | 0.20,0.10,0.08,0.02 | 10,1,8,12 |
| -0.0537+j8.4600 | 0.24,0.12,0.11 | 10,8,1 |
| -0.0617+j9.7802 | 0.22,0.20,0.03,0.025 | 8,1,7,6 |
| -0.0716+j9.7820 | 0.21,0.17,0.03,0.03,0.03 | 7,6,4,8,1 |
| -0.0701+9.8349 | 0.30,0.12,0.05 | 4,5,7 |
| -0.1161+j11.467 | 0.47 | 11 |

The participation factors for angle and speed for a particular machine are the same. We arrange them in descending order and show a few of them. The entries in the last column show the corresponding machines. It is seen that in low frequency electromechanical modes machines from different areas participate. These are known as inter-area modes [4]. At relatively high frequencies (>1.0 Hz), significant contribution is from one or two machines in a power plant. These are known as local and or intra-plant modes [4]. This is very useful when studying the behaviour of one machine with respect to the rest of the system. Participation factors reveal vital information for controlling low frequency oscillations of a system. Machines with higher participations are very effective for dampening oscillations and hence are the candidate machines to equip with power system stabiliser (PSS). For inter-area modes, PSS for many machines are needed and hence control design becomes a co-ordinated multivariable control design problem. One PSS is theoretically sufficient to control a local mode and obviously it is placed in the machine having the highest participation factor.

## 12.3 Modal Controllability, Observability and Residue

One drawback of the participation factor approach described in the previous section is that it only deals with the states and so it does not consider input and output parameters. It can not effectively identify controller site and appropriate feedback signal in the absence of information on input and output, which is more important when output feedback is employed. The effectiveness of control can, however, be indicated through controllability and observability indices. This is important as control cost is influenced to a great deal by the controllability and observability of the plant. These issues are addressed through modal controllability, observability and residue. This is briefly described next. The transfer function equivalent of equations (12.58) and (12.59) is:

$$G = C(sI - A)^{-1} B + D \qquad (12.74)$$

Let us drop the direct transmission term $D$, as it does not influence the mode (exclusion does not affect our conclusion but simplifies the explanation) and rewrite the first part of the right hand side as $G_r(s)$. We also make use of the orthogonal relationship between $V$ and $W$ i.e. $VW = I$:

$$\begin{aligned} G_r(s) &= C(sI - A)^{-1} B \\ &= CVW(sI - A)^{-1} VWB \\ &= CV\left[V^{-1}(sI - A)W^{-1}\right]^{-1} WB \\ &= CV(sI - \Lambda)^{-1} WB \\ &= \sum_{i=1}^{n} \frac{Cv_i w_i B}{s - \lambda_i} \\ &= \sum_{i=1}^{n} \frac{R_i}{s - \lambda_i} \end{aligned} \qquad (12.75)$$

$R_i$ is known as the modal residue, being the product of modal observability $Cv_i$ and modal controllability $w_iB$. It is seen from (12.75) that modes with poor damping i.e. $\lambda_i$ with small absolute real part, will significantly influence the magnitude of the transfer function $G_r$ if it is scaled up by the residue $R_i$ at and around the frequency corresponding to the imaginary part of $\lambda_i$. This means that controllability of the input signal and observability of the feedback signal become very important. The choice of feedback signal should be made after careful consideration. The feedback signal must have a high degree of sensitivity at and around the swing mode to be damped out. This will show as a high peak in the bode diagram. In other words, this means that the swing mode must be observable in the feedback signal. The output signal must have little or no sensitivity to other swing modes. This is an obvious expectation from the perspective of minimum interaction amongst modes through the controller. A FACTS-device in a transmission line will only influence those modes responsible for power swings observed on that line. The expensive control effort could be wasted if it responds to local swings within an area at one end of the line. The effect of the feed forward term on the output is also very important. The feedback signal should have little or no sensitivity to its own output in the absence of a power swing. This is known as inner loop sensitivity [13] and does not involve swing mode dynamics. It results from feed forward effect of a signal by-passing the swing mode loop. In single-input-single-output (SISO) design, the output matrix $C$ and input matrix $B$ in equation (12.75) are row and column vectors respectively and hence the residue would be a complex scalar. As the residue is a complex variable, both magnitude and phase become important. The higher magnitude of the residue implies reduced control effort (gain) whereas higher phase lag requires multiple phase compensation blocks in the feedback path.

The FACTS-devices are never sited in a location of the system with highest modal controllability. Steady state power flow and dynamic voltage support dictate the criteria for locating. Nevertheless, if the devices are smaller in size and the sole purpose of having them installed in the system is to enhance small signal stability margin, the modal controllability can be used to find the most effective location. The modal controllability index ($w_iB$) at bus location (shunt device) or line (series device) can provide valuable information about the potential locations. We have computed the modal controllability indices of an SVC in all bus locations of the study system, which is further normalized with respect to the highest modal controllability vector. The absolute values are displayed in Table 12.8.

**Table 12.8.** Normalized modal controllability indices at various bus locations

| bus location | cont indices | bus location | cont indices | bus location | cont indices | bus location | cont indices |
|---|---|---|---|---|---|---|---|
| 01 | 0.37 | 18 | 1.00 | 35 | 0.11 | 52 | 0.75 |
| 02 | 0.58 | 19 | 0.09 | 36 | 0.17 | 53 | 0.27 |
| 03 | 0.62 | 20 | 0.97 | 37 | 0.79 | 54 | 0.52 |
| 04 | 0.85 | 21 | 0.93 | 38 | 0.18 | 55 | 0.67 |
| 05 | 0.99 | 22 | 0.95 | 39 | 0.11 | 56 | 0.69 |
| 06 | 0.89 | 23 | 0.95 | 40 | 0.11 | 57 | 0.65 |
| 07 | 0.89 | 24 | 0.92 | 41 | 0.03 | 58 | 0.66 |
| 08 | 0.54 | 25 | 0.56 | 42 | 0.04 | 59 | 0.62 |
| 09 | 0.20 | 26 | 0.58 | 43 | 0.12 | 60 | 0.60 |
| 10 | 0.13 | 27 | 0.68 | 44 | 0.12 | 61 | 0.25 |
| 11 | 0.10 | 28 | 0.38 | 45 | 0.05 | 62 | 0.70 |
| 12 | 0.13 | 29 | 0.30 | 46 | 0.12 | 63 | 0.69 |
| 13 | 0.06 | 30 | 0.25 | 47 | 0.23 | 64 | 0.78 |
| 14 | 0.02 | 31 | 0.22 | 48 | 0.19 | 65 | 0.71 |
| 15 | 0.03 | 32 | 0.16 | 49 | 0.06 | 66 | 0.73 |
| 16 | 0.05 | 33 | 0.16 | 50 | 0.08 | 67 | 0.85 |
| 17 | 0.14 | 34 | 0.15 | 51 | 0.01 | 68 | 0.89 |

It is seen that bus 18 is the most effective location for SVC to offer effective damping of the first inter-area mode (0.39 Hz). This is for a particular power flow, specific to a type of load and network conditions. Changes in any of theses conditions may produce a different bus location with the highest controllability. The modal observability $(Cv_i)$ indices on the other hand relate to feedback signals. Once the location is selected, the modal controllability is fixed. One particular type of signal such as power or line current or speed can be taken and the modal observability indices would be computed. The comparison of modal observability of two types of signal such as bus voltage magnitude and power in a line must not be done. Even though they are expressed in p.u., one p.u. voltage does not necessarily ensure one p.u. of line power. We have computed modal observability of line real power for SVC located at bus 18. The 700 MW power with constant impedance load and full tie line strength was considered i.e. operating condition was similar to that used for computing the modal observability. The results are shown in Table 12.9.

It is seen that the active power in the line between bus 13 and bus 17 has the highest normalized modal observability magnitude (in this case 1.00). There are many other signals having high modal observability too. The results in Table 12.9 reveal an interesting fact. The power signal in the line 13-17 with modal observability of 1.00 is the most effective signal to dampen first inter-area mode with least control effort. However, the signal needs to be transmitted from a remote location, making it less reliable. The observability of power signals in the lines originating from bus 18 are 0.14 (line 18-42), 0.30 (line 18-50) and 0.41 (line 16-18). These are local signals and so they are reliable, but need higher control effort when compared to remote signals. Synchronised phasor measurement technology coupled with dedicated high speed fibre optic network is available to use remote signal for damping of oscillations [7].

**Table 12.9.** Normalized modal indices for various line power signal

| line between | obserb indices | line between | obserb indices | line between | obserb indices | line between | obserb indices |
|---|---|---|---|---|---|---|---|
| 13-17 | 1.00 | 18-49 | 0.32 | 25-54 | 0.18 | 56-57 | 0.12 |
| 45-51 | 0.72 | 15-42 | 0.32 | 66-67 | 0.17 | 21-68 | 0.12 |
| 50-51 | 0.60 | 31-38 | 0.31 | 12-36 | 0.17 | 39-44 | 0.12 |
| 34-35 | 0.59 | 43-44 | 0.31 | 30-61 | 0.17 | 19-20 | 0.12 |
| 35-45 | 0.58 | 17-43 | 0.30 | 59-60 | 0.17 | 31-53 | 0.12 |
| 18-50 | 0.56 | 18-50 | 0.30 | 58-59 | 0.16 | 56-66 | 0.11 |
| 34-36 | 0.52 | 37-68 | 0.28 | 67-68 | 0.16 | 21-22 | 0.10 |
| 47-53 | 0.52 | 47-48 | 0.24 | 58-63 | 0.15 | 06-22 | 0.09 |
| 53-54 | 0.49 | 14-41 | 0.21 | 62-63 | 0.14 | 24-68 | 0.09 |
| 40-41 | 0.44 | 54-55 | 0.21 | 30-31 | 0.14 | 01-54 | 0.09 |
| 40-48 | 0.43 | 52-55 | 0.20 | 18-42 | 0.14 | 27-37 | 0.09 |
| 16-18 | 0.41 | 37-52 | 0.20 | 44-45 | 0.14 | 30-32 | 0.09 |
| 38-46 | 0.38 | 19-68 | 0.19 | 32-33 | 0.13 | 03-62 | 0.09 |
| 60-61 | 0.37 | 37-52 | 0.19 | 30-53 | 0.13 | 09-29 | 0.09 |
| 17-36 | 0.37 | 41-42 | 0.19 | 25-26 | 0.13 | 30-32 | 0.09 |
| 46-49 | 0.35 | 57-60 | 0.19 | 21-68 | 0.12 | 09-29 | 0.09 |

In this chapter, the modelling of various components of power systems is discussed. A power injection and small signal model of various FACTS-devices are described. The eigenvalue analysis on a 16-machine and 68-bus study system model is carried out using various modelling complexities. The influence of various system operating conditions on the damping and frequencies of electromechanical modes are analysed. It is concluded that power flow, load characteristics and network topologies affect the damping and frequency of inter-area mode significantly. The method of participation factor is applied to compute relative participation of machine in a particular mode to identify most effective machine for installing power system stabilizer. The methods of modal controllability and observability are explained. The modal controllability indices of a static var compensator at various bus locations of the study system are computed as an effective approach to identify best location. The computed modal observability indices in power signals from various lines provide useful information on the most effective stabilising signal.

## References

[1]  Foud AA, Vittal V (1992), Power System Transient Stability Analysis Using the Transient Energy Function Method. Prentice-Hall, USA
[2]  Pai MA (1989). Energy Function Analysis for Power System Stability. Kluwer Academic Publishers, USA
[3]  Pavella M, Muthy PG (1994): Transient Stability of Power Systems: Theory and Practice. John Wiley and Sons, Chichester
[4]  Kundur P (1994). Power System Stability and Control. McGraw Hill, USA
[5]  Sauer PW, Pai MA (1998). Power System Dynamics and Stability, Prentice Hall, USA
[6]  Concordia C (1969). IEEE committee report on recommended phasor diagram for synchronous machines. IEEE Transactions on Power Apparatus and Systems, vol 88, no 11, pp 1593-1610
[7]  Pal B, Chaudhuri B (2005), Robust Control in Power Systems. Springer USA
[8]  Pereira L, Undrill J, Kosterev D, Davies D, Patterson S (2003). A New Thermal governor modeling approach in the WECC. IEEE Transactions on Power Systems, vol 18, no 2, pp 819-829
[9]  Hill D (1993), Non linear dynamic load models with recovery for voltage stability studies. IEEE Transactions on Power Systems, vol 8, no 1, pp 166-176.
[10] Walve K (1986), Modelling of power system components under severe disturbances. CIGRE paper 38-18
[11] Noroozian M, Anguist L, Gandhari M. Andersson G (1997), Improving power system dynamics by series connected FACTS devices, IEEE Transactions on Power Delivery, vol 12, no 4, pp 1635-1641
[12] Verghese G, Perez-Arriaga IJ, Scheweppe FC (1982), Selective modal analysis with applications to electric power systems, Part-I & II. IEEE Transactions on Power Apparatus and Systems, vol 101, no 9, pp 3117-3134
[13] Larsen EV, Sanchez-Gasca JJ, Chow JH (1995), Concepts for design of FACTS controllers to damp power swings. IEEE Transactions on Power Systems, vol 10, no 2, pp 948-956

# 13 Linear Control Design and Simulation of Power System Stability with FACTS

Inter-area oscillations in power systems are triggered by, for example, disturbances such as variation in load demand or the action of voltage regulators due to a short circuit. The primary function of the damping controllers is to minimize the impact of these disturbances on the system within the limited dynamic rating of the actuator devices (excitation systems, FACTS-devices). In $H_\infty$ control term, this is equivalent to designing a controller that minimizes the infinity norm of a chosen mix of closed-loop quantities.

The concept of $H_\infty$ techniques for power system damping control design is about ten years old [1]-[5]. An interesting comparison between various techniques is made in [6]. There are two approaches for solving a standard $H_\infty$ optimization problem: analytical and numerical. While the analytical approach seeks a positive semi definite solution to the Riccati equation [7], the numerical approach is to solve the Riccati inequality to optimize the relevant performance index. Although the Riccati inequality is non-linear, there are linearization techniques to convert it into a set of linear matrix inequalities (LMIs) [8][7], which simplifies the computational process.

The analytical approach is relatively straightforward but generally produces a controller that suffers from pole-zero cancellations between the plant and the controller [9]. The closed-loop damping ratio, which is very important in power system control design, can not be captured in a straight forward manner in a Riccati based design [10]. The numerical approach to the solution, using the linear matrix inequality (LMI) approach, has a distinct advantage as these design specifications can be addressed as additional constraints. Moreover, the controllers obtained using a numerical approach do not, in general, suffer from the problem of pole-zero cancellation [11].

Application of the $H_\infty$ approach using LMIs has been reported in [12][13] for design of power system stabilizers (PSS). A mixed-sensitivity approach with an LMI based solution was applied for the design of damping control for superconducting magnetic energy storage (SMES) devices [10][14][15]. Recently this approach has been extended for the design of damping control provided by different FACTS-devices [16][17][5]. This chapter describes the basic concept of mixed-sensitivity design formulation with the problem translated into a generalized $H_\infty$ problem [8][7]. The entire control design methodology is illustrated by a couple of case studies on a study power system model. The damping control performance is validated in both frequency domain and time domain.

The second half of this chapter focuses on extending these design techniques to a time delayed system. We assume that in centralized design remote signals are instantly available. However, in reality, depending on signal transmission protocol, a delay is introduced. This would transform the system into a delayed system, which the control algorithm must take into consideration. We have applied a predictor based approach [5]. An SVC is used to damp oscillations through a delayed remote signal. The performance of the control has been validated on the same study system and conclusions are made.

## 13.1 H-Infinity Mixed-Sensitivity Formulation

The standard mixed-sensitivity formulation for output disturbance rejection and control effort optimization is shown in Fig. 13.1, where $G(s)$ is the open loop system model and $K(s)$ is the controller to be designed. The sensitivity $S = (I-GK)^{-1}$ represents the transfer function between the disturbance input $w(s)$ and the measured output $y(s)$. In the case of a power system, typically the sensitivity $S$ is the impact of load changes on the oscillations of angular or machine speed. So it is required to minimize $\|S\|_\infty$. It is also required to minimize $H_\infty$ norm of the transfer function between the disturbance and the control output to optimize the control effort within a limited bandwidth. This is equivalent to minimizing $\|KS\|_\infty$. Thus, the minimization problem can be summarized as follows:

$$\min_{K \in \mathbf{S}} \left\| \begin{matrix} S \\ KS \end{matrix} \right\|_\infty \tag{13.1}$$

where $\mathbf{S}$ is the set of all internally stabilizing controllers $K$.

It is, however, not possible to simultaneously minimize both $S$ and $KS$ over the whole frequency spectrum. This is not required in practice either. The disturbance rejection is usually required at low frequencies. Thus $S$ can be minimized over the low frequency range where as, $KS$ can be minimized at higher frequencies where limited control action is required.

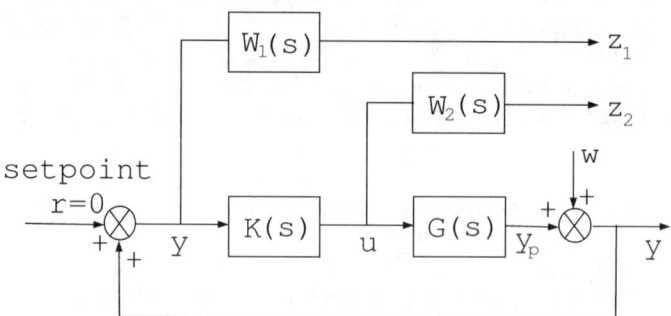

**Fig. 13.1.** Mixed-sensitivity formulation

Appropriate weighting filters $W_1(s)$ and $W_2(s)$ are used to emphasize the minimization of each individual transfer function at the different frequency ranges of interest. The minimization problem is formulated such that $S$ is less than $W_1(s)^{-1}$ and $KS$ is less than $W_2(s)^{-1}$. The standard practice, therefore, is to select $W_1(s)$ as an appropriate low pass filter for output disturbance rejection and to select $W_2(s)$ as a high-pass filter to reduce the control effort over the high frequency range. The problem can be restated as follows:

*find a stabilizing controller, such that*:

$$\min_{K \in S} \left\| \begin{bmatrix} W_1 S \\ W_2 KS \end{bmatrix} \right\|_\infty < 1 \tag{13.2}$$

## 13.2 Generalized H-Infinity Problem with Pole Placement

The mixed-sensitivity design problem is translated into a generalized $H_\infty$ problem. The first step is to set up a generalized regulator $P$ corresponding to the mixed-sensitivity formulation. For simplicity, it is assumed that the weights $W_1$ and $W_2$ are not present but will be taken care of later. Without the weights, the mixed-sensitivity formulation in Fig. 13.1 can be redrawn in terms of the $A$, $B$ and $C$ matrices of the system, as shown in Fig. 13.2. Without any loss of generality, it can be assumed that $D=0$.

From Fig. 13.2, it can be readily seen that:

$$\dot{x} = Ax + Bu \tag{13.3}$$

$$z_1 = Cx + w \tag{13.4}$$

$$z_2 = u \tag{13.5}$$

$$y = Cx + w \tag{13.6}$$

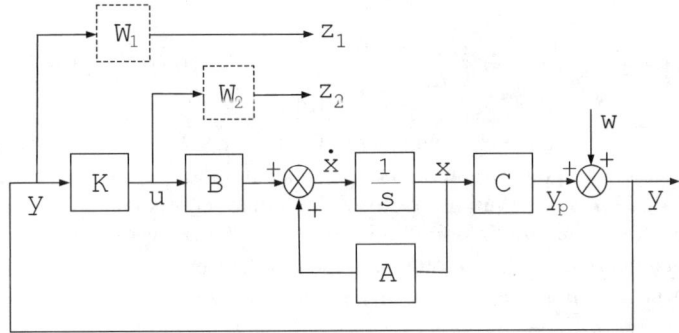

**Fig. 13.2.** Generalized regulator set-up for mixed-sensitivity formulation

The state-space representation of a generalized regulator $P$ is given as:

$$\begin{bmatrix} \dot{x} \\ z_1 \\ z_2 \\ y \end{bmatrix} = \begin{bmatrix} A & 0 & B \\ C & I & 0 \\ 0 & 0 & I \\ C & I & 0 \end{bmatrix} \begin{bmatrix} x \\ w \\ u \end{bmatrix} \quad (13.7)$$

$x$:     state variable vector of the power system (e.g. machine angle, machine speed etc),
$w$:     disturbance input (e.g. a step change in excitation system reference),
$u$:     control input (e.g. output of PSS or FACTS-devices),
$y$:     measured output (e.g. power flow, line current, bus voltage etc ),
$z$:     regulated output.

For the weighting filters of the generalized regulator, i.e. the state-space representations of $W_1$ and $W_2$, are placed in a diagonal form using the *sdiag* function available in Matlab [18]. The result is multiplied with $P$ (without the weights) using the *smult* function also available in Matlab.

The task now is to find an LTI control law $u = Ky$ for some $H_\infty$ performance index $\gamma > 0$, such that $\| T_{wz} \|_\infty < \gamma$ where, $T_{wz}$ denotes the closed-loop transfer function from $w$ to $z$. If the state-space representation of the LTI controller is given by:

$$\begin{aligned} \dot{x}_k &= A_k x_k + B_k y \\ u &= C_k x_k + D_k y \end{aligned} \quad (13.8)$$

then the closed-loop transfer function $T_{wz}$ from $w$ to $z$ is given by $T_{wz}(s) = D_{cl} + C_{cl}(sI - A_{cl})^{-1} B_{cl}$ where,

$$A_{cl} = \begin{bmatrix} A + B_2 D_k C_2 & B_2 C_k \\ B_k C_2 & A_k \end{bmatrix} \quad (13.9)$$

$$B_{cl} = \begin{bmatrix} B_1 + B_2 D_k D_{21} \\ B_k D_{21} \end{bmatrix} \quad (13.10)$$

$$C_{cl} = [C_1 + D_{12} D_k C_2 \quad D_{12} C_k] \quad (13.11)$$

$$D_{cl} = D_{11} + D_{12} D_k D_{21} \quad (13.12)$$

In addition to guaranteeing robustness by satisfying $\| T_{wz} \|_\infty < \gamma$, another design requirement in power systems is to ensure that the oscillations settle within 10-15 s [14]. This is achieved if the closed-loop poles corresponding to the critical modes have the minimum damping ratio. In consideration of this, the above problem statement can be modified to include the pole-placement constraint so that the problem is now: *find an LTI control law $u = Ky$ such that:*

- $\| T_{wz} \|_\infty < \gamma$
- Poles of the closed-loop system lie in $D$

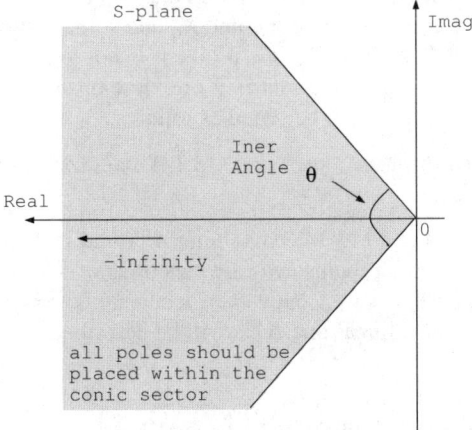

**Fig. 13.3.** Conic sector region for pole-placement

$D$ defines a region in the complex plane having certain geometric shapes like disks, conic sectors, vertical/horizontal strips, etc. or intersections of these. A 'conic sector', with inner angle $\theta$ and apex at the origin is an appropriate region for power system applications as it ensures a minimum damping ratio $\varsigma_{min} = \cos^{-1}\frac{\theta}{2}$ for the closed-loop poles.

## 13.3 Matrix Inequality Formulation

The bounded real lemma [11] and Schur's formula for the determinant of a partitioned matrix [7], enable one to conclude that the $H_\infty$ constraint $/T_{wz}/_\infty < \gamma$ is equivalent to the existence of a solution $X_\infty = X_\infty^T > 0$ to the following matrix inequality:

$$\begin{pmatrix} X_\infty A_{cl} + A_{cl}^T X_\infty & B_{cl} & X_\infty C_{cl}^T \\ B_{cl}^T & -\gamma I & D_{cl}^T \\ C_{cl} X_\infty & D_{cl} & -\gamma I \end{pmatrix} < 0 \qquad (13.13)$$

A 'conic sector' with inner angle $\theta$ and apex at the origin is chosen as the region $D$ within which the pole-placements are confined to. The closed-loop system matrix $A_{cl}$ has all its poles inside the conical sector $D$ if and only if there exists $X_D = X_D^T > 0$, such that the following matrix inequality is satisfied [21].

$$\begin{pmatrix} \sin\theta(A_{cl}X_D + X_D A_{cl}^T) & \cos\theta(A_{cl}X_D - X_D A_{cl}^T) \\ \cos\theta(X_D A_{cl}^T - A_{cl}X_D) & \sin\theta(X_D A_{cl}^T + A_{cl}X_D) \end{pmatrix} < 0 \qquad (13.14)$$

The design specifications are feasible if and only if (13.13) and (13.14) hold for some positive semi-definite matrices $X_\infty$ and $X_D$ and some controller K with state-space ($A_k$, $B_k$, $C_k$, and $D_k$). However, the problem is not jointly convex in $X_\infty$ and $X_D$ unless it is solved for the same matrix X. In view of this, the sub-optimal $H_\infty$ problem with pole-placement can be stated as follows:

*find X >0 and a controller K, such that (13.13) and (13.14) are satisfied with X = $X_\infty$ = $X_D$* [20][21].

The inequalities (13.13) and (13.14) containing $A_{cl}X$ and $C_{cl}X$ are functions of the controller parameters, which themselves are functions of X. This makes the products $A_{cl}X$ and $C_{cl}X$ non-linear in X. However, a change of controller variables can convert the problem into a linear one. This will be described in the next section.

## 13.4 Linearization of Matrix Inequalities

The controller variables are implicitly defined in terms of the (unknown) matrix X. Let X and $X^{-1}$ be partitioned as:

$$X = \begin{pmatrix} R & M \\ M^T & U \end{pmatrix}, X^{-1} = \begin{pmatrix} S & N \\ N^T & V \end{pmatrix} \quad (13.15)$$

For $\Pi_1 = \begin{pmatrix} R & I \\ M^T & 0 \end{pmatrix}$ and $\Pi_2 = \begin{pmatrix} I & S \\ 0 & N^T \end{pmatrix}$, X satisfies the identity $X\Pi_1 = \Pi_2$. The new controller variables are defined in (13.16) to (13.19).

$$\hat{A} = NA_kM^T + NB_kC_2R + SB_2C_kM^T + S(A + B_2D_kC_2)R \quad (13.16)$$

$$\hat{B} = NB_k + SB_2D_k \quad (13.17)$$

$$\hat{C} = C_kM^T + D_kC_2R \quad (13.18)$$

$$\hat{D} = D_k \quad (13.19)$$

The identity $XX^{-1} = I$ together with (13.15) gives:

$$MN^T = I - RS \quad (13.20)$$

If M and N have full row rank, then the controller matrices $A_k$, $B_k$, $C_k$, and $D_k$ can always be computed from $\hat{A}, \hat{B}, \hat{C}, \hat{D}, R, S, M$ and N. Moreover, the controller matrices can be determined uniquely if the controller order is chosen to be equal to that of the generalized regulator [21].

Pre- and post-multiplying the inequality $X > 0$ by $\Pi_2^T$ and $\Pi_2$ respectively, and carrying out appropriate change of variables according to (13.16), (13.17), (13.18) and (13.19) allows obtaining the following linear matrix inequality (LMI):

## 13.4 Linearization of Matrix Inequalities

$$\begin{pmatrix} R & I \\ I & S \end{pmatrix} > 0 \qquad (13.21)$$

Similarly, by pre- and post-multiplying the inequality (13.13) by diag $(\Pi_2^T, I, I)$ and diag $(\Pi_2, I, I)$ respectively and carrying out appropriate change of variables according to (13.16), (13.17), (13.18) and (13.19), the following LMI is obtained.

$$\begin{bmatrix} \Psi_{11} & \Psi_{21}^T \\ \Psi_{21} & \Psi_{22} \end{bmatrix} < 0 \qquad (13.22)$$

where

$$\Psi_{11} = \begin{bmatrix} AR + RA^T + B_2\hat{C} + \hat{C}^T B_2^T & B_1 + B_2\hat{D}D_{21} \\ (B_1 + B_2\hat{D}D_{21})^T & -\gamma I \end{bmatrix} \qquad (13.23)$$

$$\Psi_{21} = \begin{bmatrix} \hat{A} + (A + B_2\hat{D}C_2)^T & SB_1 + \hat{B}D_{21} \\ C_1 R + D_{12}\hat{C} & D_{11} + D_{12}\hat{D}D_{21} \end{bmatrix} \qquad (13.24)$$

$$\Psi_{22} = \begin{bmatrix} A^T S + SA + \hat{B}C_2 + C_2^T\hat{B} & (C_1 + D_{12}\hat{D}C_2)^T \\ C_1 + D_{12}\hat{D}C_2 & -\gamma I \end{bmatrix} \qquad (13.25)$$

Proceeding in a similar fashion by pre- and post-multiplying the inequality (13.14) by $\Pi_2^T$ and $\Pi_2$ respectively, and carrying out the change of variables according to (13.16), (13.17), (13.18) and (13.19), the following LMI is obtained. A detailed explanation of this process can be found in [20][21].

$$\begin{pmatrix} \sin\theta(\phi+\phi^T) & \cos\theta(\phi-\phi^T) \\ \cos\theta(\phi^T-\phi) & \sin\theta(\phi^T+\phi) \end{pmatrix} < 0 \qquad (13.26)$$

where

$$\phi = \begin{pmatrix} AR + B_2\hat{C} & A + B_2\hat{D}D_{21} \\ \hat{A} & SA + \hat{B}C_2 \end{pmatrix} \qquad (13.27)$$

The system of LMIs in (13.21), (13.22) and (13.26) are solved for $R, S, \hat{A}, \hat{B}, \hat{C}$ and $\hat{D}$. A full-rank factorization of the matrix $I-RS$ is computed via singular value decomposition (SVD) such that $MN^T = I-RS$ holds for $M$ and $N$ being square and invertible. With known values of $R, S, \hat{A}, \hat{B}, \hat{C}, \hat{D}, M$ and $N$, the system of linear equations (13.16), (13.17), (13.18) and (13.19) can be solved for $D_k, B_k, C_k$ and $A_k$ in that order. The controller $K$ is obtained and the resultant controller places the closed-loop poles in D and satisfies $\|T_{wz}\|_\infty < \gamma$.

## 13.5 Case Study

In this section, the prototype power system model, described in chapter 12, is used to illustrate the control design methodology in detail. The performance and robustness of the design is also validated.

### 13.5.1 Weight Selection

As mentioned earlier, the standard practice in $H_\infty$ mixed-sensitivity design is to choose the weight $W_1(s)$ as an appropriate low pass filter for output disturbance rejection and chose $W_2(s)$ as a high-pass filter to reduce the control effort in the high frequency range. For the prototype system, the weights are accordingly chosen as follows:

$$W_1(s) = \frac{30}{s+30}$$
$$W_2(s) = \frac{10s}{s+100}$$
(13.28)

The frequency responses of these weight functions are shown in Fig. 13.4. It can be seen that the two weights intersect at around 10 rad/s, noting that the critical modes to be controlled are below the frequency of 10 rad/s. Thus, the minimization of the sensitivity is emphasized up to this frequency and the control is constrained soon after.

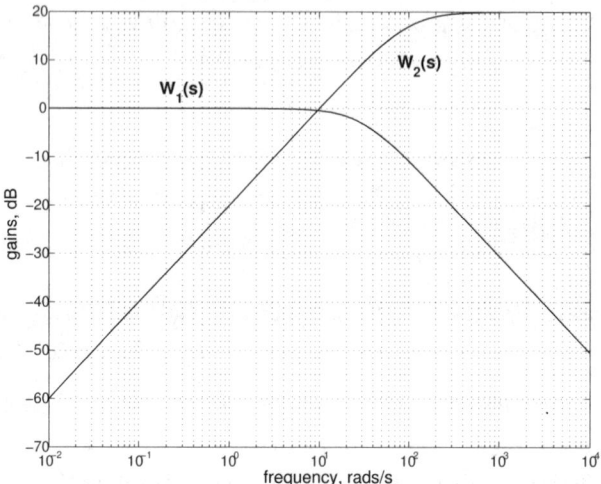

**Fig. 13.4.** Frequency response of weighting filter

## 13.5.2 Control Design

To facilitate the control design, and to reduce the complexity of the designed controller, the nominal system model was reduced to a $7^{th}$ order equivalent as described in chapter 12. The generalized regulator problem was formulated according to (13.7) using the simplified system model and the weights in (13.28). The control design problem is to minimize $\gamma$ such that (13.21), (13.22) and (13.26) are satisfied.

A series of functions which, are available with the *LMI toolbox* [22] in *Matlab* [18], is used to formulate and solve the optimization problem. The first step is to define the solution variables (also called the LMI variables). The variables $R, S, \hat{A}, \hat{B}, \hat{C}$ are defined using the appropriate Matlab toolbox function. The size of the variables and their structure is specified through this function. Having defined the solution variables, the next step is to set up the LMIs (13.21), (13.22) and (13.26) in terms of these variables. Each of the terms of an LMI and their respective positions are specified using the function from LMI toolbox. In this design, a 'conic sector' of inner angle $2\cos^{-1}0.15$ with apex at the origin was chosen as the pole-placement region to ensure a minimum damping ratio of 0.15 for the closed-loop system. To achieve this, the value of $\theta$ in (13.26) was set to $\cos^{-1}0.15$. The three sets of LMIs are combined in a system of LMI. The optimal values of $R, S, \hat{A}, \hat{B}, \hat{C}$ are retrieved from the output of the optimization function by using the suitable function from the toolbox. The control variables $A_k$, $B_k$, $C_k$, and $D_k$ are computed accordingly with the help of equations (13.16), (13.17), (13.18) and (13.19).

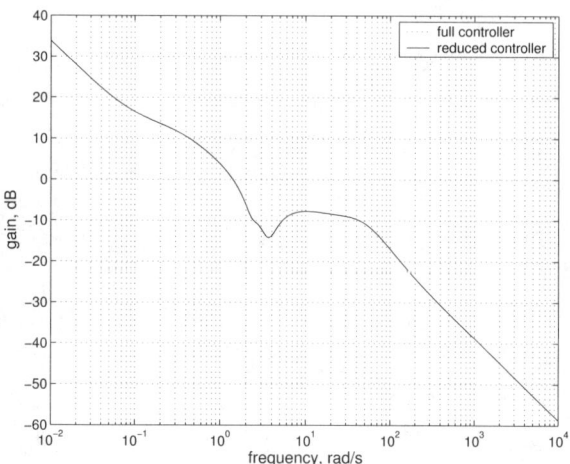

**Fig. 13.5.** Frequency response of full and reduced order controller

The order of the controller obtained from this design routine is equal to the order of the reduced system order plus the order of the weights. As there are three weights associated with the three measured outputs and one with the control input, the size of the designed controller is 14 (9+3+1). The designed controller was simplified further to a 10th order equivalent without affecting the frequency response, as shown in Fig. 13.5. The frequency response of the sensitivity $S$ and the controller sensitivity product $KS$ are plotted in Figs. 13.6 and 13.7.

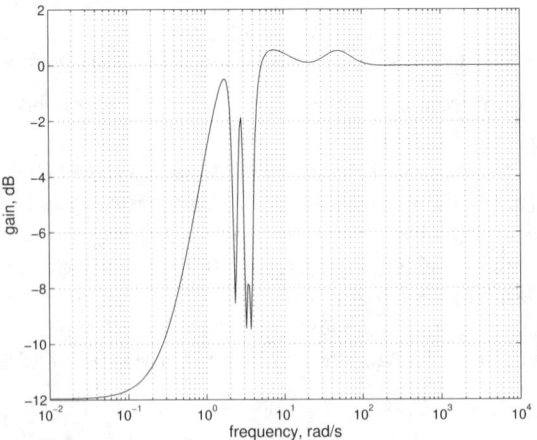

**Fig. 13.6.** Frequency response of sensitivity (S)

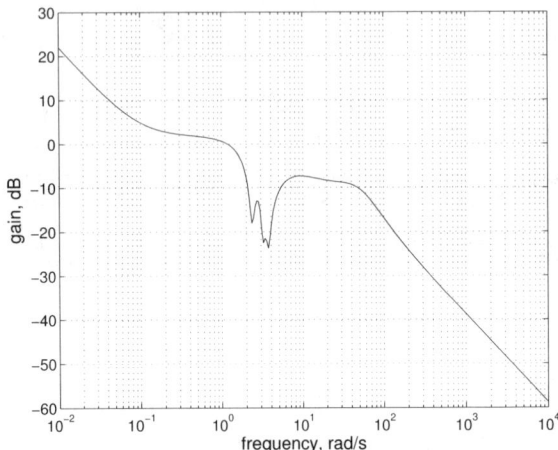

**Fig. 13.7.** Frequency response of control time sensitivity (KS)

As discussed before, $S$ should be low at the lower frequencies to achieve disturbance rejection but comparatively high values can be tolerated at higher frequencies. This is achieved in the designed controller as seen from Fig. 13.6. In contrast, to ensure satisfactory performance, $KS$ should be low at high frequencies to reduce the control effort.

The design steps can be summarized as follows:

- Simplify the system model;
- Formulate the generalized regulator using the simplified system model and the mixed-sensitivity weights;
- Define the LMI variables using the *lmivar* function;
- Construct the terms of the LMIs using the *lmiterm* function;
- Assemble the individual LMIs into a set of LMIs employing the *getlmis* function;
- Solve the $\gamma$ optimization problem with the set of LMI constraints using the *mincx* function;
- Retrieve the optimum value of the solution variables through the *dec2mat* function;
- Determine the controller using the optimum value of the solution variables; and
- Simplify the designed controller.

Alternatively, the design problem can be solved by suitably defining the objectives in the argument of the function *hinfmix*, available with the *LMI Toolbox* [22] for Matlab [18]. The pole-placement constraint can be imposed by using the *lmireg* function, which is an interactive interface for specifying different LMI regions.

**Table 13.1.** Damping ratios and frequencies of the inter-area modes

| Mode No. | Without Control | | With Control | |
|---|---|---|---|---|
| | $\xi$ | $f$(Hz) | $\xi$ | $f$(Hz) |
| 1 | **0.0626** | 0.3913 | **0.2336** | 0.3590 |
| 2 | **0.0435** | 0.5080 | **0.1316** | 0.5094 |
| 3 | **0.0554** | 0.6232 | **0.1456** | 0.6384 |
| 4 | 0.0499 | 0.7915 | 0.0550 | 0.7843 |

### 13.5.3 Performance Evaluation

The eigenvalues of the closed-loop system were computed to examine the performance of the designed controller in terms of improving the damping ratios of the inter-area modes. The results are summarized in Tables 13.1. It can be seen that the damping ratios of the three critical inter-area modes, shown in boldface, are improved in the closed loop.

It is to be noted that by imposing the pole-placement constraint, as described earlier, a minimum damping ratio of 0.15 could be ensured for the simplified closed-loop system. However, the results shown here are based on the full and original system model. Therefore the damping ratios under certain situations are less than 0.15. Nonetheless, they are still adequate enough to ensure that oscillations settle within 12-15 s.

The damping action of the designed controller was examined under different types of disturbances in the system. These included, amongst others, changes in power flow levels over key transmission corridors and change in type of loads. Table 13.2 displays the damping ratios of the inter-area modes for a range of power flows across the interconnection between the areas NETS and NYPS in the study system.

The performance of the controller was tested with various load models including a constant impedance (CI), a mixture of constant current and constant impedance (CC+CI), a mixture of constant power and constant impedance (CP+CI) and with dynamic load characteristics. The damping ratios of the inter-area modes are listed in Table 13.3 for different types of load characteristics. From the damping ratios displayed in the tables below it can be concluded that the action of the designed controller is robust against widely varying operating conditions.

**Table 13.2.** Damping ratios and frequencies of the critical inter-area modes at different levels of power flow between NETS and NYPS

| Power Flow (MW) | Mode 1 $\xi$ | Mode 1 $f$(Hz) | Mode 2 $\xi$ | Mode 2 $f$(Hz) | Mode 3 $\xi$ | Mode 3 $f$(Hz) |
|---|---|---|---|---|---|---|
| 100 | 0.2420 | 0.3566 | 0.1374 | 0.5106 | 0.1351 | 0.6640 |
| 500 | 0.2371 | 0.3578 | 0.1338 | 0.5097 | 0.1419 | 0.6451 |
| 700 | 0.2336 | 0.3590 | 0.1316 | 0.5094 | 0.1456 | 0.6384 |
| 900 | 0.2300 | 0.3609 | 0.1289 | 0.5093 | 0.1491 | 0.6251 |

**Table 13.3.** Damping ratios and frequencies of the critical inter-area modes for different load models

| Type of Load | Mode 1 $\xi$ | Mode 1 f(Hz) | Mode 2 $\xi$ | Mode 2 f(Hz) | Mode 3 $\xi$ | Mode 3 f(Hz) |
|---|---|---|---|---|---|---|
| CI | 0.2336 | 0.3590 | 0.1316 | 0.5094 | 0.1456 | 0.6384 |
| CI+CC | 0.2313 | 0.3608 | 0.1308 | 0.5175 | 0.1314 | 0.6353 |
| CI+CP | 0.2251 | 0.3621 | 0.1309 | 0.5260 | 0.1175 | 0.6351 |
| Dynamic | 0.2304 | 0.3582 | 0.1399 | 0.5135 | 0.1456 | 0.6381 |

### 13.5.4 Simulation Results

One of the most severe disturbances, in terms of producing poorly damped inter-area oscillations, is a three-phase fault in one of the key transmission circuits. For temporary faults, the circuit breaker re-closes after few cycles and normal operation is restored. For a permanent fault, the line is tripped out of operation. The other types of disturbances in the system, such as change of load characteristics and sudden change in power flow, are less severe compared to three faults and are not considered here.

The simulations were carried out to determine system performance during probable fault scenarios in the NETS and NYPS inter-connection. There are three inter-connections between NETS and NYPS connecting buses #60-#61, #53-#54 and #27-#53, respectively. Each of these inter-connections consists of two lines and an outage of one of these lines weakens the interconnection considerably. To examine the effect of such disturbances, a series of solid three-phase solid faults, each of about 80 ms (about 5 cycles) in duration, were simulated in the following locations:

(a) bus #60 followed by auto-reclosing of the circuit breaker
(b) bus #53 followed by outage of one of the tie-lines between buses #53-#54
(c) bus #53 followed by outage of one of the tie-lines between buses #27-#53
(d) bus #60 followed by outage of one of the tie-lines between buses #60-#61

Simulations were carried out in Matlab *Simulink* [23] for 25 s employing the *trapezoidal integration* method with a variable step size. The disturbance was created 1 s after the start of the simulation. The dynamic response of the system following the disturbance is shown in Figs. 13.8, 13.9 and 13.10. These figures illustrate the relative angular separation between the generators located in separate geographical regions. Inter-area oscillations are mostly manifested in these angular differences and are, therefore, chosen for display. It can be seen that inter-area oscillations settle within the desired performance specification of 12-15 s for a range of post-fault operating conditions. The TCSC was limited to provide between 0.1 to 0.8 p.u. of compensation. The variation in the percentage compensation provided by the TCSC is shown in Fig. 13.10.

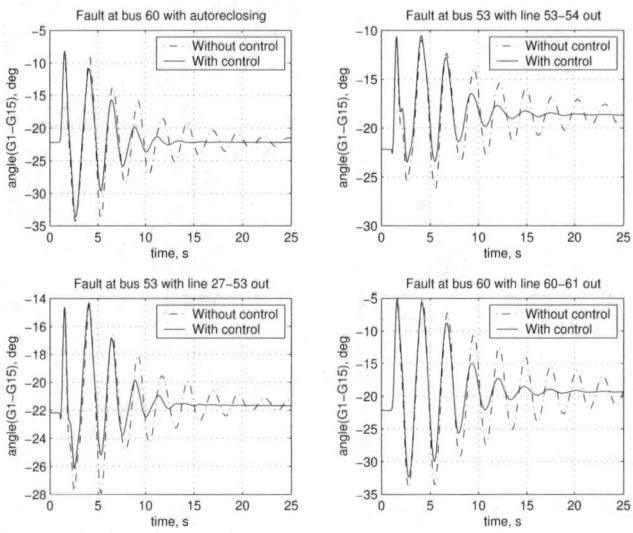

**Fig. 13.8.** Dynamic response of the system

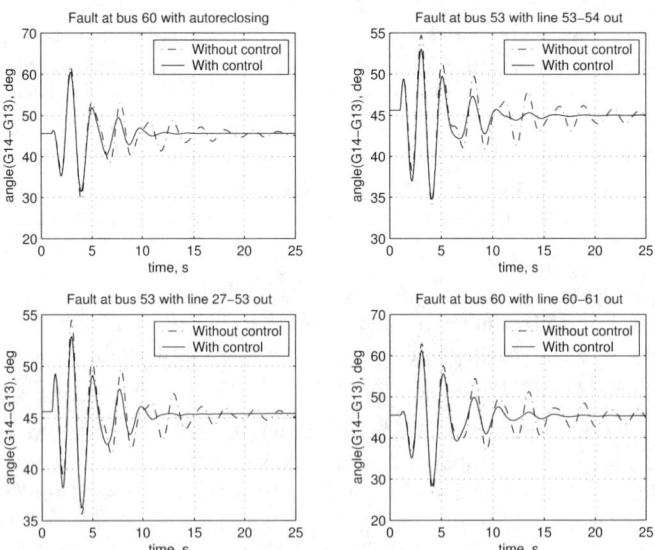

**Fig. 13.9.** Dynamic response of the system

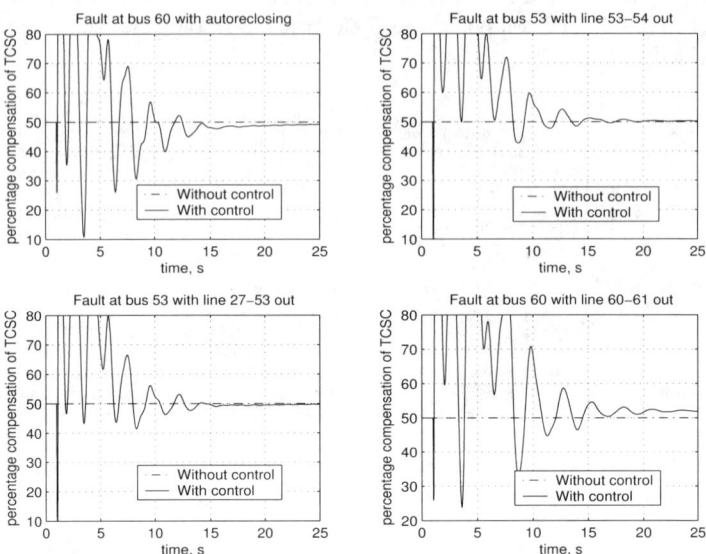

**Fig. 13.10.** Dynamic response of the system

## 13.6 Case Study on Sequential Design

In this section, a case study considering sequential design of damping controllers for multiple FACTS-devices is presented. The basic control design formulation is exactly the same as in the previous section. However, a separate controller is designed for each of the FACTS-devices sequentially. The feedback signals are chosen appropriately out of those locally available.

### 13.6.1 Test System

The study system described in chapter 12 is used again. Three FACTS-devices are considered to be installed as shown in Fig. 13.11. The TCSC is installed in the line between buses #18 and #50 to provide compensation ($k_c$) of 50%. An SVC is present at bus #18 to provide voltage support in the presence of the 1500 MW power transfer between area #5 and NYPS. The SVC is set to provide 117 MVAr to ensure nominal voltage at bus #18. A TCPS with a steady state phase angle ($\varphi$) setting of 10 degrees is installed in the line connecting buses #13 and #17 to facilitate 3000 MW power transfer from the equivalent generation G13 to the rest of the NYPS. The aim of this exercise is to design three separate damping controllers $K_1$, $K_2$ and $K_3$, which use locally available signals only, such that inter-area oscillations are damped. The location of the FACTS-devices and the corresponding damping controllers are shown in Fig. 13.11, where $y_1$, $y_2$ and $y_3$ are the measured feedback signals and $u_1$, $u_2$ and $u_3$ are the derived control signals.

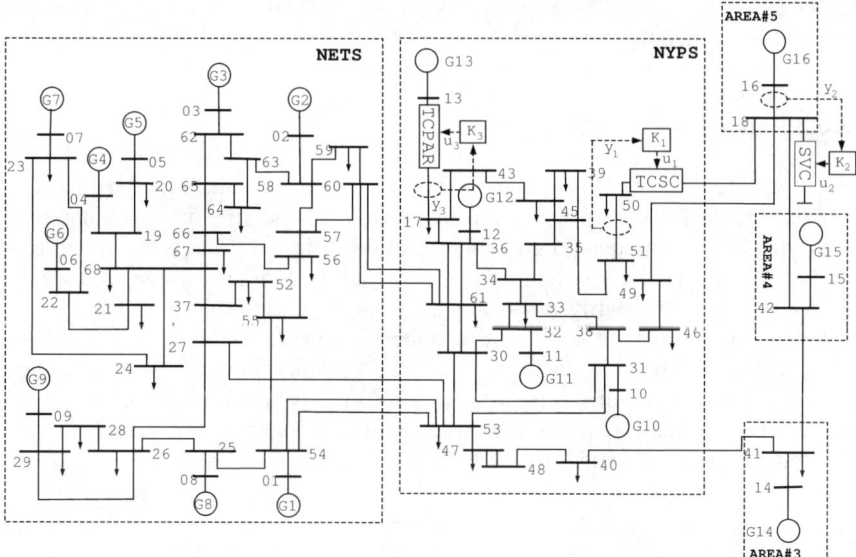

**Fig. 13.11.** Sixteen machine five area study system with three FACTS-devices

## 13.7.2 Control Design

The control design formulation described in Section 13.6.2 produces centralized controllers in multi-variable form. The design is now to be carried out in a sequential manner. The basic idea is to design a damping controller for one device to start with. The closed-loop system using this controller is used to design the controller for the second device. Exactly the same procedure is repeated for the third device. At each stage of this sequential design, the system model is updated with the designed controller model. In this process, the order of the system increases as each loop is closed depending on the number of states associated with the controllers of the individual FACTS-devices.

The sequential design of the controllers $K_1$, $K_2$ and $K_3$ for the TCSC, SVC and TCPS has been carried out in sequence. The choice of this sequence improves the damping of modes #1, #2 and #3 in that order. Other sequences were tested and found to produce slightly different controllers but, in each case, similar performance was achieved. The same set of weights given in (13.29) and (13.30) has been found to work well for the design of all three controllers.

$$W_1(s) = 0.8475 \frac{99s + 11400}{s^2 + 156s + 12504} \tag{13.29}$$

$$W_2(s) = 0.8475 \frac{0.1055s^2 + 0.037s + 0.0094}{s(s + 0.0020)^2} \tag{13.30}$$

Each of the controllers was reduced to lower order by balanced truncation without significantly affecting the frequency response. The gains of the controllers (but not the controller structure) were scaled slightly to produce a damping ratio that ensured the oscillations settled in 10-12 seconds.

## 13.6.3 Performance evaluation

The eigenvalues of the closed-loop system produced by sequential loop closure were examined. Table 13.4 shows the eigenvalues considering only the controller $K_1$ for the TCSC. The damping of mode #1 shown in boldface, is improved primarily with very little effect on modes #2, #3 and #4. Similarly Table 13.5 shows that the controller primarily associated with the SVC improves the damping of mode #2, shown in boldface, besides improving mode #1 slightly. Finally, as shown in Table 13.6, the controller for the TCPS mainly improves the damping of mode #3, shown in boldface, besides adding to the damping ratios of modes #1 and #2. The combined action of the three controllers improves the damping of all three critical inter-area modes to adequate level.

**Table 13.4.** Damping ratios and frequencies of inter-area modes with the controller of the TCSC (Control loops of the SVC and TCPS open)

| Mode No | Open-loop $\xi$ | $f$(Hz) | Closed-loop $\xi$ | $f$(Hz) |
|---|---|---|---|---|
| 1 | **0.0626** | 0.3945 | **0.1544** | 0.3434 |
| 2 | 0.0434 | 0.5105 | 0.0545 | 0.4991 |
| 3 | 0.0560 | 0.6269 | 0.0656 | 0.6191 |
| 4 | 0.0499 | 0.7923 | 0.0502 | 0.7918 |

**Table 13.5.** Damping ratios and frequencies of inter-area modes with the controllers of the TCSC and SVC (control loop of the TCPS open)

| Mode No | Open-loop $\xi$ | $f$(Hz) | Closed-loop $\xi$ | $f$(Hz) |
|---|---|---|---|---|
| 1 | 0.1544 | 0.3434 | 0.1795 | 0.3158 |
| 2 | **0.0545** | 0.4991 | **0.1031** | 0.4549 |
| 3 | 0.0656 | 0.6191 | 0.0643 | 0.6184 |
| 4 | 0.0502 | 0.7918 | 0.0603 | 0.7864 |

**Table 13.6.** Damping ratios and frequencies of inter-area modes with the controllers of the TCSC, SVC and TCPS (all the control loops closed)

| Mode No | Open-loop $\xi$ | $f$(Hz) | Closed-loop $\xi$ | $f$(Hz) |
|---|---|---|---|---|
| 1 | 0.1795 | 0.3158 | 0.3140 | 0.2682 |
| 2 | 0.1031 | 0.4549 | 0.2266 | 0.4444 |
| 3 | **0.0643** | 0.6184 | **0.1105** | 0.4585 |
| 4 | 0.0603 | 0.7864 | 0.0600 | 0.7858 |

### 13.6.4 Simulation Results

As a part of the performance testing and validation exercise, a non-linear simulation was carried out under the same set of operating conditions described in section 13.5.4. Again, a series of disturbances, consisting of a solid three-phase fault lasting for 80 ms (5 cycles) followed by the contingency conditions depicted in Figures 13.8 to 13.10, were applied to the system.

The angular separation between machines G1 and G15 located in different areas is shown in Fig. 13.12 under different operating scenarios. In each case, the designed controllers of the TCSC, SVC and TCPS are able to settle the oscillations within 12-15 s. The outputs of the individual FACTS-devices are shown in Fig. 13.13, 13.14 and 13.15 for the same operating conditions. Appropriate limits were imposed on the variation of the control variables, as seen from their output response. The limit imposed on the TCSC is the same as before. For the SVC, the output variation limit was set to -150 (inductive) to 200 (capacitive) MVAr. The limit on the phase angle of the TCPS was set to between 0 and 20 degrees.

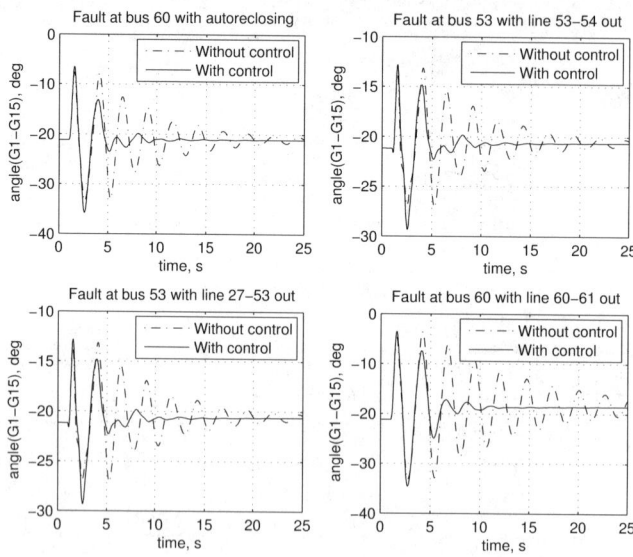

**Fig. 13.12.** Dynamic response of the system: angle between G1 and G15

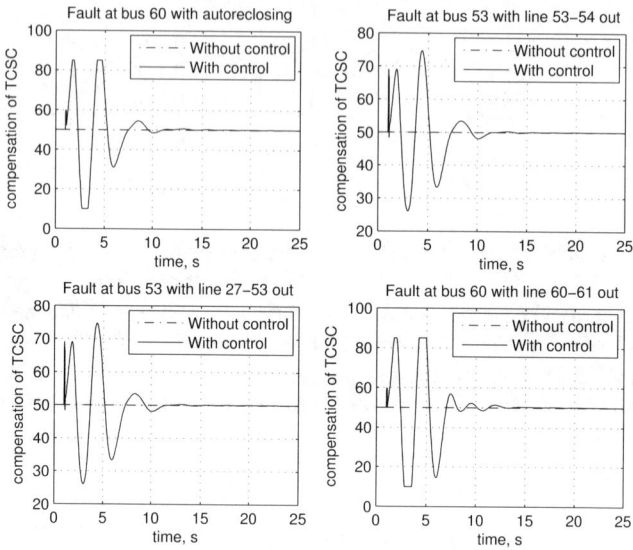

**Fig. 13.13.** Percentage compensation of the TCSC

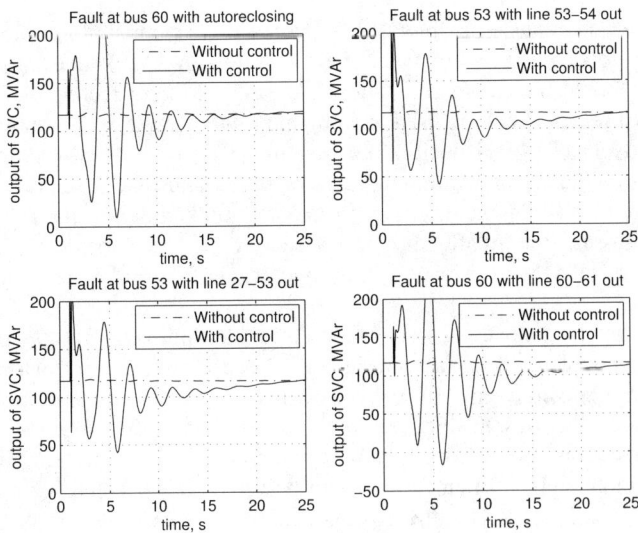

**Fig. 13.14.** Output of the SVC

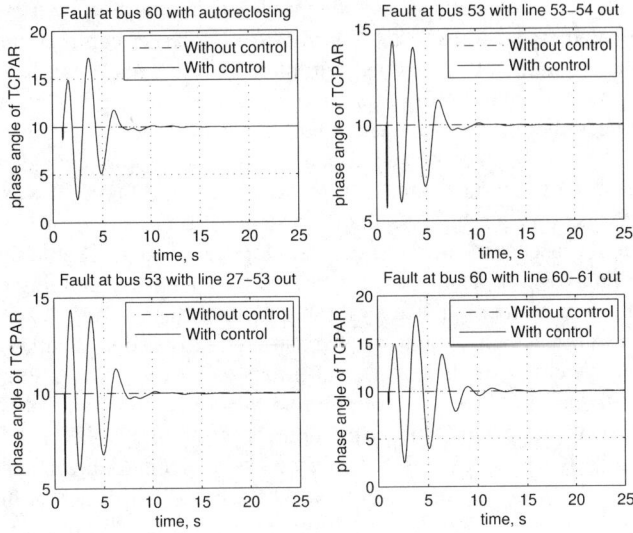

**Fig. 13.15.** Phase angle of the TCPS

## 13.7 H-Infinity Control for Time Delayed Systems

In large scale power systems, inter-area response can be damped more effectively through the use of wide-area measurement systems (WAMS), like introduced in chapter 11. Optical fiber communication is the right kind of technology for sensing and measurement systems used to implement the wide-area measurement systems. The advent of global positioning system (GPS) technology has made time stamping fairly routine and measurements of phase and other temporal information are, therefore, attainable through the use of commercially available equipment [24]. The information architecture, proposed in [25] is capable of providing timely, secure, reliable information exchange among various entities in a power system.

With the rapid advancements in Wide Area Measurement System (WAMS-) Technologies coupled with a fast and reliable data transmission infrastructure, the prospect of centralized control of power systems has gained momentum. Damping of inter-area oscillations through remote measurements, therefore, is becoming more feasible. From an economic viewpoint, implementation of centralized control using remote signals often turns out to be more cost effective than installing new control devices [26]. The obvious question is, at what speed are these remote measurements available to the control site? Employing phasor measurement units (PMUs), it is possible to deliver the signals at a speed as high as a 30 Hz sampling rate [3],[24]. It is possible to deploy the PMUs at strategic locations of the grid and obtain a coherent picture of the entire network in real time [24]. Many utilities in the USA and Europe have already installed demonstration projects.

The infrastructure cost and associated complexities, however, restrict the use of such sophisticated signal transmission hardware on a larger commercial scale. As a more viable alternative, the existing communication channels can be effectively used to transmit the signals from remote locations. The major problem is the delay involved between the instant of measurement and that of the signal being available to the controller. In the previous sections, no time-delay was considered as fast transmission of the necessary signals (typically within 0.02-0.05 s) was assumed. As the delay was considerably less than the smallest time period of the inter-area modes, it was not necessary to consider it in the design stage. A conservative estimate of the delay can typically be in the range of 0.5-1.0 s depending on the distance, transmission protocol and several other factors. Such a long delay should be accounted for in the design stage itself, to ensure effective control action.

The power system is treated as a dead-time system involving a long delay in transmitting the measured signals from remote locations to the controller site. It is not straightforward to control such time-delayed systems [27]. Normal $H_\infty$ controllers are unlikely to guarantee satisfactory control of time-delayed systems. The Smith predictor [28][29] approach, proposed in the early fifties, was the first effective tool for handling such control problems. The difficulties associated with the design of $H_\infty$ controllers for time-delayed systems and potential solutions using the predictor based approach is discussed later.

Since its arrival in the 1950s, a number of variations of the Smith predictor has been proposed in the literature. One of the drawbacks of the classical Smith predictor (CSP) approach is that it is very difficult to ensure a minimum damping ratio of the closed loop poles, when the open-loop system model has lightly damped poles (as often encountered in power systems). A modified Smith predictor (MSP) approach proposed in [30], was used to overcome the drawbacks of the CSP. However, the control design using the MSP approach might run into numerical problems for systems with fast stable poles. To overcome the drawbacks of CSP and MSP, a unified Smith predictor (USP) approach was proposed very recently by Zhong [31]. The USP approach effectively combines the advantageous features of both the CSP and MSP.

In this section, we illustrate the application of the USP approach for designing a damping controller for a power system with time-delayed signals. The predictor based control design methodology is illustrated by applying a case study to the same power system model described earlier. The performance and robustness of the designed controller is validated using frequency domain analysis and time domain simulations.

## 13.8 Smith Predictor for Time-Delayed Systems

In a time-delayed or dead-time system, either the measured output takes a certain time before it affects the control input or the action of the control input takes a certain time before it influences the measured outputs. Typical dead time systems consist of input and/or output delays. The general control setup for a system having an output delay is shown in Fig. 13.16, where,

$$P(s) = \begin{bmatrix} P_{11}(s) & P_{12}(s) \\ P_{21}(s) & P_{22}(s) \end{bmatrix} \tag{13.31}$$

The closed-loop transfer matrix from $d$ to $z$ is: $T_{zd}=P_{11}+P_{12}Ke^{-sh}(I-P_{22}Ke^{-sh})^{-1}P_{21}$. An equivalent structure is shown in Fig. 13.17. This suggests that there exists an instantaneous response through the path $P_{11}$ (path 1 in Fig. 13.17) without any delay.

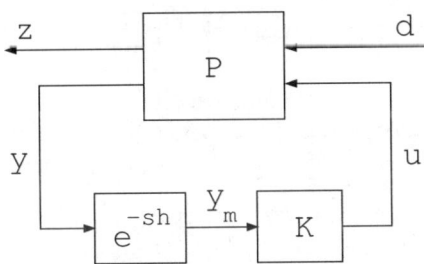

**Fig. 13.16.** Control setup for dead-time systems

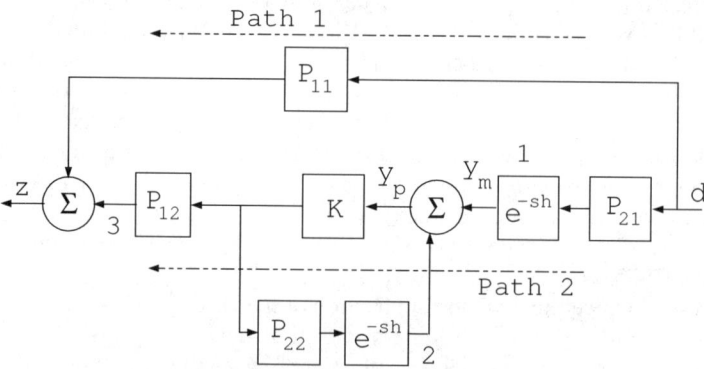

**Fig. 13.17.** An equivalent representation of dead-time systems

It can be seen that during the period $t = 0 \sim h$ after $d$ is applied, the output $z$ is not controllable, since it is only determined by $P_{11}$ and $d$, with no response coming through the controlled path (path 2). This means that the $H_\infty$ performance index $\| T_{zd} \|_\infty$ is likely to be dominated by a response that cannot be controlled, which is not desirable. It is extremely difficult to design a controller for such systems [27]. The Smith predictor (SP) represents the first effective tool for tackling such control problems. The primary idea is to eliminate any uncontrollable response that is likely to govern the $H_\infty$ performance index. One possible way of achieving this is to introduce a uniform delay in both paths (path 1 and path 2), as shown in Fig. 13.19. There are two steps towards achieving this. Firstly, the delay blocks $e^{-sh}$ at points 1 and 2 need to be shifted to point 3 by introducing a suitable predictor block in parallel with $K$. Secondly, a delay block needs to be introduced into path 1. The first step is achieved by introducing a Smith predictor block $Z = P_{22}(s) - P_{22}(s)e^{-sh}$, as shown by the dotted box in Fig. 13.18.

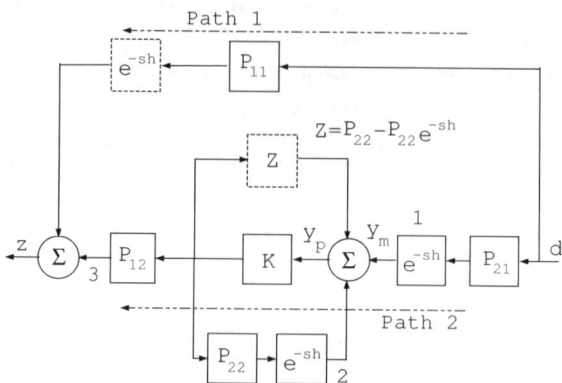

**Fig. 13.18.** Introduction of Smith predictor and delay block

## 13.8 Smith Predictor for Time-Delayed Systems

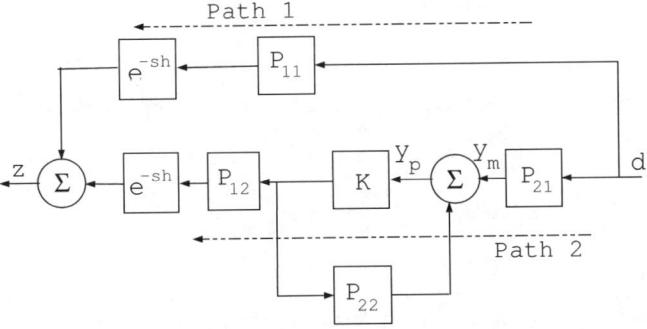

**Fig. 13.19.** Uniform delay in both paths

The second task of bringing a delay in path 1 is done while forming the generalized regulator prior to control design. The presence of the predictor block $Z$ and the delay in path 1 ensures that the responses (through path 1 and path 2) governing the performance index are delayed uniformly, as shown in Fig. 13.19.
A predictor-based controller for the dead-time system $P_h(s) = P_{22}(s)\,e^{-sh}$ consists of a predictor $Z = P_{22}(s) - P_{22}(s)e^{-sh}$ and a stabilizing compensator $K$, as shown in Fig. 13.20. The predictor $Z$ is an exponentially stable system such that $P_h + Z$ is rational i.e. it does not involve any uncontrollable response governing the $H_\infty$ performance index.

The shortcomings of the CSP approach for systems having poorly damped open-loop poles are overcome with the MSP approach [27]. Let us consider a generalized delay-free system given by equation 13.32.

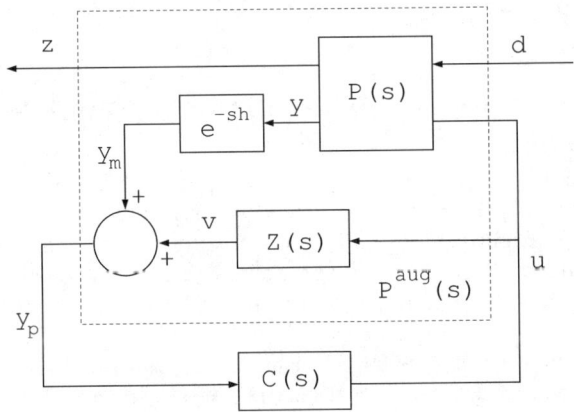

**Fig. 13.20.** Smith predictor formulation

$$P(s) = \begin{bmatrix} A & B_1 & B_2 \\ C_1 & D_{11} & D_{12} \\ C_2 & D_{21} & D_{22} \end{bmatrix} = \begin{bmatrix} P_{11} & P_{12} \\ P_{21} & P_{22} \end{bmatrix} \quad (13.32)$$

For a delay of $h$ seconds, the generalized regulator formulation using the MSP approach [27] is as follows:

$$\tilde{P}(s) = \begin{bmatrix} A & e^{Ah}B_1 & B_2 \\ C_1 & 0 & D_{12} \\ C_2 e^{-Ah} & D_{21} & 0 \end{bmatrix} \quad (13.33)$$

where $\tilde{P}(s)$ is $P_{aug}(s)$ including the effect of the delay block between $d$ and $z$, as shown in Fig. 13.20.

The computation of matrix exponential $e^{-Ah}$ in (13.33) suffers from numerical problems, especially for systems having fast stable eigen-values. In the worst case it might well be non-computable. This problem can even arise with reasonably small amount of delay, if some of the stable eigenvalues are very fast.

In $H_\infty$ mixed-sensitivity formulation for power system damping control design, the presence of fast stable eigenvalues in the augmented system cannot be ruled out. Possible sources of fast stable eigenvalues include the fast sensing circuits ($T \sim 0.02s$), fast damper circuits ($T \sim 0.05s$) and even the weighting filters. These often lead to numerical instability when solving the problem using LMIs. These problems are overcome through the use of the USP [31] formulation, achieved by decomposing the delay free system model $P$ into a critical part $P_c$ and a non-critical part $P_{nc}$. The critical part contains the poorly damped poles of the system, whereas the non-critical part consists of poles with sufficiently large negative real values. The next section describes the generalized problem formulation using this approach.

## 13.9 Problem Formulation using Unified Smith Predictor

As indicated in the previous section, the first step towards formulating the control problem using the USP approach is to decompose the delay-free system model into critical and non-critical parts. This is normally done by applying a suitable linear coordinate transformation on the state space representation of the system. In this work, a suitable transformation matrix $V$ is chosen such that the transformed matrix $J = V^{-1}AV$ is in the Jordan canonical form and is free from complex entries. The transformation matrix $V$ is chosen using the *eig* function available in Matlab [18]. The elements of the transformed matrix $J$ were converted from complex diagonal form to a real diagonal form using the *cdf2rdf* function in Matlab [18]. The transformed augmented delay-free system model $P_{22}'$ is given by:

## 13.9 Problem Formulation using Unified Smith Predictor

$$P'_{22}(s) = \begin{bmatrix} V^{-1}AV & V^{-1}B_2 \\ C_2V & D_{22} \end{bmatrix} = \begin{bmatrix} A_c & 0 & B_c \\ 0 & A_{nc} & B_{nc} \\ C_c & C_{nc} & D_{22} \end{bmatrix} \quad (13.34)$$

where $A_c$ is the critical and $A_{nc}$ is the non-critical part of $A$. The augmented system model $P'_{22}$ can be split as $P'_{22} = P_c + P_{nc}$, where:

$$P_c(s) = \begin{bmatrix} A_c & B_c \\ C_c & D_{22} \end{bmatrix} \quad (13.35)$$

$$P_{nc}(s) = \begin{bmatrix} A_{nc} & B_{nc} \\ C_{nc} & 0 \end{bmatrix} \quad (13.36)$$

The predictor for the critical part is formulated using the MSP approach by applying a completion operator [27]. On a rational transfer matrix $G = C(sI-A)^{-1}B+D$, the completion operator $\pi_h\{e^{-sh}G\}$ is defined as follows:

$$\pi_h\{e^{-sh}G\} = \begin{bmatrix} A & B \\ C_c e^{-Ah} & 0 \end{bmatrix} - \begin{bmatrix} A & B \\ C & D \end{bmatrix} e^{-sh} \quad (13.37)$$

Using (13.37), the predictor for the critical part $P_c$ is given by (13.38). More details are provided in reference [27].

$$Z_c(s) = \pi_h\{e^{-sh}P_c\} = \begin{bmatrix} A_c & B_c \\ C_c e^{-A_c h} & 0 \end{bmatrix} - \begin{bmatrix} A_c & B_c \\ C_c & D_c \end{bmatrix} e^{-sh} = P_c^{aug}(s) - P_c(s)e^{-sh} \quad (13.38)$$

The predictor for the non-critical part is constructed following the CSP formulation and is given by:

$$Z_{nc}(s) = P_{nc}(s) - P_{nc}(s)e^{-sh} \quad (13.39)$$

The USP, denoted by $Z$, is simply the sum of $Z_c$ and $Z_{nc}$, as shown in Fig. 13.21.

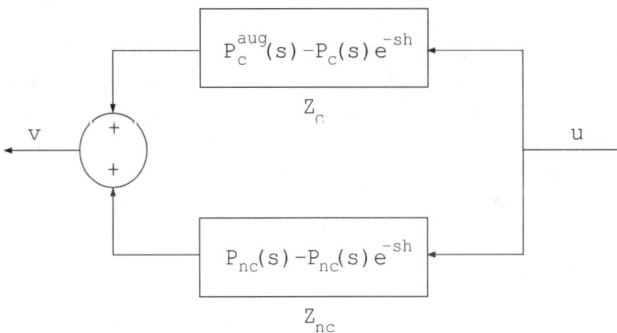

**Fig. 13.21.** Unified Smith predictor

It is given by:

$$Z(s) = P_{22}^{aug}(s) - P_{22}^{t}(s)e^{-sh} \quad (13.40)$$

where $P_{22}^{aug} = P_{nc} + P_{c}^{aug}$. Using (13.34), (13.36) and (13.38), the realization for $P_{22}^{aug}$ can be expressed in the form:

$$P_{22}^{aug} = \begin{bmatrix} A & B_2 \\ C_2 E_h & 0 \end{bmatrix} \quad (13.41)$$

where

$$E_h = V \begin{bmatrix} e^{-A_c h} & 0 \\ 0 & I_{nc} \end{bmatrix} V^{-1} \quad (13.42)$$

The augmented system model $P^{aug}$ is obtained by connecting the original dead-time system and the USP in parallel, as shown in Fig.13.21. The new set of measured outputs is $y_p$. The expression for $P^{aug}$ is given by:

$$P^{aug} = \begin{bmatrix} P_{11}(s) & P_{12}(s)e^{-sh} \\ P_{21}(s) & P_{22}(s)^{aug} \end{bmatrix} \quad (13.43)$$

The generalized regulator $\tilde{P}$ can be formulated from $P^{aug}$ after inserting the delay block $e^{-sh}$ in between $d$ and $z$, as shown by a dotted box in Fig. 13.21. The steps to arriving at the final expression for $\tilde{P}$ are detailed in [31]. The final form of the generalized regulator is as follows:

$$P = \begin{bmatrix} A & 0 & E_h^{-1}B_1 & B_2 \\ 0 & A_{nc} & \begin{bmatrix} 0 & e^{A_{nc}h} - I_{nc} \end{bmatrix} V^{-1}B_1 & 0 \\ C_1 & C_1 V \begin{bmatrix} 0 \\ I_{nc} \end{bmatrix} & 0 & D_{12} \\ C_2 E_h & 0 & D_{21} & 0 \end{bmatrix} \quad (13.44)$$

Having formulated the generalized regulator $\tilde{P}$, following the USP approach, the objective is to design a controller $K$ to meet the desired performance specifications. If $K$ ensures the desired performance for $\tilde{P}$ then the controller predictor combination $K_e = K(I-ZK)^{-1}$ is guaranteed to achieve the same for the original dead-time system [27].

## 13.10 Case Study

In this section, the prototype power system model described earlier is considered to illustrate the control design methodology in details. The performance and robustness of the design is also validated. A conservative estimate of 0.75 s was considered as the signal transmission delay, with a view to cover the worst case scenario. We have used an SVC as the FACTS-device in the system.

### 13.10.1 Control Design

The control design problem was formulated using the standard mixed-sensitivity approach [16][17] with modifications to include the effect of delay. The overall control setup is shown in Fig. 13.22, where:

$G_p(s)$: power system model,
$G_{svc}(s)$: SVC model,
$W_1(s), W_2(s)$: weighting filters,
$K(s)$: controller to be designed,
$Z(s)$: Smith predictor,
$d$: disturbance at the system output,
$z$: weighted exogenous outputs,
$y_m$: measured output and
$u$: control input.

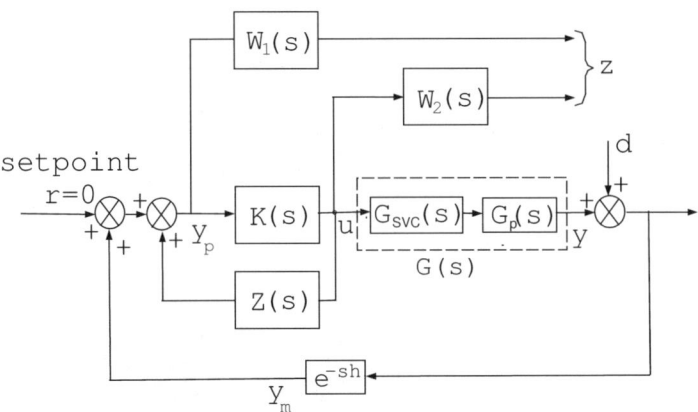

**Fig. 13.22.** Control setup with mixed-sensitivity design formulation

The design objective is as follows:

*Find a controller K from the set of internally stabilizing controllers S such that:*

$$\min_{K \in S} \left\| \begin{array}{c} W_1 S \\ W_2 KS \end{array} \right\|_\infty < 1 \qquad (13.45)$$

where $S=(I-GK)^{-1}$ is the sensitivity. The solution to the problem was sought numerically using the LMI solver with an additional pole-placement constraint.

Once the generalized regulator is formulated following the steps given in section 13.3, the basic steps for control design are exactly similar to those outlined earlier. The weights were chosen as follows:

$$W_1(s) = \frac{100}{s+100}, W_2(s) = \frac{100s}{s+100} \qquad (13.46)$$

The frequency responses for the weighting filters are shown in Fig. 13.23. They are in accordance with the basic requirement of mixed-sensitivity design.

To facilitate the control design and to reduce the complexity of the designed controller, the nominal system model was reduced to a lower order. Using the simplified system model and the above-mentioned weights, the generalized problem was formulated according to (13.44). The solution was sought numerically using suitably defined objectives in the argument of the function *hinfmix* of the *LMI Toolbox* in Matlab [18]. The pole-placement constraint was imposed by using a 'conic sector' of inner angle $2cos^{-1}(0.175)$ with apex at the origin. The order of the controller obtained from the LMI solution was equal to the reduced order augmented system order plus the order of the weights. The designed controller was reduced to an $8^{th}$ order equivalent using Schur's method without affecting the frequency response as shown in Fig. 13.24.

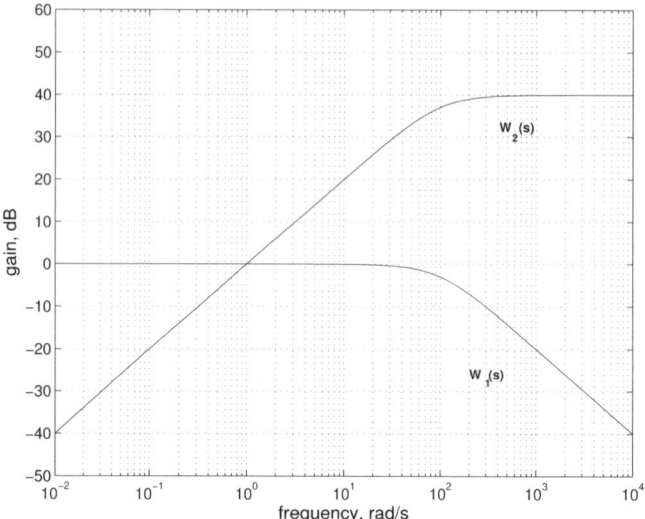

**Fig. 13.23.** Frequency response of the weighting filters

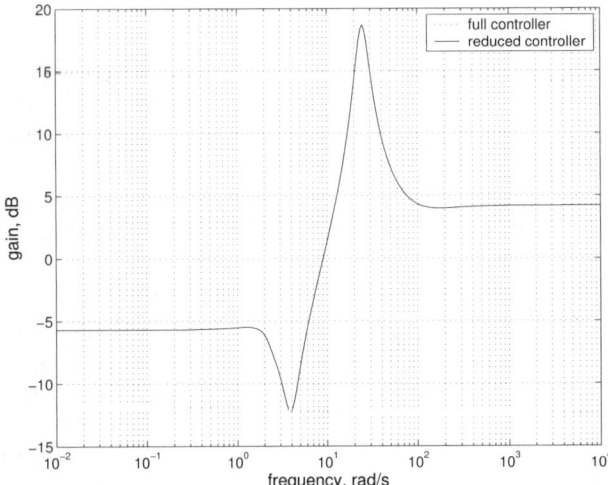

**Fig. 13.24.** Frequency response of the full and reduced controller

### 13.10.2 Performance Evaluation

The eigenvalues of the closed-loop system were computed to examine the performance of the designed controller in terms of improving the damping ratios of the inter-area modes. A $4^{th}$ order Pade approximation was used to represent the delay (= 0.75 s) in the frequency domain.

### 13.10.3 Simulation Results

The controller for the SVC was designed following exactly the same procedure as used for the TCSC earlier in this chapter. The same disturbance, as considered in the previous section, was examined. The dynamic response of the system following this disturbance is shown in Fig. 13.25. This figure illustrates the relative angular separation between the generators G1 and G15. It is clear that the inter-area oscillation is damped out in 12-15 s, even though the feedback signals arrive at the control location after a finite time delay of 0.75 s. The variation of the output of the SVC is shown in Fig. 13.26. It is within a range of -1.5 p.u. to 2.0 p.u.

The performance of the controller for different delays is shown in Figs. 13.27 and 13.28. The simulation results in Fig. 13.29 demonstrate the detrimental effect of not considering the delay at the design stage, if there is one in practice. The results highlight a potential application of the USP approach for power system. It could be used for the design of damping control with different types of FACTS-devices, where the transmission of remote feedback signals involves a finite amount of time delay.

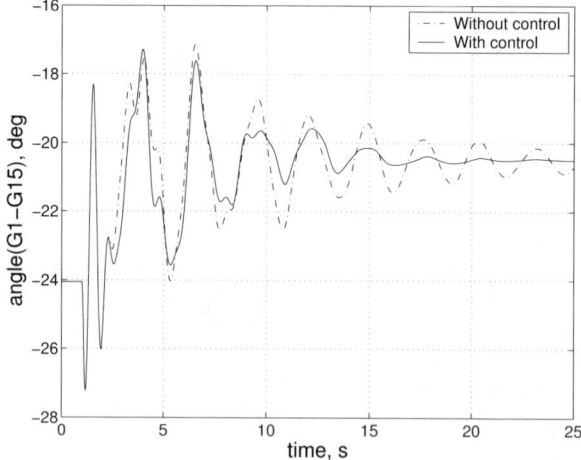

**Fig. 13.25.** Dynamic response of the system with SVC; controller designed considering delay

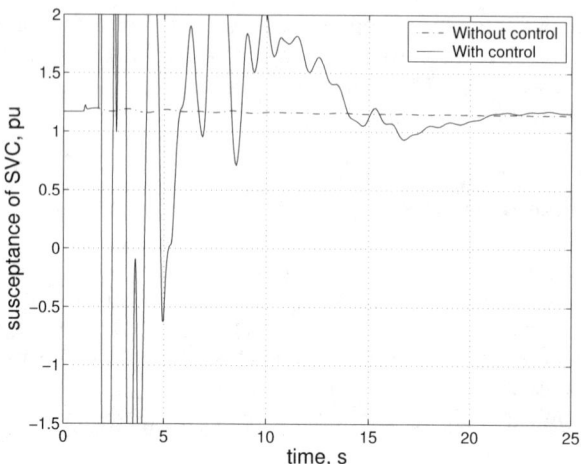

**Fig. 13.26.** Output of the SVC

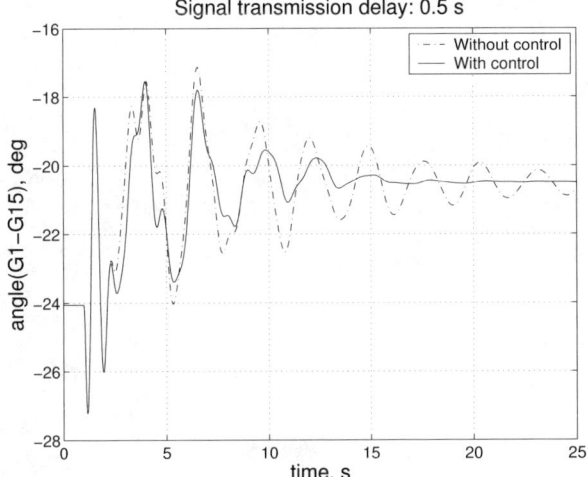

**Fig. 13.27.** Dynamic response of the system with a delay of 0.5 s

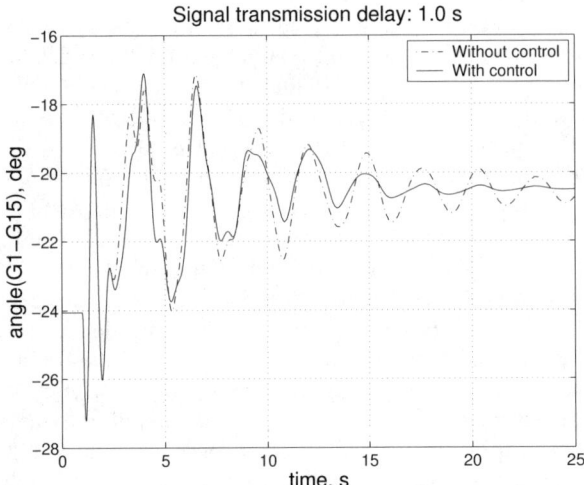

**Fig. 13.28.** Dynamic response of the system with a delay of 1.0 s

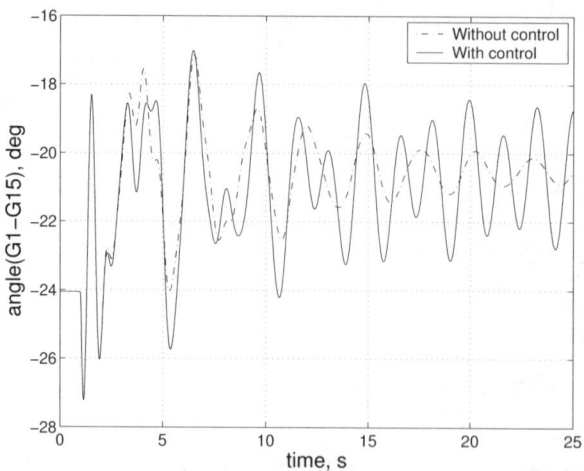

**Fig. 13.29.** Dynamic response of the system; controller designed without considering delay

In this chapter, the basic concept of mixed-sensitivity design formulation has been elaborated. The problem has been translated into a generalized $H_\infty$ problem. The solution to the problem was sought numerically using LMIs with pole-placement. The control design methodology was illustrated by two case studies. In the first case, a centralized controller was designed for a single FACTS-device. Feedback signals from three remote locations were employed for the controller. In the second case, a sequential design methodology was adopted for multiple FACTS-devices. Local feedback signals were used to design decentralized controllers for individual FACTS-devices. The performance and robustness of the design was validated using frequency domain analysis and non-linear simulations.

A methodology for power system damping control design that accounts for delayed arrival of feedback signals from remote locations is also described. A predictor based $H_\infty$ control design strategy has been presented for such time-delayed systems. The design procedure based on the USP approach has been applied for the centralized design of a power system damping controller for FACTS-devices such as TCSC and SVC. A combination of the USP and the designed controller was found to work satisfactorily under different operating scenarios, even though the stabilizing signals could reach the controller site only after a finite time.

In this chapter, a fixed time delay has been considered for all the communication channels. In practice, this might not always be the case as the distances from the measurement sites differ. Therefore, different amount of delay for each signal needs to be considered during the design. Further research is currently being conducted in this area.

# References

[1] Klein M, Le LX, Rogers GJ, Farrokpay S, and Balu NJ (1995) H∞ damping controller design in large power system. IEEE Transactions on Power Systems, vol 10, no 1, pp 158-166

[2] Zhao Q, Jiang J (1995) Robust SVC controller design for improving power system damping. IEEE Transactions on Power Systems, vol 10, no 4, pp 1927-1932

[3] Kamwa I, Trudel G, Gerin-Lajoie L (2000) Robust design and coordination of multiple damping controllers using non-linear constrained optimization. IEEE Transactions on Power Systems, vol 15, no 3, pp 1084-1092

[4] Taranto GN, Chow JH (1995) A robust frequency domain optimization technique for tuning series compensation damping controllers. IEEE Transactions on Power Systems, vol 10 no 3, pp 1219-1225

[5] Pal BC, Chaudhuri B (2005) Robust Control in Power in Power Systems. Springer, USA

[6] Boukarim GE, Wang S, Chow JH, Taranto GN, Martins N (2000) A comparison of classical, robust and decentralized control designs for multiple power system stabilizer. IEEE Transactions on Power Systems, vol 15, no 4, pp 1287-1292

[7] Skogestad S, Postlethwaite I (2001) Multivariable Feedback Control. John Wiley and Sons, UK

[8] Zhou K, Doyle J, Glover K (1995) Robust and Optimal Control. Prentice Hall, USA.

[9] Sefton J, Glover K (1990) Pole/zero cancellations in the general $H_\infty$ problem with reference to a two block design. Systems and Control Letters, vol 14, pp 295-306

[10] Pal BC, Coonick AH, Jaimoukha IM, Zobaidi H (2000) A linear matrix inequality approach to robust damping control design in power systems with superconducting magnetic energy storage device. IEEE Transaction on Power Systems, vol 15, no 1, pp 356-362

[11] Gahinet P, Apkarian P (1994) A linear matrix inequality approach to $H_1$ control. International Journal of Robust and Non-linear Control, vol 4, no 4, pp 421-448

[12] Rao PS, Sen I (2000) Robust pole placement stabilizer design using linear matrix inequalities. IEEE Transactions on Power Systems, vol 15, no 1, pp 313-319

[13] Taranto GN, Wang S, Chow JH, Martins N (1998) Decentralized design of power system damping controllers using a linear matrix inequality algorithm. Proceedings of VI SEPOPE

[14] Pal BC, Coonick AH, Cory BJ (1999) Robust damping of inter area oscillations in power systems with superconducting magnetic energy storage devices. IEE Proceedings on Generation Transmission and Distribution, vol 146, no 6, pp 633-639

[15] Pal BC, Coonick AH, Cory BJ (2001) Linear matrix inequality versus root-locus approach for damping inter-area oscillations in power systems. International Journal on Electrical Power Energy Systems, vol 23, no 6, pp 481-489

[16] Chaudhuri B, Pal BC, Zolotas AC, Jaimoukha IM, Green TC (2003) Mixed-sensitivity approach to H1 control of power system oscillations employing multiple facts devices. IEEE Transactions on Power Systems, vol 18, no 3, pp 1149-1156

[17] Chaudhuri B, Pal BC (2004) Robust damping of multiple swing modes employing global stabilizing signals with a TCSC. IEEE Transactions on Power Systems, vol 19 no 1, pp 499-506

[18] (1998) Matlab Users Guide. The Math Works Inc., USA

[19] Paserba J (1996) Analysis and control of power system oscillation. CIGRE Special Publication 38.01.07, Technical Brochure 111

[20] Chilali M, Gahinet P (1997) Multi-objective output feedback control via LMI optimization. IEEE Transactions on Automatic Control, vol 42, no 7, pp 896-911
[21] Scherer C, Gahinet P, Chilali M (1996) H1 design with pole placement constraints: An LMI approach. IEEE Transactions on Automatic Control, vol 41, no 3, pp 358-367
[22] Gahinet P, Nemirovski A, Laub AJ, Chilali M (1995) LMI Control Toolbox for use with Matlab. The Math Works Inc, USA
[23] (2002) Using Simulink. The Math Works Inc., USA
[24] Heydt GT, Liu CC, Phadke AG, Vittal V (2001) Solutions for the crisis in electric power supply. IEEE Computer Applications in Power, vol 14, no 3, pp 22-30
[25] Xie Z, Manimaran G, Vittal V, Phadke AG, Centeno V (2002) An information architecture for future power systems and its reliability analysis. IEEE Transactions on Power Systems, vol 17, no 3, pp 857-863
[26] Chow JH, Sanchez-Gasca JJ, Ren H, Wang S (2000) Power system damping controller design using multiple input signals. IEEE Control Systems Magazine, vol 20, no 4, pp 82-90
[27] Zhong QC (2003) $H_1$ control of dead time systems based on a transformation. Automatica, no 39, pp 361-366
[28] Smith, OJM (1957) Closer control of loops with dead time. Chem. Eng. Progress, vol 53, no 5, pp 217–219
[29] Smith, OJM (1958) Feedback Control Systems. McGraw-Hill Book Company Inc., USA
[30] Wantanable K, Ito M (1981) A process-model control for linear systems with delay. IEEE Transactions on Automatic Control, vol 26, no 6, pp 1261-1269
[31] Zhong, Q-C, Weiss G (2004) A unified smith predictor based on the spectral decomposition of the plant. International Journal of Control, vol 77, no 15, pp 1362-137

# Index

active power flow control, 46, 62
adaptation rules, 277
adaptive control, 281
adverse control interactions, 260
architecture of a WAMS, 290
ATC, 240
augmented Lagrangian function, 118
automatic voltage regulator, 320
autonomous control, 269
autonomous system, 270
auto-regressive (AR) model, 297

back-to-back, 24
balanced system, 224
bay control level, 270, 275
building blocks, 261
bus voltage control, 47
bus-impedance method, 139

capacitive compensation, 31
centralized control, 366
classical Smith predictor, 367
communication delay, 366
computational intelligence, 269
concrete rules, 278
congestion management, 239
constraint enforcement, 36, 49
contingency control, 263, 266
continuation power flow, 189
continuation three-phase power flow, 192, 218, 222
control architecture, 259
control modes, 46, 62
control path, 272
controllable transmission line, 261
controlled Lyapunov function, 265
coordinating control, 272
corridor voltage stability, 292
cross-border transmission capability, 246
CSP, 367

current constraints, 47, 67

damping controller, 281
damping ratio, 335
decoupling-compensation bus-admittance method, 139
design process, 259
DFC, 4, 19
direct voltage injection, 63
DVR, 4, 19
Dynamic Power Flow Controller, 4, 19, 239
Dynamic Voltage Restorer, 4, 19

economic benefit, 248
eigenvalues, 335
emergency control, 303
energy function, 265
equivalent impedance, 31
equivalent injected voltage magnitude, 32
excitation system, 322

fast controllability, 254
fast-decoupled method, 139
functional specification, 259
fuzzy adaptation, 277
fuzzy modules, 278

Gauss-Seidel Method, 139
Generalized Unified Power Flow Controller, 4, 22, 113
generation cost, 254
generation shift distribution factors, 278
generic rules, 272
governor model, 324
gradient method, 102
GSDF, 278
GUPFC, 4, 22, 70

$H_\infty$ control, 347

half-bridge, 8
harmonics, 9
HVDC, 4
hybrid control model, 176

IEEE 30-bus system, 40
IGBT, 6
IGCT, 6
impedance (reactance) control, 47
impedance compensation, 95
Implicit Bus-Impedance Method, 139
increase of transmission capacity, 248
induction motor load, 326
inductive compensation, 32
initialization, 78
Insulated Gate Bipolar Transistor, 6
Insulated Gate Commutated Thyristors, 6
interior point method, 103
Interline Power Flow Controller, 4, 22, 113
IPFC, 22, 70

Kalman filter, 297

line impedance compensation, 65
line temperature, 296
linear frequency, 335
linear programming, 102
linear system analysis, 319
Load Flow Controller, 239
load model, 324
loadability, 198, 302
loadability calculation, 292
loss optimal, 254
loss reduction, 239
LTI control law, 350

marginal cost, 254
market simulation, 249
matrix inequality formulation, 351
mixed-sensitivity formulation, 348
modal controllability, 343
modal observability, 320, 343
mode shape, 341
modified Smith predictor, 367
modular controller design, 262
MSP, 367
multi-control capabilities, 40
multi-control functions, 29
multi-converter, 59

multi-functional, 59
multi-terminal VSC-HVDC, 123
M-VSC-HVDC, 82, 123

N-1-Criterion, 242
network control level, 270, 277
Newton equation, 143
Newton's method, 103
Newton-Raphson Method, 139
NISC, 259
Non-Intrusive System Control, 259
non-linear control, 319
Nonlinear Interior Point OPF, 106
nonlinear programming, 101

OPF, 101
OPF-Formulation, 104
optimal power flow, 101
optimized power flow, 307
oscillation monitoring, 292, 297
overload, 273
overvoltage protection, 16

packaging, 7
parallel path, 272
participation factor, 342
Phase Shifting Regulation, 65
Phase Shifting Transformer, 4, 239
Phasor Measurement Units, 289
PMU, 289
polar coordinates, 28
pole placement, 349
power converters, 8
power electronics, 5
power flow calculations, 27
power mismatch equations, 142
Power Transfer Distribution Factor, 246
PQ bus, 28, 141
PQ control, 85
PQ machine, 142
preventive coordination, 279
primary converters, 85
PST, 4
PV bus, 28, 140
PV control, 85
PV machine, 143

quadratic programming, 103

reactive power control, 31

# Index

reactive power flow control, 46, 62
remote control of (maximum) apparent power, 33
remote feedback control, 309
remote reactive power flow control, 33
remote voltage magnitude control, 33
robustness, 262

secondary converter, 86
security constrained transfer capability, 211
semiconductors, 5
sequential design, 361
Series Compensation, 3, 16
series devices, 15
short circuit, 274
shunt devices, 10
singular decomposition, 189
slack bus, 27, 140
small disturbances, 319
small impedance, 95
small signal stability, 319
small signal stability control, 281
Smith predictor, 367
SSSC, 4, 18, 152
stacking, 7
STATCOM, 3, 13, 152
state calculation, 292, 301
STATic COMpensator, 13
Static Synchronous Compensator, 152
Static Var Compensator, 11
Static Voltage Restorer, 4
steady state voltage stability, 232
Substation Control Level, 270, 277
Sub-Synchronous Resonance, 17
SVC, 3, 11, 328
SVR, 4
switching patern, 9
symmetrical components control, 172
synchronous generator, 320

tap position, 242, 256
TCPS, 329
TCSC, 3, 15, 17
thermal constraints, 34
thermal monitoring, 292, 296
three-phase OPF, 184
three-phase power flow, 139, 220
Thyristor, 5
Thyristor Controlled Phase Shifter, 329
Thyristor Controlled Reactor, 328

Thyristor Controlled Series Capacitors, 17
Thyristor converter, 8
time-delayed systems, 366
time-synchronized, 290
topology analysis, 277
topology calculation, 292
topology changes, 274
topology detection, 301
Total Transfer Capability, 197, 240
Transistor, 5
transmission functions, 260
TTC, 240
turbine model, 324
TWIN converter, 9

UCTE, 249
unbalanced three-phase voltage stability, 217
unbalanced three-phase power system, 232
Unified Power Flow Controller, 4, 21, 152, 239
unified Smith predictor, 367
unified transfer capability, 205, 206
UPFC, 4, 21, 60, 166
USP, 367, 370

voltage constraints, 34, 47, 67
voltage control, 31
voltage shifting, 62
voltage stability analysis, 189
voltage stability limit, 197
voltage stability monitoring, 292
voltage stability prediction, 303

WAMS, 289
wide-area control, 306
wide-area measurement system, 289

zonal balance, 248